Engineering Materials

This series provides topical information on innovative, structural and functional materials and composites with applications in optical, electrical, mechanical, civil, aeronautical, medical, bio- and nano-engineering. The individual volumes are complete, comprehensive monographs covering the structure, properties, manufacturing process and applications of these materials. This multidisciplinary series is devoted to professionals, students and all those interested in the latest developments in the Materials Science field, that look for a carefully selected collection of high quality review articles on their respective field of expertise.

Indexed at Compendex (2021) and Scopus (2022)

Elkhattabi El Mehdi · Mourad Boutahir
Editors

Advanced Materials for Sustainable Energy and Engineering

Volume 2: Advancements in Sustainable Energy Technologies and the Green Revolution

Springer

Editors
Elkhattabi El Mehdi
Ecole Normale Supérieure
Moulay Ismail University
Meknes, Morocco

Mourad Boutahir
Ecole Normale Supérieure
Moulay Ismail University
Meknes, Morocco

ISSN 1612-1317 ISSN 1868-1212 (electronic)
Engineering Materials
ISBN 978-3-032-10068-9 ISBN 978-3-032-10069-6 (eBook)
https://doi.org/10.1007/978-3-032-10069-6

This Springer imprint is published by the registered company Springer Nature Switzerland AG
The registered company address is: Gewerbestrasse 11, 6330 Cham, Switzerland

In memory of my beloved parents, Ahmed Boutahir and Saida Belkhyate.

and

To my wonderful wife, Dr. Lamiae El Alaoui, and our son, Ahmed Ghayth Boutahir: your love and support are the essential foundations of my success in turning ideas into reality.

Preface

Dear Colleagues,

In a world that is constantly evolving, the significance of scientific research and innovation is paramount. Our relentless pursuit of knowledge drives us to explore new frontiers, uncover hidden truths, and devise solutions to the challenges we encounter. It is within this context that we present this collection of research articles, encompassing a diverse array of topics in engineering, materials science, chemistry, and environmental science.

This book embodies our steadfast commitment to advancing knowledge and our desire to share the outcomes of our efforts with the scientific community. Each chapter reflects the results of thorough research and analysis conducted by dedicated scholars. The topics span a wide spectrum, showcasing the varied interests and expertise of authors from various academic and research institutions. From seismic fragility analysis to the processing of volatile organic compounds, and from 3D printing parameters to the dynamics of metal-insulator transitions, these chapters offer insights into the latest advancements in their fields.

We aspire for this book to not only inform but also inspire new ideas and collaborations that will shape the future of materials science in energy applications. Our heartfelt gratitude goes to the authors for their invaluable contributions and to the readers for their interest in the vibrant and evolving realm of materials science for energy.

Meknes, Morocco

Prof. Elkhattabi El Mehdi
Prof. Mourad Boutahir

Contents

Modeling and Calculation of Reactivity Feedback Coefficients and Neutron Flux Using Monte Carlo Code OpenMC for SLOWPOKE-2 Reactor

A. El Mtili, O. El Hajjaji, and M. El Hadri

Abstract SLOWPOKE-2 is a pool-type research reactor difficult to model due to the importance of neutron leakage. The aim of this paper is to evaluate a three-dimensional continuous energy Monte Carlo code OpenMC of the reactor and then it is used to compute the reactivity feedback coefficients and the neutron flux. The OpenMC geometry model was validated by criticality calculation, where good agreement is observed between the calculated core reactivity and excess reactivity measurement. The reactivity feedback coefficients are accounted for variation in fuel temperature, moderator temperature and void coefficients of the SLOWPOKE-2 reactor. Only one parameter was changed, while all other parameters remain constant. To perform this study and evaluate cross-section data, the new version VIII.0 of the ENDF/B nuclear data library was used to generate a data set in ACE (A Compact ENDF) format at various temperatures (313, 333, 353 and 373 K), utilizing NJOY2016 code to process the data at the particular temperatures. It was observed that all reactivity feedback coefficients were found negative which meets the design safety criteria. The neutron flux in outer and inner irradiation sites is also presented. The results of calculations are analyzed and discussed.

Keywords SLOWPOKE-2 · Open MC code · Criticality calculation · Reactivity coefficients · $LEU - UO_2$ · Neutron flux

A. E. Mtili (✉) · O. E. Hajjaji
Radiations and Nuclear Systems Laboratory, Faculty of Sciences, Abdelmalek Essaadi University, Tetouan, Morocco
e-mail: elmtili.afaf-etu@uae.ac.ma

M. E. Hadri
Faculty of Sciences, Abdelmalek Essaadi University, Tetouan, Morocco

E. El Mehdi and M. Boutahir (eds.), *Advanced Materials for Sustainable Energy and Engineering*, Engineering Materials, https://doi.org/10.1007/978-3-032-10069-6_1

1 Introduction

The Safe LOW POwer Critical Experiment reactor (SLOWPOKE-2) is one of the low power research reactors. Many parameters during the reactor operation, are usually managed and controlled to ensure safe and stable operation of the reactor [1] like temperature, reactivity feedback, power level, pressure, and cooler flow. The various effects of changes in these parameters differ greatly depending upon the reactor design. During reactor operation, one of the important effects is the temperature: a change in temperature will cause a change in reactivity [2] and the value of multiplication factor (k_{eff}). The significant effect is caused by water density changes with variations of temperature. The direction of the change, even if it is positive or negative influence on the reactor safety and control. Various materials in a reactor are at different temperatures and make various effects on reactivity. So, we must make a distinction between different effects contributing to the global temperature effect such as the fuel and moderator. In the case of LEU fuels, the fuel temperature coefficient is usually considered even more important than the moderator temperature coefficient. It will be the first and the most important feedback that will compensate for the inserted positive reactivity. This fuel temperature coefficient also responds faster to an increasing power transient than moderator temperature coefficient [3], because reactor power causes an immediate increase in fuel temperature. The moderator delays due to the time for the transfer of heat from the fuel to the moderator. Also, this is true for decreasing power (fuel temperature decrease). The exception to this is when a change in steam demand initiates.

The objective of this research activity is to evaluate a 3-D model for the reactor using the code of Monte Carlo OpenMC code. Then this model is used to illustrate the ability to simulate our reactor by computing the reactivity feedback coefficients and neutron flux which account for variation in fuel temperature, moderator temperature as well as the void fraction. In order to achieve this study, one parameter was changed and the others were kept constant, also the temperature was increased and the reactivity was measured. All the neutronics parameters are determined by Monte Carlo transport code OpenMC. As far as our reading, no such analysis was already done using OpenMC. Monte Carlo particle transport methods are being considered as a viable option for high-fidelity simulation of nuclear reactors. Our calculations were performed using the ENDF/B-VIII.0.

2 Reactor Description and Computer Codes

2.1 Reactor Description

The SLOWPOKE-2 reactor is a pool-type research reactor developed by Atomic Energy of Canada, with a maximum nominal thermal power of 20 kW [4]. This reactor has a number of specific features including low critical mass, unattended operation and inherent self-limiting feedback. The SLOWPOKE-2 reactor uses Low Enriched Uranium (LEU) of UO_2 fuel pellets, with an enrichment of ≈ 20 % (19.88 % ^{235}U), light water as a coolant and moderator, and beryllium as a reflector. Major aspects of the LEU-fueled SLOWPOKE-2 reactor are the lack of symmetry and its small size.

The core consists of total 198 fuel pins cladded in zircaloy, disposed on a hexagonal lattice to maximize the neutron flux on the periphery where the five irradiation sites are located. It was also designed to have a relatively high flux at the center where a single control rod is situated. The core of the reactor is ringed by an annulus reflector, and the reactor has also a bottom reflector beryllium.

In addition to the five internal irradiation sites mentioned previously, four outer irradiation sites are located between the radial reflector and the vessel wall. At RMC, one of the outer sites is cadmium lined for fast neutron irradiation purposes, and another outer sites adaptable to large.

The reactor is located in a light water filled pool that measures 6.1 m deep with a diameter of 2.5 m.

The reactor at RMC is also supplied with a thermal column of heavy water, which replaces one of the outer sites [5], to provide a pathway for thermal neutrons to travel from the core region radially though the reactor container.

2.2 *Computer Codes*

OpenMC code

OpenMC is a young open-source Monte Carlo particle transport code developed in 2011 at the Massachusetts Institute of Technology. It is a powerful simulation platform for neutron criticality calculation and particle transport simulation based on a 3D simulated model of solid geometry construction [6]. The code supports multigroup cross sections and continuous energy.

Similar to other Monte Carlo codes, OpenMC reads continuous-energy cross sections from Compact ENDF (ACE) data libraries, which can be generated with the NJOY nuclear data processing code. Since this data format is shared with other Monte Carlo codes (such as MCNP and SERPENT), any ACE format data library generated for these codes can be used with OpenMC [7]. Two types of neutronic problems can be resolved the external fixed fission source problems during Monte Carlo calculations and the k-eigenvalue criticality fission source.

ERSN-OpenMC

In the present work, we used ERSN-OpenMC, which is a Graphical User Interface for OpenMC Monte Carlo particle transport simulation code, originally developed by Jaafar EL Bakkali and Tarek EL Bardouni, members of Radiation and Nuclear Systems Group (ERSN; Equipe Radiations et Syst‘emes Nucl´eaires) at Faculty of Sciences Tetouan (Morocco).This java-based application has been created to provide a user-friendly human-computer interface using OpenMC Monte Carlo code as a neutrons transport calculation engine. This java-based application allows the creation/edition of separate mandatory XML files describing the geometry, the materials involved and the settings of a given OpenMC simulation [7]. On-line Doppler broadening is not employed in OpenMC code. Thus, continuous-energy neutron data library in ACE format and thermal $S(\alpha, \beta)$ cross-sections at given temperature are generated using NJOY code based on ENDF/B-VIII.0 before OpenMC calculation.

Njoy2016 code

Njoy2016 is a modular computer program, developed in the Los Alamos National Laboratory of the United States [8]. It is started from the original ENDF-formatted file and ended with an ACE file suitable for use in an OpenMC calculation. The NJOY code consists of a set of main modules, each performing a well-defined processing task. Each of these main modules is essentially a separate computer program. They are linked to one another by input and output files. The task of NJOY2016 is to obtain the basic data from the nuclear database (NDL) and transform it into the form required by the application. For thermal neutron scattering data generation, the THERMR module expects that the requested temperature to be processed should be one of the temperatures contained in the ENDF/B thermal file, or be within a few degrees of that value (ie, 296 K is utilized if 300 K is requested) [9]. Data for other temperatures should be obtained by interpolation between the cross-sections [6]. Therefore, the user Intermediate temperatures can be obtained by interpolating between the resulting cross-sections. The current version of NJOY2016 is freely available, and can be downloaded from http://njoy.lanl.gov.

3 OpenMC Modelling of SLOWPOKE-2 Reactor

Monte Carlo methods solve transport equations differently than deterministic codes. They generate results by simulating individual particles and recording some scores (tallies) of their average behavior. MC codes are widely used because of their ability to model complex geometries and the high fidelity of the resulting solutions when using continuous energy cross-sections data. The OpenMC model of the SLOWPOKE-2 reactor encapsulated all components of the reactor within the reactor pool. This included the fuel, the reflector, the control rod, irradiation sites, the thermal column, the reactor container, and the surrounding pool. A vacuum boundary condition has been applied to the surface, it is mean that the particle is killed and any surface current tallies are scored to as needed.

The reactor was modelled in full 3-D details using OpenMC code, which is considered the first for the SLOWPOKE-2 research reactor. It was created based on previous MCNP 4A model of RMC SLOWPOKE-2 by Pierre [5].

The images created using the ParaView application of OpenMC are presented to aid in the description of the MCNP 4A model. It is important to note that these images visualize the newly created input, not the original input from Pierre's work.

This model is extended axially from z = − 53.29575 to z = + 510.70425, with its mid plane being at z = 0. At the core level, the fuel rods occupy the center of cells in a hexagonal lattice with a pitch of 1.10363 cm.

Figures 1 and 2 represent respectively the horizontal and vertical cuts of the reactor vessel and the 3-D view of SLOWPOKE-2 reactor compiled with the code ParaView 5.10.1, color by material.

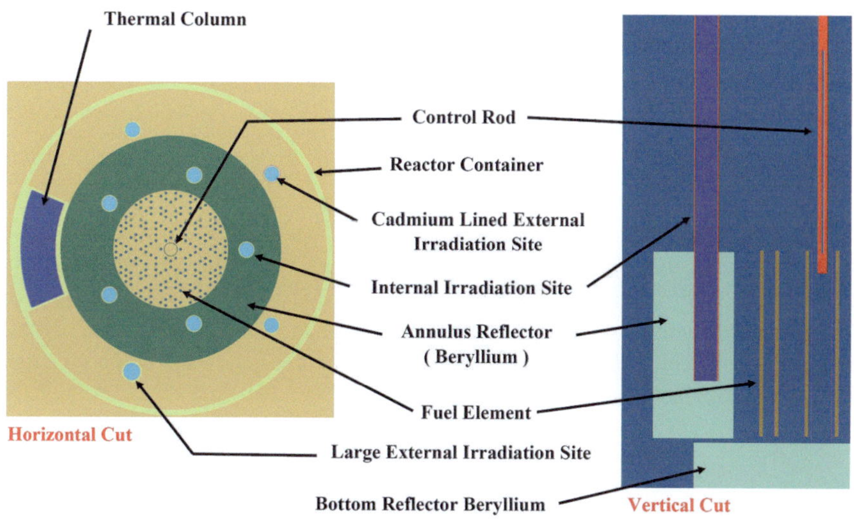

Fig. 1 Horizontal and vertical cuts of the reactor vessel

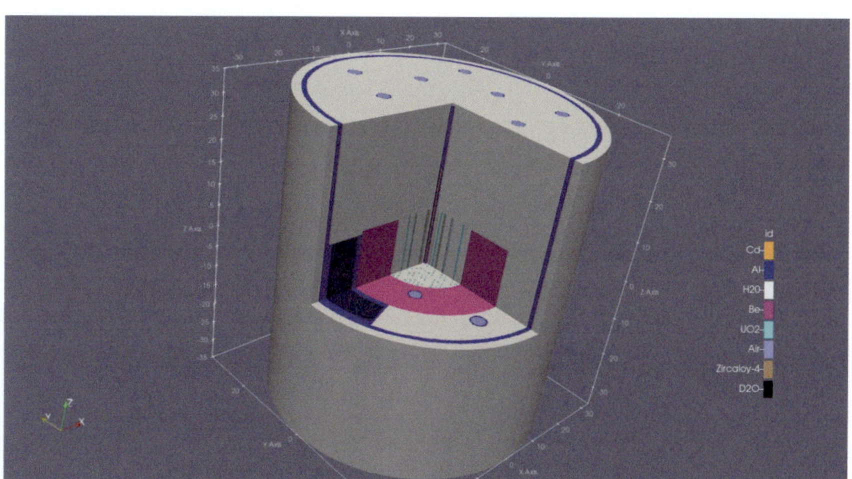

Fig. 2 3-D view of SLOWPOKE-2 reactor compiled with the code ParaView 5.10.1, color by material

4 Calculation of Reactivity Coefficients

Reactivity ρ is a measurement that represents the deviation of the reactor from critical state ($k_{eff} = 1$):

$$\rho = \frac{k_{eff} - 1}{k_{eff}} \tag{1}$$

Noting that no delayed neutrons were used. Generally, reactivity coefficient quantifies the effect that a variation in a reactor parameter (i.e., a control rod position, change in temperature, etc.) has on the overall reactivity of the core [2]. It is expressed mathematically as:

$$\gamma\xi = \frac{\Delta\rho}{\Delta\xi}$$

where γ is the reactor parameter that impacts the reactivity and $\Delta\rho$ is corresponding of the change in reactivity. In the current research paper, the temperature coefficients of reactivity have been calculated, so ξ represent T_f, T_m and V_m, and $\gamma\xi$ defines the fuel temperature of reactivity, moderator temperature of reactivity and void coefficient of reactivity respectively.

5 Results and Discussion

5.1 Criticality Calculation

Based on the OpenMC model, the criticality is calculated and compared against the experimen- tal commissioning data value. The generation of k_{eff} of the reactor was determined at room temperature for the control rod in its fully out position. These calculations were run using 500 inactive and 11000 active batches of 100000 source neutrons, with the ENDF/B-VIII.0 nuclear cross-section libraries.

From Table 1, it can be concluded that the excess reactivity of the SLOWPOKE-2 reactor as calculated from the OpenMC simulation with the control rod in the fully out position, is in close agreement with the experimental data. Indeed, the ratio between the calculated value and the experimental value in 0.977 mk or only about 1%, which is less than one percent.

Table 1 Comparison of SLOWPOKE-2 model with experimental commissioning data [5]

	Calculated	Experiment	C/E*
k_{eff}	1.00309 ± 0.00003	1.00316	0.999
Reactivity value (mk)	3.08	3.15	0.977

* C/E = Calculated value/Experimental value

5.2 *Reactivity Coefficients*

Effect of fuel temperature

Doppler broadening and spectrum hardening are the two primary factors that affect the fuel temperature reactivity coefficient. The former effect is common in reactors containing 238U [10]. An increase in fuel temperature causes a shift and deformation of Maxwell's spectrum in fuel, whereas the spectrum in water is only slightly affected because the water temperature remains almost constant. The values of the effective multiplication factor were determined at fuel temperatures of 293, 313, 333, 353 and 373 K. For all calculations, the moderator temperature and the other regions were constant 293 K. Table 2 and Fig. 3 show the change in reactivity with increasing fuel temperature.

Table 2 and Fig. 3 show the general trend in reactivity that by increasing fuel temperature, the core reactivity decreases. When the fuel temperature increases, the target core has more energy. As a result, the nucleus absorbs a wider band of neutrons (off-peak neutrons). Peak broadening occurs and makes resonance detection more likely. Therefore, the resonance exit probability is reduced, resulting in a decrease in k_{eff}.

Effect of moderator temperature

The moderator temperature is varied from 293 to 573 K in steps of 20 K, and the corresponding k_{eff} value is calculated by the OpenMC code. Next, the reactivity was calculated for each value of moderator temperature, given in Table 3 and plotted in Fig. 4.

Table 3 and Fig. 4 shows that the reactivity is almost linearly related to the moderator temperature. This is expected, an increase in water temperature reduces the

Table 2 Effect in reactivity due to change in fuel temperature of SLOWPOKE-2 reactor

Temperature (K)	k_{eff}	$\rho = 1\text{–}1/k_{eff}$	$\Delta\rho = \rho\text{-}\rho$ (293 K)*1000	
293	1.00309	0.003080481	0	
313	1.00285	0.002841901	– 0.238580729	
333	1.00261	0.002553206	– 0.477275679	
353	1.00238	0.002334349	– 0.706132263	
373	1.00217	0.002165301	– 0.915180017	

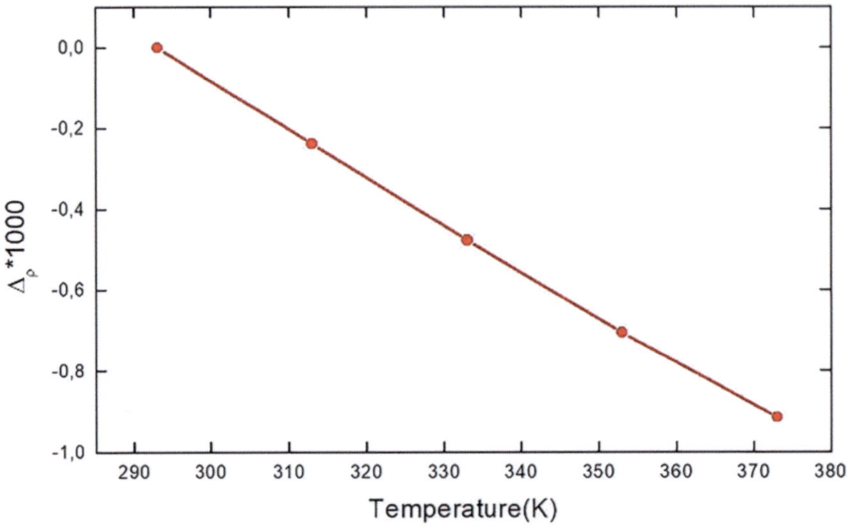

Fig. 3 Variation of reactivity with fuel temperature

Table 3 Effect in reactivity due to change in moderator temperature of SLOWPOKE- 2 reactor

Temperature (K)	k_{eff}	$\rho = 1{-}1/k_{eff}$	$\Delta\rho = \rho - \rho$ (293 K)*1000
293	1.00309	0.003080481	0
313	1.00260	0.002593258	– 0.487223782
333	1.00198	0.001976087	– 1.104393966
353	1.00140	0.001398043	– 1.682438573
373	1.00028	0.000279922	– 2.800559691

density of water, negatively affecting the reactivity due to the reduction in the density of hydrogen atoms [10]. It can be concluded that the temperature feedback coefficient becomes more negative as the water temperature increases, so there is no adverse effect on the abnormal operation of the reactor.

Void fraction of reactivity

Void is a term used to describe the presence of air bubbles in a coolant/moderator near or near its boiling point. In order to calculate the void reactivity coefficient, the reactivity change due to void fraction is simulated only by reducing the water density.

We calculate the reactivity coefficient using the value of k_{eff}. Table 4 and Fig. 5 show that as the coolant void increases, the reactivity of the core decreases. This is because the density of mixture (water and vapors) decreases as the void fraction increases. Thus, in the case of a water reactor, the effect of void space on reactivity must always be negative if the reactor is not over-moderated. Nevertheless, it was

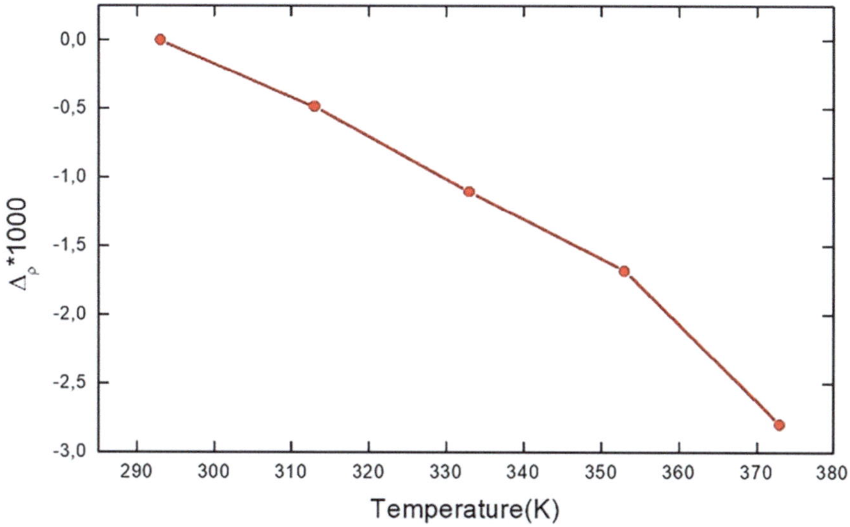

Fig. 4 Variation of reactivity with moderator temperature

shown that the SLOWPOKE-2 reactor does not form voids during normal operation or even during a slope positive reactivity insertion accident.

Overall reactivity

The combined effect of fuel, moderator temperature and water density are called the system overall reactivity coefficient. All coefficients are considered simultaneously in order to account for the mutual coupling and to understand the impact of the interconnection. Because it is clear that an increase in fuel temperature will also lead to an increase in the temperature of the moderator, thereby reducing its density.

In this analysis, both fuel temperature and moderator temperature were varied from 293 to 373 K in steps of 20 K. The corresponding moderator densities are

Table 4 Effect in reactivity due to change in coolant void of SLOWPOKE-2 reactor

Void fraction (%)	k_{eff}	$\rho = 1{-}1/k_{eff}$	$\Delta\rho = \rho\text{-}\rho\ (0\%)*1000$
0	1.00309	0.003080481312744	0
1	1.00105983	0.001058707949554	– 2.02177336318985
3	0.999302589	– 0.000697897721548	– 3.77837903429152
5	0.998209093	– 0.00179412010225	– 4.87460141499357
7	0.997205469	– 0.002802362288288	– 5.88284360103155
9	0.996201794	– 0.003812687372053	– 6.89316868479634
11	0.995108118	– 0.004915930150215	– 7.99641146295877
13	0.994104676	– 0.0059302849512	– 9.01076626394334
15	0.993101959	– 0.006945954478779	– 10.0264357915224

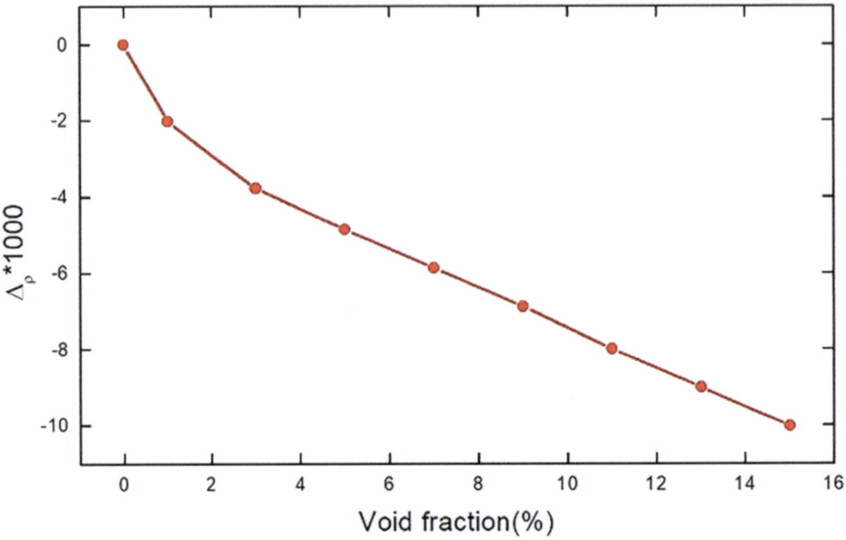

Fig. 5 Reactivity as a function of void fraction

assumed to be 0.998, 0.992, 0.983, 0.972, and 0.958 g/cm^3. The values are listed in Table 5 and the coupled reactivity coefficient as a function of temperature is plotted in Fig. 6.

According to the current calculation, it was observed that the value of coupled reactivity coefficient is larger than the individual values of fuel temperature, moderator temperature and void reactivity coefficients. The values of fuel, moderator and void temperature reactivity coefficients were calculated according to the gradient of the reaction curve, and the results were all negative, which met the design criteria [3]. The values are listed in Table 6.

Also, the overall reactivity was found to be – 13.70 pcm K^{-1}, while the measured value for the SLOWPOKE-2 reactor as mentioned by [5] is – 10.11 pcm K^{-1}. The relative difference between the calculated and the measured results is 30%. Good agreement between the measured and calculated of OpenMC.

Table 5 Combine effect of fuel, moderator and density on reactivity of SLOWPOKE- 2 reactor

Temperature (K)	Dens	g/cm3	^{k}eff	$\rho = 1–1/k_{eff}$	$\Delta\rho = \rho$-ρ (293 K)*1000
293	0.998		1.00309	0.003080481	0
313	0.992		1.000286	0.000285918	– 2.794563085
333	0.983		0.997124	– 0.002884295	– 5.964776546
353	0.972		0.994238	– 0.005795393	– 8.875874368
373	0.958		0.992179	– 0.007882650	– 10.963131520

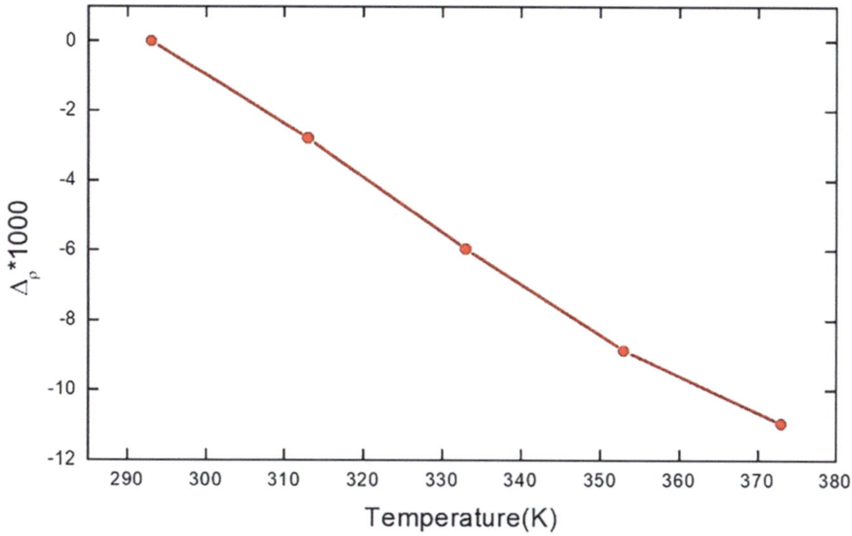

Fig. 6 System overall reactivity as a function of temperature

Table 6 Average values of reactivity coefficients

Fuel temperature coefficient	$-$ 1.144 pcm K^{-1}
Moderator temperature coefficient	$-$ 3.501 pcm K^{-1}
Void reactivity coefficient	$-$ 66.84 pcm%V K_M^{-1}
System overall temperature coefficient	$-$ 13.70 pcm K^{-1}
Experimental value of overall temperature coefficient [5]	$-$ 10.11 pcm K^{-1}
Relative difference, %	30%

5.3 Neutron Spectrum

OpenMC is employed to calculate the static neutronics parameters of the system e.g., k eigen- value, neutron flux and power distribution. The general process goes as follows. It starts by taking an estimate of the total reaction rate and dividing it by the total macroscopic cross-section. In this paper, we had worked a comparison between the neutron flux spectra in the inner and the outer irradiation sites of LEU. The flux in the OpenMC model was tallied by using 69 energy groups. Comparison between flux per unit lethargy spectra in the inner and outer irradiation sites are shown in Fig. 7.

As shown, the thermal neutron spectrum in the inner irradiation sites is 0.3% higher than in the outer irradiation sites, the epithermal neutron spectrum is almost the same (less than 0.03% difference), and the fast neutron spectrum is 0.2% higher in outer irradiation sites than in inner irradiation sites (see Fig. 7). These results

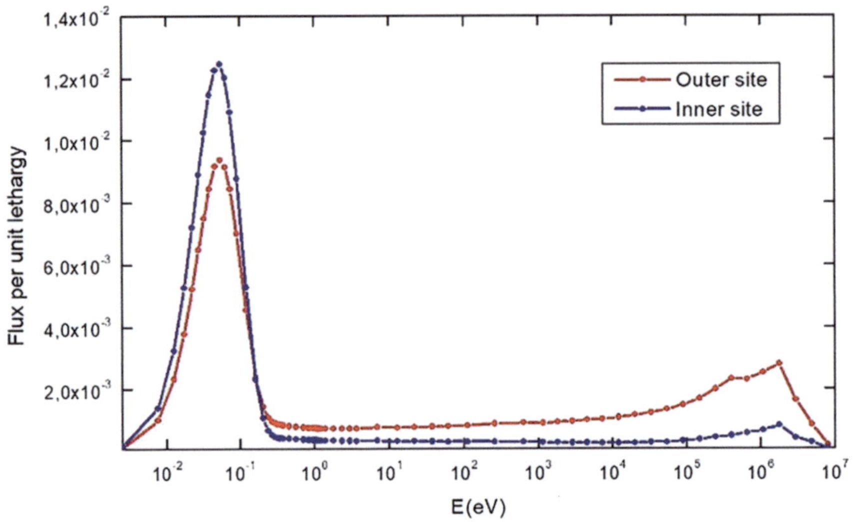

Fig. 7 The neutron flux spectrum in the inner and outer irradiation sites calculated by OpenMC

are expected because the outer irradiation sites in water exist outside the annular beryllium reflector, while the inner irradiation sites exist in the center of the annulus beryllium reflector, and beryllium is a better reflector than water.

6 Conclusion

In the presented study, a three-dimensional model of the SLOWPOKE-2 research reactor was created using OpenMC Monte Carlo code. Some practical calculations such as core excess reactivity was performed and compared with experimental. Good agreement between calculations and experiments ensures model validity. The calculations of reactivity feedback coefficients were successfully accomplished, using the ENDF/B-VIII.0. Nuclear data Libraries. The calculated values of fuel, moderator, void and overall reactivity coefficients were found -1.144 pcm K^{-1}, -3.501 pcm K^{-1}, -66.84 pcm% V K_M^{-1} and -13.70 pcm K^{-1} respectively. In the work, good agreement between the measured and calculated overall reactivity was obtained. These results are a good initial indication of inherent safety for this reactor concept. In studying the spectral effect, the neutron flux spectra in the inner irradiation site is greater than the neutron flux spectra in the outer irradiation site, according to our findings.

As a result, all this makes SLOWPOKE-2 a very suitable tool for employing the NAA technique to analyse the unknown samples and for the production of radioisotopes with medium or short half-lives. Also, a coupled neutronics-thermal hydraulics code can be used in the future to evaluate the overall temperature coefficient for the transient situations.

References

1. El Hajjaji, O.: La physique du r´eacteur SLOWPOKE-2 de l'E´cole Polytechnique. E´cole Polytechnique de Montr´eal (1999)
2. do Prado Souza, R.M.G., Mesquita, A.Z.: Measurements of the isothermal, power and temperature reactivity coefficients of the IPR-R1 TRIGA reactor. Prog. Nucl. Energy **53**(8), 1126–1131 (2011)
3. Trkov, A., Brown, D.A.: Endf-6 Formats Manual: Data Formats and Procedures for the Evaluated Nuclear Data Files. Technical Report, Brookhaven National Lab (BNL), Upton, NY (United States) (2018)
4. Pierre, J., Low Enrichment Uranium (LEU)-fueled SLOWPOKE-2 Nuclear Reactor Simulation with the Monte-Carlo Based MCNP 4A Code. MASc, Royal Military College of Canada (1996)
5. Pierre, M.J., Bonin, H.W.: Monte Carlo simulation of the LEU-fueled SLOWPOKE-2 nuclear reactor using MCNP 4A. Nucl. Technol. **125**(1), 1–12 (1999)
6. Xoubi, N., Darda, S.A., Soliman, A.Y., Abulfaraj, T.: An investigative study of enrichment reduction impact on the neutron flux in the in-core flux-trap facility of MTR research reactors. Nucl. Eng. Technol. **52**(3), 469–476 (2020)
7. Boukhal, H., Kaddour, M., Benaalilou, K., Chham, E.: ERSN-OpenMC, a Java-based GUI for OpenMC Monte Carlo code. J. Radiat. Res. Appl. Sci. **30**(1), e8 (2015)
8. Brown, D.A., Chadwick, M., Capote, R., Kahler, A., Trkov, A., Herman, M., Sonzogni, A., Danon, Y., Carlson, A., Dunn, M., et al.: ENDF/B-VIII. 0: the 8th major release of the nuclear reaction data library with CIELO-project cross sections, new standards and thermal scattering data. Nucl. Data Sheets **148**, 1–142 (2018)
9. Raffuzzi, V., Sublet, J.-C., Jouanne, C., Koning, A., Rochman, D.: An Iron Evaluation Story: From TALYS Model Parameters to Validation on the ASPIS Benchmark with the Monte Carlo Code TRIPOLI-4®
10. El Mtili, A., et al.: Measurement of the reactivity coefficients and neutron flux of the SLOWPOKE-2 reactor using OpenMC. Int. J. Eng. Appl. **11**(4) (2023)
11. Hilborn, J.W., Kay, R., Stevens-Guille, P., Jervis, R.: SLOWPOKE at the University of Toronto: A Laboratory Reactor for Neutron Irradiation. Technical Report, Atomic Energy of Canada Ltd. (1972)
12. Stacey, W.M.: Nuclear Reactor Physics. John Wiley & Sons (2018)

Design and Evaluation Performance of Single-Layer and Double-Layer Radiative Cooling Coatings Using Silicon Dioxide and Titanium Dioxide. A Case Study in North Africa: Rabat, Morocco

Cherqi Oumayma, Malha Mustapha, Said Kardellass, and Abdessamad idouanaou

Abstract Radiative cooling (RC) has emerged as a significant advancement in passive cooling technologies, presenting a sustainable and energy-efficient strategy to address the issue of the growing demand for air conditioning. This promising approach utilizes natural process to dissipate infrared radiation to outer space cold, without required input consumption. A study focuses designing several single layer and double layers configurations of RC samples. An experimental investigation aimed to understanding the thermal behavior RC samples under real time conditions. The study providing considerable insights which the temperature of each RC sample drops below ambient temperature during nighttime. The results of this study contribute significantly in advancing research in the field of radiative cooling.

Keywords Radiative cooling · Atmospheric window · Sustainable and energy efficient strategy · Thermal behavior of RC samples

1 Introduction

By 2050, The global cooling demand is expected to surpass heating needs as temperatures continue to rise [1]. According to the International Energy Agency, the number of air conditioning units is projected to increase dramatically from 1.6 billion today to 5.7 billion [2]. The demand for space cooling could grow even faster than current estimates suggest, with installed capacity expected to surge from 11,670 GW in 2016

C. Oumayma (✉) · M. Mustapha · S. Kardellass · A. idouanaou
Thermal and Energy Research Team (ERTE), ENSAM Rabat, University Mohamed V, Rabat, Morocco
e-mail: Oumayma_cherqi@um5.ac.ma

C. Oumayma
Center for Doctoral Studies in Information Technology and Engineering Sciences (ST2I), ENSIAS Rabat, Rabat, Morocco

E. El Mehdi and M. Boutahir (eds.), *Advanced Materials for Sustainable Energy and Engineering*, Engineering Materials, https://doi.org/10.1007/978-3-032-10069-6_2

to over 36,500 GW by 2050 [3]. In parallel, global electricity consumption for space cooling is forecasted to escalate from 2020 TWh in 2016 to approximately 6200 TWh in 2050 [4]. Moreover, the anticipated rise in stationary air conditioning and refrigeration could lead to cumulative climate-related emissions ranging between 230 and 430 gigatonnes of CO_2 from 2020 to 2050. This corresponds to the equivalent of 7 to 13 years of global energy-related CO_2 emissions at 2018 levels [5]. While cooling is vital for climate adaptation, it paradoxically contributes to further warming through the release of high-GWP hydrofluorocarbon (HFC) refrigerants and emissions from inefficient systems, including CO_2. This represents a major challenge not only for energy systems but more importantly for climate mitigation efforts. Nonetheless, a wide array of existing technologies and strategies can substantially reduce emissions from the cooling sector while addressing the increasing demand. Consequently, in recent years, radiative sky cooling has garnered significant attention within the engineering community, particularly regarding the development of materials aimed at enhancing energy efficiency and mitigating carbon emissions [6].

Radiative cooling is a passive thermal management that uses specially designed RC samples to emit infrared thermal radiation, enabling samples to cool by escaping heat in the form of longwave radiation into the cold outer space especially through the atmospheric window [8–13 μm] [7]. This technique is particularly effective at night when the absence of solar radiation eliminates heat gain from the sun, allowing surfaces to achieve sub-ambient cooling are called a nighttime radiative cooling material. Furthermore, Surfaces engineered with high solar reflectivity and high infrared emissivity, enabling all-day self-cooling, are referred to as passive daytime radiative cooling materials [8]. The selection of radiative cooling (RC) materials is a critical factor in optimizing the performance of RC samples for daytime and nighttime radiative cooling. Silicon dioxide (SiO_2) is chosen due to the coating's ability to emit thermal energy efficiently, which is essential for cooling. On the other hand, Titanium dioxide (TiO_2) is selected due to its high reflectivity, allowing the coating to effectively reflect a large amount of solar energy. Cheng et al. [9] presented a numerical study of a single-layer radiative cooling coating made of titanium dioxide (TiO_2) and silicon dioxide (SiO_2) particles, achieving a solar reflectivity of 95.6% and emissivity of 94.9% and it is highlighted that the best cooling performance is achieved with TiO_2 of 0.4 μm and SiO_2 of 5 μm in diameter. Bao et al. [10] examined a double-layer coating using TiO_2, SiO_2, and SiC nanoparticles to improve cooling on Earth. These coatings were created for different surfaces. It emphasized the necessity for surfaces to have high solar reflectivity and emissivity in the 8–13 μm 'sky window' for effective thermal management. Theoretical models predict temperature drops of 17 °C at night and 5 °C during the day, with experiments showing good cooling on aluminum and black surfaces. Yin et al. [11] presented a double-layer radiative cooling emitter made of SiO_2 and TiO_2, achieving an emissivity of 96.42% and a net cooling power of 164.39 W/m^2, resulting in a temperature reduction of 9.6 °C. Cherqi et al. [12] conducted a combined numerical and experimental study to evaluate the radiative cooling performance of single-layer SiO_2 coatings on aluminum substrates under humid climatic conditions in Rabat, Morocco. Using COMSOL Multiphysics simulations validated by outdoor measurements, they tested four coating thicknesses and observed nighttime surface temperature reductions of 2.4–3.15 °C below ambient air and 1.4–4.23 °C compared to uncoated aluminum.

In this paper, different RC samples was constructed on single layer and double layers configurations for testing for daytime and nighttime under real time conditions in the ENSAM Rabat-Morocco. The objective of this study is to understanding the thermal behavior RC samples and evaluating the temperature of RC sample in comparison of the air ambient and the uncoated aluminum temperature.

2 Overview of RSC

2.1 Principles of Radiative Cooling

Theoretically, any thermal entity possessing a temperature exceeding absolute zero (0 K) will inherently generate thermal radiation when subjected to electromagnetic radiation. According to Kirchhoff's law of thermal radiation, the emissivity and absorptivity of all bodies reach parity in a state of thermal equilibrium, functioning as thermal emitters and absorbers with identical efficacy. Radiative cooling exemplifies a technological advancement that is predicated upon these well-established principles [7].

To realize the phenomenon of radiative cooling, two principal concepts warrant careful examination. Firstly, in accordance with the second law of thermodynamics, thermal energy is inherently transferred from bodies of elevated temperature to those of diminished temperature. Consequently, thermal energy from terrestrial entities exhibiting temperatures around 300 K can invariably dissipate to environments characterized by lower temperatures, with outer space representing the coldest region at approximately 3 K. Secondly, there exists a transparent atmospheric window (8–13 μm) within the atmosphere that facilitates the exchange of thermal energy between the Earth and outer space via the mechanism of infrared radiation [7, 12, 13].

Figure 1 presents the thermal energy exchange between a radiator on Earth, its immediate surroundings, and the cold sink of outer space.

2.2 Previous Studies

Many researchers have dedicated their efforts to creating radiative cooling designs by exploring various structures such as multilayer films, microspheres, and coatings made from nanoparticles [14–16], with the goal of providing uninterrupted cooling throughout the day and night. Table 1 provides an overview of significant studies on materials developed for nocturnal and daytime radiative cooling.

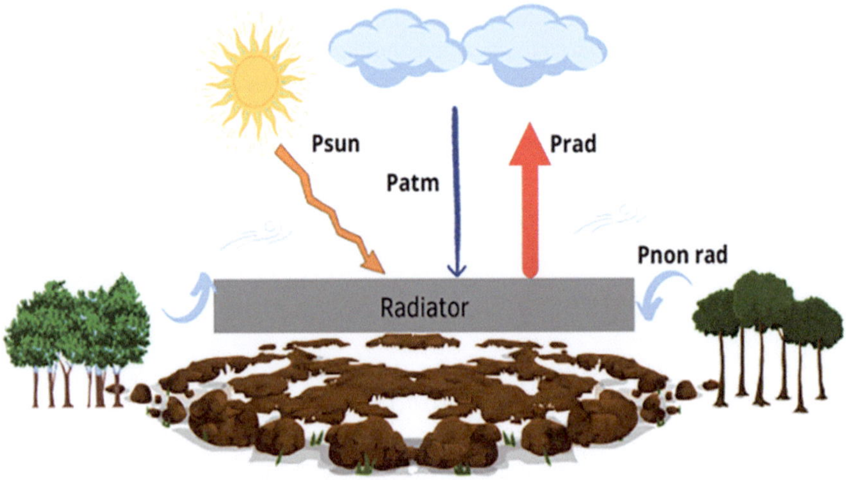

Fig. 1 Heat flux balance of radiative cooling [12]

3 Methodology

3.1 Selection of Materials and Coating Fabrication Technique

Figure 2 Shows RC samples are designed with the both single-and double-layer coatings. These coatings are used to apply on aluminum substrates. The aluminum substrates with dimensions of 11 × 11 cm. However, most previous studies have focused on using reflective substrates, such as aluminum, to attain particular radiative characteristics [10, 12, 27, 34]. The coating (1) is consisted of 0.2 g Silicon dioxide single layer. The Coating (2) is made from 0.3 g of Titanium dioxide single layer. The Coating (3) is composited of double layers: The top layer is formed from 0.3 g of titanium dioxide and the bottom layer from 0.2 g of Silicon dioxide. The last coating (4) is constructed from layer of varnish spray. The choice of these materials is crucial because of good optical properties allowing for efficient heat dissipation through thermal radiation in the infrared spectrum especially in the atmospheric window, while reflecting solar radiation in the visible spectrum. In addition, the spraying coating technique [10, 12, 35] was chosen for our experiment due to its ease of use and suitability for potential home applications. During the spraying process, a consistent spray distance of 15–30 cm was maintained, which ensured the formation of densely packed particles on the substrates. After coating, the samples were left to dry at room temperature (25 °C) for 24 h before undergoing further testing [12] as shown Fig. 3

Table 1 Summary of key studies on nocturnal and daytime radiative cooling

References	Cooler structure	Performance data	Cooling performance notes
Catalanotti et al. (1975) [17]	Thin film (12.5 μm) of TEDLAR (polyvinyl fluoride plastic) deposited on an evaporated aluminum substrate	$\Delta T_{min} = -12$ °C	They tested their structures during the day and found a 15 °C drop in temperature compared with the uncovered substrate. uncovered substrate
Granqvist and Hjortsberg (1981) [18]	Film mince (1 μm) de SiO évaporé sur un substrat d'aluminium	$\Delta T_{min} = -14$ °C $P_{tot} = 61$ W/m^2	
Grenier(1979) [19]	Thin film (340 μm) of TPX (polymethylpentene) deposited on an aluminum substrate		
Granqvist et al. (1982) [20]	Si_3N_4 thin film deposited on an aluminum substrate		
Lushiku et al. (1984) [21]	Gas plates (0.1–50 cm thick) ammonia (NH_3), ethylene (C_2H_4) and ethylene oxide (C_2H_4O) supported by a reflective aluminum plate		Temperature drops of 8.5 °C were observed between heat exchanger inlet and outlet
Eriksson et al. (1984) [22]	Thin film (1.3 μm) of silicon oxynitride ($SiO_{0.6}N_{0.2}$) evaporated on an aluminum substrate	$\Delta T_{min} = -16$ °C	
Kimball(1985) [23]	White TiO_2 paint and black paint	$\Delta T_{min} = -11$ °C for $TiO2_2$paint $\Delta T_{min} = -6$ °C for black paint	White TiO_2 paint has an emissivity close to that of black body paint throughout the thermal IR range
Eriksson et al. (1985) [24]	An aluminum substrate coated with layers of SiO_2 and $SiO_{0.25}N_{1.52}$ (each layer 0.7 μm thick)		
Diatezua et al. (1996) [25]	An aluminum substrate covered with multilayer structures of silicon oxynitride and SiO_2		Maximum cooling heat flux of 118–125 W/m^2 is calculated
Tazawa et al. (1998) [26]	An aluminum substrate coated with SiO and doped with tungsten vanadium dioxide ($V_{1-X}\,W_X\,O_2$) bilayers		

(continued)

Table 1 (continued)

References	Cooler structure	Performance data	Cooling performance notes
Gentle et Smith (2010) [27]	An aluminum substrate covered with a 25μm-thick fim of doped polyethylene. 25 μm-thick doped polyethylene. The film contains a mixture of SiO_2 and SiC nanoparticles (approx. 50 nm in diameter)	$\Delta T_{min} = -45$ °C $P_{tot} = 75$ W/m ^{2}	Performance indicators are simulation results, not actual experimental results
Hossain et al. (2015) [28]	Two-dimensional periodic conical metal-dielectric nanostructure consisting of 7 alternating layers of aluminum (30 nm thick) and germanium (110 nm thick) on an aluminum substrate		Although the reflection profiles presented show a high potential for night-time radiative cooling, the idea of adding a broadband mirror to the top of the structure and using it for daytime cooling is not cost-effective
Miyazaki et al. (2016) [29]	An aluminum substrate coated with a thin layer of Si_2N_2O	$\Delta T_{min} > -1$ °C	For top layer thicknesses ranging from 6.5 to 35 μm, the structure was tested for night-time cooling. The results did not show effective cooling
Zou et al. (2017) [30]	A plasmonic emitter consisting of an array of phosphorus-doped N-type silicon cubes coated with a thin layer of silver on a silver back reflector	$\Delta T_{min} = -10$ °C	Although the simulation results are close to the ideal IR emitter characteristics, the measurement results show a high emissivity outside the atmospheric window
Bao et al. (2017) [10]	Double-layer covering of SiO_2, TiO_2 and SiC nanoparticles,	$\Delta T_{min} = -17$ °C at night $\Delta T_{min} = -5$ °C	The RC structure demonstrating solar reflectivity of 90.7% while its emissivity was higher than 90%
Perrakis et al. (2020) [31]	Surface patterning with TiO_2 in crystalline silicon-based photovoltaic systems to improve cooling efficiency		The structure leading to high emissivity (between 95.85% and 96.42%), independent of incidence angles or metal substrates

(continued)

Table 1 (continued)

References	Cooler structure	Performance data	Cooling performance notes
Benlattar et al. (2021) [32]	A double-layer coating with an Al_2O_3 bottom layer embedded with Ni nanoparticles to give high mid-infrared emissivity and a top layer of TiO_2 and SiO_2 to block solar radiation, further improving surface cleanability	$\Delta T_{min} = -30$ °C, during the day $\Delta T_{min} = -62.3$ °C, during the night	Daytime and nighttime cooling power densities were more than 75 W/m^2 and 129 W/m^2
Mishra et al. (2023) [33]	Radiative cooling (RC) paint coating developed from titanium dioxide (TiO_2) and polydimethylsiloxane (PDMS), focusing on its cooling capabilities under varying environmental conditions	$\Delta T_{min} = -11.5$ °C	

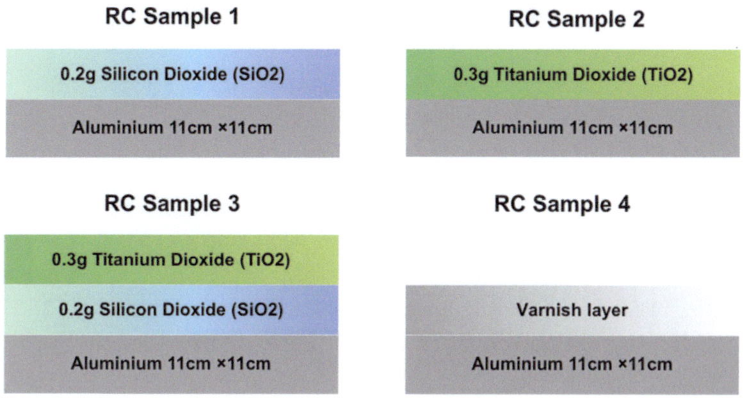

Fig. 2 RC samples with the both single-and double-layer coating

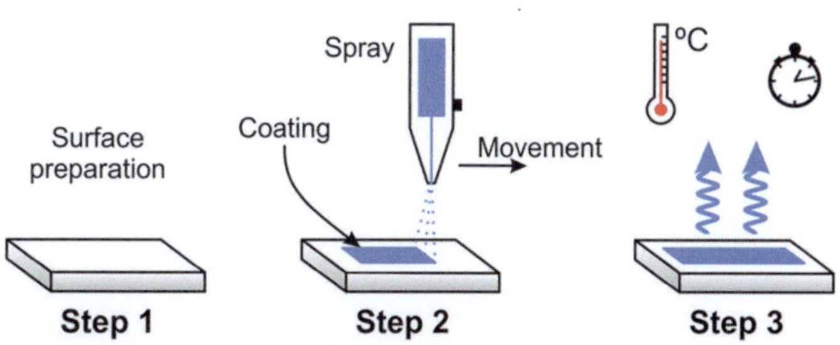

Fig. 3 Spraying coatings method [36]

Fig. 4 Test setup of RC Sample

3.2 *Experiment Configuration and Testing Protocol*

To measure the temperature of the radiative cooling RC samples, PT100 thermocouples were placed beneath the aluminum substrate, with thermal paste applied to improve heat transfer between the RC sample surface and the thermocouples. These PT100 sensors were connected to HOBO U12-600 data loggers, which have a temperature measurement accuracy of ± 0.4 °C for temperatures above 25 °C. The HOBO devices were calibrated and programmed to record temperature readings every 15 min over 24-h periods using HOBOware software. Additionally, a weather station was installed at ENSAM Rabat to monitor real-time climatic conditions [12]. The RC samples were tested in Rabat, Morocco (latitude 34°00′47″ N, longitude 6°49′57″ W), to evaluate their performance under a temperate Mediterranean climate. The testing spanned a 48-h period from 4:00 PM on June 18 to 4:00 PM on June 20, allowing for comprehensive data collection across varying environmental conditions as shown in Fig. 4.

The experiments were conducted during the summer, when there is a greater need for cooling. The RC samples' thermocouple temperature readings were continually recorded and contrasted with the uncoated aluminum sample and the ambient temperature [12].

3.3 Data Collection and Analysis

By calculating the temperature difference between the samples' surface and the surrounding environment, the coating's cooling performance was determined. The analysis was conducted using (1):

$$\Delta T = T_{sample} - T_{amb} \tag{1}$$

where T_{amb} is ambient temperature, and T_{sample} is the temperature of the coated or uncoated sample.

A negative ΔT indicated effective cooling, where the surface sample temperature dropped below the ambient temperature due to radiative cooling effects. The temperature profiles were plotted from the collected data, showing the cooling achieved over time. To determine whether the differences in cooling performance among the various coatings were statistically significant, a statistical analysis was performed [12].

4 Results and Discussion

In this test, the RC Samples are compared under the same climatic conditions Rabat, Morocco (L: 34°00′47″ N, λ: 6°49′57″ W) during the day/night of the 18 to 20 June 2021. The first sample has 0.2 g of silicon dioxide single layer coated on aluminum substrate, the second sample has 0.3 g of titanium dioxide single layer on exposed surface of aluminum, the third sample has double layers of 0.2 g SiO_2 and 0.3 g TiO_2 on exposed surface of aluminum and the last sample has a varnish layer coated on the top of aluminum substrate. This test is designed to analyze the temperature at the base of each sample. Figure 5 shows the comparison the temperature of RC samples with the ambient temperature and aluminum uncoated. The 0 h in the X axis refers to the starting time of the test and 48 h corresponds the ending time of the test.

During the daytime, particularly during peak solar radiation hours, our observations indicate that all RC samples exhibit surface temperatures slightly higher than the ambient temperature see Table 1. This suggests that while the RC coatings possess some reflective capacity, their relatively high absorption in the visible solar spectrum limits their effectiveness in reflecting solar radiation. At night and the absence of solar radiation, the temperature of RC samples is below the ambient temperature. This implies that the RC samples can radiate heat efficiently to outer space at night.

For instance, the RC sample 1 compared with ambient environment had a maximum cooling differential temperature of 3.04 °C. Moreover, the RC Sample 2 achieved a maximum cooling differential temperature of 3.32 °C below ambient temperature. Thereafter, the RC Sample 3 exhibited a significant performance, achieving a maximum cooling differential temperature of 4.2 °C. Furthermore, the RC Sample 4 demonstrated even greater performance, achieving a maximum cooling

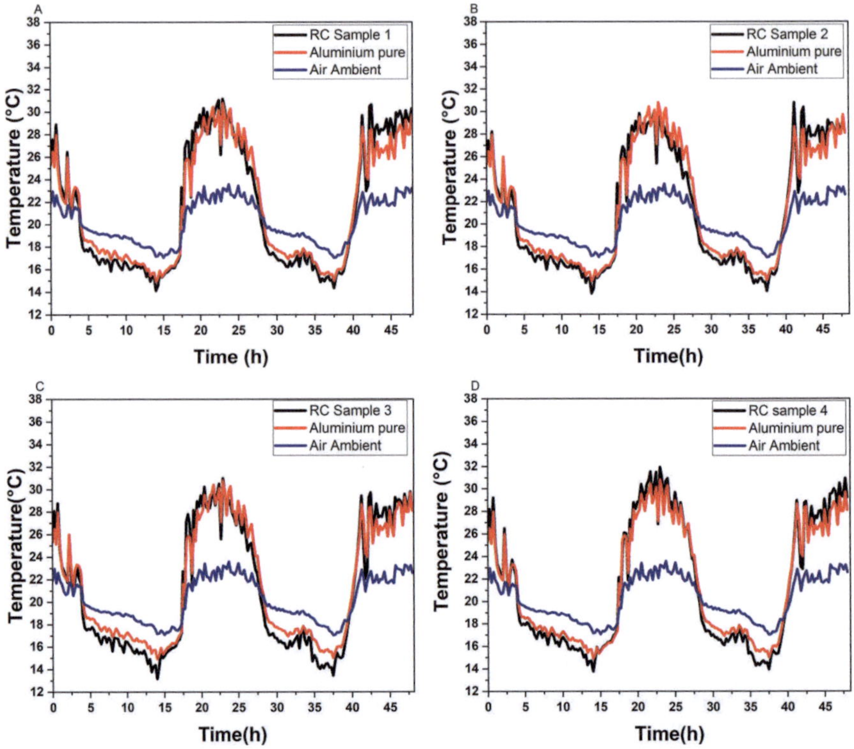

Fig. 5 Temperature evolution of the samples

differential of 3.65 °C under ambient temperature Lastly, the uncoated aluminum compared with ambient environment had a maximum cooling differential of 2. 6 °C. It can be seen clearly that RC samples achieve only nighttime radiative cooling as shown in Fig. 3 and reported in Table 2.

Table 3 presents temperature difference between the sample and the aluminum uncoated temperature during night, the temperature RC sample1 decreased by 2 °C compared to aluminum pure. Though the temperature RC sample 2 recorded a drop of 2.4 °C compared to the aluminum pure. While RC the temperature sample 3 showed a diminution temperature of 2.2 °C relative to the aluminum pure. In contrast, RC sample 4 demonstrated a reduction of 1.3 °C in comparison to the aluminum pure. This analysis emphasizes the potential of various coatings to enhance the cooling performance of RC samples surface.

Table 2 Temperature difference between the sample and the ambient temperature during peak solar hours

Samples	RC sample 1	RC sample 2	RC sample 3	RC sample 4	Aluminum pure
ΔT (°C)	8.2	8	7.8	7.64	8.7

Table 3 Temperature difference between the sample and the ambient temperature during nighttime

Samples	RC sample 1	RC sample 2	RC sample 3	RC sample 4	Aluminum pure
ΔT (°C)	– 3.04	– 3.32	– 4.2	– 3.65	– 2.6

Table 4 Temperature difference between the sample and the aluminum uncoated temperature during nighttime

Samples	RC sample 1	RC sample 2	RC sample 3	RC sample 4
ΔT (°C)	– 2	– 2.4	– 2.2	– 1.3

Table 5 Temperature difference between the sample and the aluminum uncoated temperature during peak solar hours

Samples	RC sample 1	RC sample 2	RC sample 3	RC sample 4
ΔT (°C)	2.4	2.1	1.9	1.7

During peak solar hours, the temperature difference between the RC samples and the uncoated aluminum surface was clearly observed. RC Sample 1 exhibited a temperature increase of 2.4 °C compared to pure aluminum. Similarly, RC Sample 2 recorded a rise of 2.1 °C, while RC Sample 3 showed an increase of 1.9 °C relative to the uncoated aluminum. In contrast, RC Sample 4 demonstrated a smaller temperature rise of 1.7 °C. This analysis highlights the influence of solar absorption on the surface temperature of the RC samples and reflects the limited capacity of the coatings to mitigate heat gain under direct sunlight. Table 4 shows temperature difference between the sample and the aluminum uncoated temperature during nighttime (Table 5).

5 Conclusion

The RC (radiative cooling) samples were tested under real-time environmental conditions in Rabat, Morocco, with various surface coatings applied. The main conclusions are as follows:

- All RC samples exhibited radiative cooling exclusively during nighttime.
- During nighttime, both single-layer and double-layer coated RC samples achieved lower sub-ambient temperatures compared to the uncoated aluminum substrate and ambient air.
- The study demonstrated the potential of different coatings to enhance the radiative cooling performance of RC surfaces.
- The TiO_2 particle size used in this study was not optimal for effective solar reflection, which limited the samples' ability to achieve daytime radiative cooling.

References

1. Elnagar, E., et al.: Framework to assess climate change impact on heating and cooling energy demands in building stock: a case study of Belgium in 2050 and 2100. Energy Build. **298**, 113547 (2023)
2. Kian Jon, C., Islam, M.R., Kim Choon, N., Shahzad, M.W.: Advances in Air Conditioning Technologies: Improving Energy Efficiency. Springer Singapore, Singapore (2021). https://doi.org/10.1007/978-981-15-8477-0
3. Castro, P.J., Aráujo, J.M.M., Martinho, G., Pereiro, A.B.: Waste management strategies to mitigate the effects of fluorinated greenhouse gases on climate change. Appl. Sci. **11**, 4367 (2021)
4. Khean, C.C., Leman, Z., Adam, N.M., Supeni, E.E.: Effects of building materials on building thermal load in Malaysian institutional library. ARFMTS **107**, 191–207 (2023)
5. Dhawan, V., et al., Low Carbon Development Pathways for Cooling: Leveraging Kigali Amendment across Residential Applications, p. 1985251. None. ark:/13030/qt4f2090jn https://www.osti.gov/servlets/purl/1985251/ (2023) https://doi.org/10.2172/1985251
6. Liu, J., et al.: Recent advances in the development of radiative sky cooling inspired from solar thermal harvesting. Nano Energy **81**, 105611 (2021)
7. Zhao, B., Hu, M., Ao, X., Chen, N., Pei, G.: Radiative cooling: a review of fundamentals, materials, applications, and prospects. Appl. Energy **236**, 489–513 (2019)
8. Aili, A., et al.: Passive daytime radiative cooling: moving beyond materials towards real-world applications. Next Energy **3**, 100121 (2024)
9. Cheng, Z., et al.: Optical properties and cooling performance analyses of single-layer radiative cooling coating with mixture of TiO_2 particles and SiO_2 particles. Sci. China Technol. Sci. **64**, 1017–1029 (2021)
10. Bao, H., et al.: Double-layer nanoparticle-based coatings for efficient terrestrial radiative cooling. Sol. Energy Mater. Sol. Cells **168**, 78–84 (2017)
11. Yin, H., Fan, C.: Realization of an efficient radiative cooling emitter with double layer inorganic SiO_2 and TiO_2 metamaterial. Results Phys. **45**, 106216 (2023)
12. Cherqi, O., Idouanaou, A., Malha, M., Kardellass, S.: Numerical modeling and experimental study of SiO_2 radiative cooling coatings on aluminum substrate in humid conditions: a case study in Morocco Mediterranean climate. Energy Build. **339**, 115771 (2025)
13. Hossain, Md.M., Gu, M.: Radiative cooling: principles, progress, and potentials. Adv. Sci. **3**, 1500360 (2016)
14. Xiang, B., et al.: 3D porous polymer film with designed pore architecture and auto-deposited SiO_2 for highly efficient passive radiative cooling. Nano Energy **81**, 105600 (2021)
15. Bijarniya, J.P., Sarkar, J., Maiti, P.: Review on passive daytime radiative cooling: fundamentals, recent researches, challenges and opportunities. Renew. Sustain. Energy Rev. **133**, 110263 (2020)
16. Sun, X., Sun, Y., Zhou, Z., Alam, M.A., Bermel, P.: Radiative sky cooling: fundamental physics, materials, structures, and applications. Nanophotonics **6**, 997–1015 (2017)
17. Catalanotti, S., et al.: The radiative cooling of selective surfaces. Sol. Energy **17**, 83–89 (1975)
18. Granqvist, C.G., Hjortsberg, A.: Radiative cooling to low temperatures: general considerations and application to selectively emitting SiO films. J. Appl. Phys. **52**, 4205–4220 (1981)
19. Grenier, Ph. Réfrigération radiative. Effet de serre inverse. Rev. Phys. Appl. (Paris) **14**, 87–90 (1979)
20. Granqvist, C.G., Hjortsberg, A., Eriksson, T.S.: Radiative cooling to low temperatures with selectivity IR-emitting surfaces. Thin Solid Films **90**(2), 187–90 (1982)
21. Lushiku, E.M., Eriksson, T.S., Hjortsberg, A., Granqvist, C.G.: Radiative cooling to low temperatures with selectively infrared-emitting gases. Sol. Wind Technol. **1**, 115–121 (1984)
22. Eriksson, T.S., Lushiku, E.M., Granqvist, C.G.: Materials for radiative cooling to low temperature. Sol. Energy Mater. **11**, 149–161 (1984)
23. Kimball, B.A.: Cooling performance and efficiency of night sky radiators. Sol. Energy **34**, 19–33 (1985)

24. Eriksson, T.S.: Surface coatings for radiative cooling applications: silicon dioxide and silicon nitride made by reactive RF-sputtering. Sol. Energy Mater. **12**(5), 319–25 (1985)
25. Diatezua, D.M., et al.: Silicon oxynitride multilayers as spectrally selective material for passive radiative cooling applications. Sol. Energy Mater. Sol. Cells **40**(3), 253–259 (1996)
26. Tazawa, M., Jin, P., Yoshimura, K., Miki, T., Tanemura, S.: NEW MATERIATLODOEBSTIGANINWTIETMHPVE1R−AxWTUxOR2EFSILTMABFILOI RTYSKYRADIATOR
27. Gentle, A.R., Smith, G.B.: Radiative heat pumping from the earth using surface phonon resonant nanoparticles. Nano Lett. **10**, 373–379 (2010)
28. Hossain, M.M., Jia, B., Gu, M.: A metamaterial emitter for highly efficient radiative cooling. Adv. Opt. Mater. **3**, 1047–1051 (2015)
29. Miyazaki, H., Okada, K., Jinno, K., Ota, T.: Fabrication of radiative cooling devices using Si_2N_2O nano-particles. J. Ceram. Soc. Japan **124**, 1185–1187 (2016)
30. Zou, C., et al.: Metal-loaded dielectric resonator metasurfaces for radiative cooling. Adv. Opt. Mater. **5**, 1700460 (2017)
31. Perrakis, G., et al.: Passive radiative cooling and other photonic approaches for the temperature control of photovoltaics: a comparative study for crystalline silicon-based architectures. Opt. Express **28** (2020)
32. Benlattar, M., Ibourk, I., Adhiri, R.: Simple double-layer coating for efficient daytime and nighttime radiative cooling. Atmosphere **12**, 1198 (2021)
33. Mishra, B.R., Sundaram, S., Sasihithlu, K.: Impact of Experimental and Environmental Factors on Performance of Radiative Cooling Coating
34. Granqvist, C.G., Hjortsberg, A., Eriksson, T.S.: Radiative cooling to low temperatures with selectivity IR-emitting surfaces. Thin Solid Films **90**, 187–190 (1982)
35. Zhao, X., He, X.D., Sun, Y., Wang, L.D.: Carbon nanotubes doped SiO_2/SiO_2–PbO double layer high emissivity coating. Mater. Lett. **65**, 2592–2594 (2011)
36. Frederichi, D., Scaliante, M.H.N.O., Bergamasco, R.: Structured photocatalytic systems: photocatalytic coatings on low-cost structures for treatment of water contaminated with micropollutants—a short review. Environ. Sci. Pollut. Res. **28**, 23610–23633 (2021)

Sustainability of Photovoltaic Materials and Smart Monitoring: Case Study in Morocco

Ali Ait Ali, Mbarek Chahboun, Mohcine Abouyaakoub, Youssef Ouhassan, and Hicham Hihi

Abstract In This study explores real-time monitoring of solar energy production via the Internet of Things (IoT), which allows for the continuous assessment and monitoring of the performance and degradation of photovoltaic materials in arid and semi-arid environments in Morocco. Analysis of the collected data, such as current, voltage, and power, highlights the influence of local climatic conditions (heat, dust, humidity) on the properties of semiconductor materials and, more specifically, encapsulate. This intelligent monitoring is an essential tool for better understanding material aging, optimizing maintenance, and guiding the choice of more resistant materials for arid or semi-arid environments. A Thing Speak platform processes the information measured by the sensors. This data, including voltage and current, is then sent in real time via a Wi-Fi module. To receive and visualize the data, we used the Thing Speak platform, which enables data to be accessed online from any location. The graphs generated showed a very satisfactory evolution of voltage, current and power parameters. The results obtained are very promising, and it has been found that our solution offers numerous advantages over traditional solutions, such as the ability to present data in graphical form. In this way, it can reduce the need for human interference in order to ensure that potential data reading-related problems from multiple devices, to reduce costs and minimize time. As a result, the suggested system can facilitate data acquisition and storage, as well as the creation and customization of re-ports that can be evaluated to improve the performance of solar power systems.

Keywords Sustainability · Monitoring system · Photovoltaic materials · Thing-speak · Internet of things

---

A. Ait Ali (✉) · M. Chahboun · M. Abouyaakoub · H. Hihi
Laboratory of Engineering, Systems and Applications, National School of Applied Sciences, Sidi Mohamed Ben Abdellah University, Fez, Morocco
e-mail: ali.aitali@usmba.ac.ma

Y. Ouhassan
Department of Physics, Faculty of Sciences, Laboratory of Electronic Systems, Information Processing, Mechanics and Energy, Ibn Tofail University of Kenitra, Kenitra, Morocco

E. El Mehdi and M. Boutahir (eds.), *Advanced Materials for Sustainable Energy and Engineering*, Engineering Materials, https://doi.org/10.1007/978-3-032-10069-6_3

1 Introduction

The energy transition is accelerating faster in Morocco, where solar energy plays a strategic role. However, the sustainability of photovoltaic systems installed in hot and dusty environments remains a challenge. In this context, the combination of advanced materials and smart monitoring technologies is becoming essential. The Internet of Things will provide a good real-time monitoring of the efficiency of photovoltaic panels, as well as better comprehension of the degradation of photovoltaic materials.

As behalf of our work in the optimization of energy efficiency, particularly photovoltaic panels, we are studying the effect of factors such as temperature, who has a big effect on the production of photovoltaic modules. Indeed, the increase in temperature to 45 °C can result in a decrease in power by around 20% [1] on photovoltaic panels. We are also exploring the integration of innovative solutions, such as cooling, to improve the efficiency of these systems. Moreover, in the field of wind energy, we analyze the application of the Back stepping technique, which makes it possible to solve robustness issues associated with changes in operating conditions and system parameters [2]. In the context of photovoltaic system monitoring, several works can be found in the literature over the last two decades proposing different monitoring and communication systems. The authors of [3, 4] A automated system for collecting data has been proposed to monitor weather conditions and the performance of a renewable energy system. This system uses a set of sensors to measure weather data (temperature, humidity, etc.) and electrical parameters (photovoltaic voltage and current, etc.). The collected data is first processed by precision electronic circuits and then transmitted to a computer via an acquisition card. This automated and cost-effective acquisition system was designed using an innovative approach based on virtual instrumentation, as presented in [5]. In the context of the PVSAT-2 research project, a new system for remote performance monitoring has been developed, including automated detection of photovoltaic system failures. This procedure is based on the use of satellite radiation measurement data, and can be used to simulate the energy efficiency of a photovoltaic system [6]. They have also developed a data processing system for monitoring solar energy and the global environment. An innovative AT mega 16 microcontrollers is used to analyze 8 sensors in parallel at programmable intervals. This article presents an in-depth study of the project, its applications and the main specifications of the prototype system and programs [7]. Some researchers have proposed a programmable logic controller (PLC) which is suitable for Home Plug software, designed to manage and monitor photovoltaic systems in smart homes [8]. The authors of [9] recommended a simple PV monitoring system. This consists of a laptop computer connected to a low-cost microcontroller board (AT MEL ATMEGA16) via a serial cable. The sensors are emulated in a prototype board that interfaces with the microcontroller system. A graphical user interface program is developed in LabVIEW to monitor the data collected on the collected on the computer monitor. In [10], system design and development for a data acquisition system based on a low-cost microcontroller (PIC 16F877) technology for photovoltaic remote photovoltaic water supply systems. The communications are based on the GSM network and, in specific, on the short text message (SMS) option. The ambient temperature was measured using an analog sensor (LM35). Voltage and current were measured using Hall-effect transducers. Solar radiation in W/m^2 was As measured by the short-circuit current of a solar cell. In [11] have proposed an industrial PLC compatible with the Home Plug software for the management and monitoring of photovoltaic systems for the smart home. The author of [12] recommended a short-range monitoring and control system used for monitoring and cooling to eliminate the problem of solar panel heating using the Beagle Bone module. In [13], The researchers developed a low-cost system for monitoring photovoltaic installations, based on an Arduino UNO board. This system enables remote data collection, even in areas with no telecommunications network, by storing data locally on an SD card. It uses a temperature sensor (DS18B20) as well as current and voltage sensors to measure electrical and environmental parameters, which are then recorded by the data acquisition device. In [14] A smart grid is a network of interconnected electrical distribution systems that can be controlled remotely. It is designed to be more efficient and reliable than traditional electrical grids. Data acquisition and display were performed using the PLC system coupled with an SCADA interface, while data ex-traction and transmission were handled by the Arduino MEGA 2560 R3 platform. In addition, other researchers have suggested a wireless solution based on the Wi-Fi 802.11 standard to develop a wireless local area network (WLAN). This technology is widely used around the world due to its popularity and high throughput, which can reach between 11 and 54 Mb/s, with a range of up to 100 meters. [15]. In [16], the authors have developed an open-source SCADA design for the implementation of a photovoltaic monitoring system. This structure included various components, such as ACS712 Hall-effect current sensors, an ESP32 Thing microcontroller, Raspberry Pi devices as well as a Wi-Fi network gateway. The ESP32 system microcontroller is designed to gather the electric data from the different devices, to transfer this data to an IoT (Internet of Things) platform via the Wi-Fi network for storage. The new development of a photovoltaic monitoring system on a small scale, providing the user with crucial information on the return on investment of the plant, was presented. The design is based on a PIC16F877a microcontroller and LabVIEW. The data acquisition system comprises a voltage sensor, a current sensor and an ambient temperature sensor. Solar radiation is observed on the basis of electrical panel measurements.

When we install a photovoltaic system, we are monitoring to find out if this system provides the maximum power, which helps us to see if the system is working efficiently or not. The IoT monitoring system we present in this paper really helps to supervise the real time data of the solar panel installed in the house or the workplace. The present study is organized into three parts. The first part presents a brief description of the project and its objectives, as well as an overview of the methods and hardware used to monitor the maximum power of the school's supply. The second part of the paper is focused on the analysis of the results received. Finally, the final section provides concluding remarks and suggestions for later work.

2 Materials and Methods

2.1 Objective of the Project and Maps Location of the Area

The school is located in an unfavorable rural area with limited access to electricity. This is the reason for installing a photovoltaic system. However, this system requires continuous, real-time monitoring. To meet this need in order to supervise the stability of the system and control its Maximum power. We have opted for a remote monitoring solution via Wi-Fi, in order to be able to intervene rapidly in the event of a malfunction.

2.2 Geographical Location of the Study Area

The study was conducted in a rural area of the region known for its hot, dry summers. A photovoltaic system with a capacity of 1 kWp is installed on the roof (Fig. 1). IoT sensors monitor the following variables: current, voltage, and power. The modules used are made of polycrystalline silicon, encapsulated in EVA (ethylene vinyl acetate) with a tempered glass front. The project has located in the rural district of Ksar Tafroukht, Rissani (31°10′5.32″N; 4°29′32.61″ W).

2.3 IoT Architecture

The IoT system is based on an ESP32 architecture coupled with a visualization platform (Thing Speak). Data is recorded every 5 minutes over a period of 1 day. We are going to implement a remote monitoring and control system via an Android platform, as well as other compatible platforms. The Bluetooth and Wi-Fi technology used in this system enables communication between different devices such as cell phones, laptops and desktop computers (Fig. 2). The Bluetooth and Wi-Fi module provides a communication link for data exchange between the hardware and the energy management unit. The system is implemented as below.

The monitoring system enables us to monitor data transmission in real time and quickly identify any problems. It also provides the opportunity to efficiently synchronize the hardware and software within the system.

In general, there are several methods for monitoring a renewable energy production system. Various communication modules are used, such as the Bluetooth communication module and the Wi-Fi module, which in fact enable (Fig. 3).

o Development of a system for instant monitoring of renewable energy production, consisting of solar panels measuring the current and voltage of each reliable source. The associated values are measured by the voltage and current sensor and processed by the Arduino or a microprocessor, which is communicated via a Module Wi-Fi to

Fig. 1 Location maps of the school

an online module (Fig. 4). They are also controlled by a computer and can be stored in a database or monitored in real time.

2.4 Model Obtained by a Test Simulation in Proteus 8

2.4.1 Components of Simulation

(See Fig. 5).

2.4.2 Ideal Model for Real Time Monitoring of Solar Panel Productivity

This system integrates solar cells which convert the light of the sun from energy, and installed in photovoltaic panels. We have one of these systems in our database. The sensor technology is used to monitor the voltage and current parameters. The values for the voltage and current are equal. All the information appearing on the

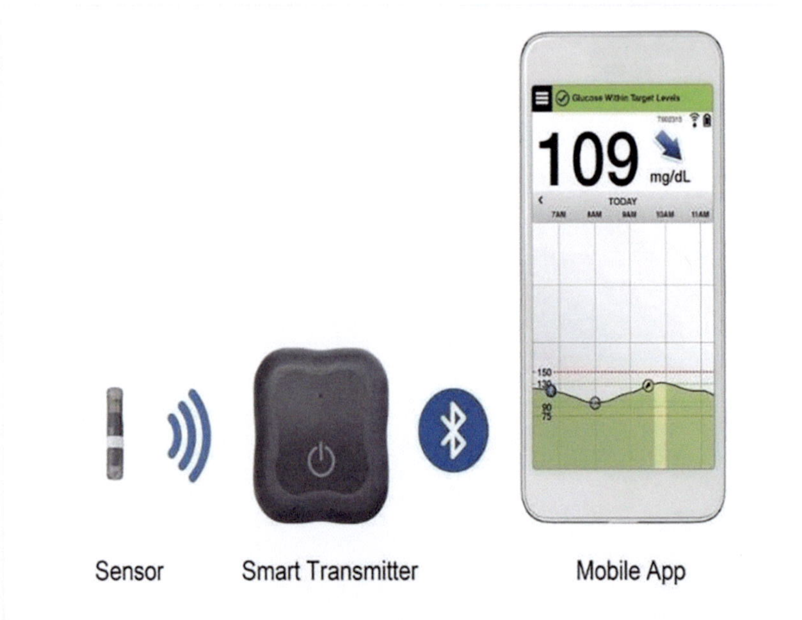

Fig. 2 Bluetooth and Wi-fi integration in monitoring

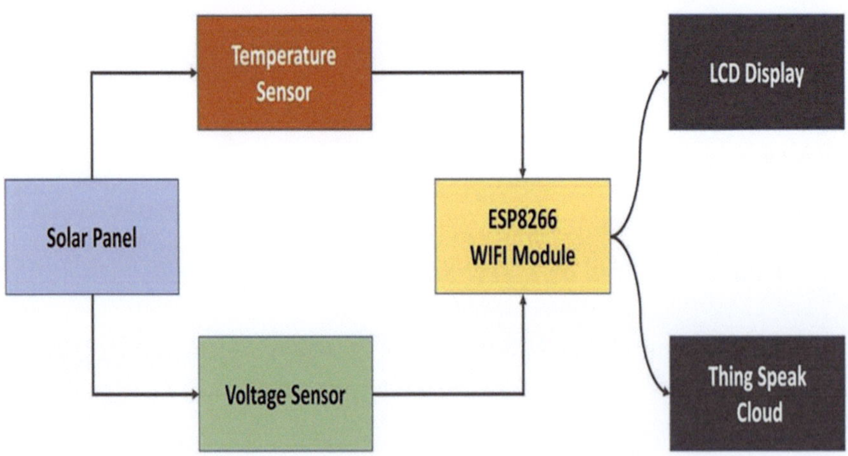

Fig. 3 Using Wi fi for visualization

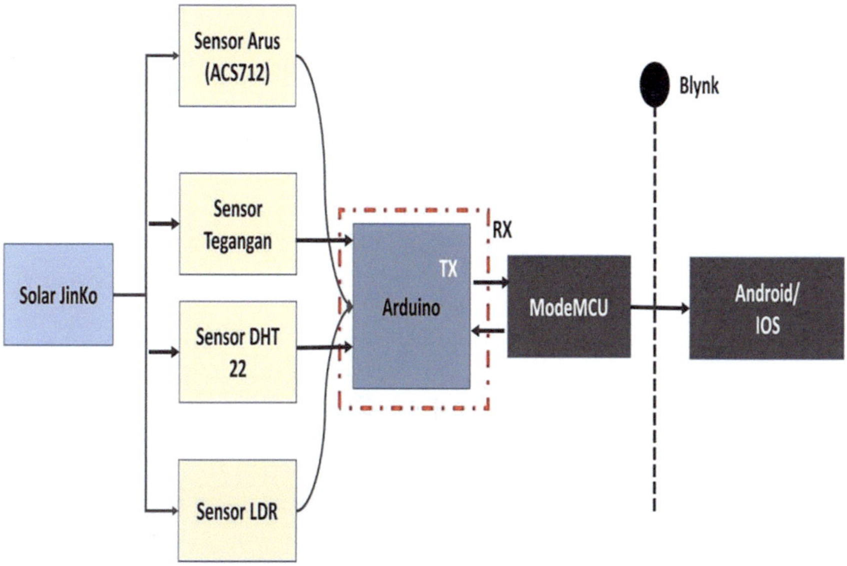

Fig. 4 Treatment by Arduino of the values measured by the sensors

LCD display is available here. Using the detectors to set up, when connected, an IoT device is also able to visualize a setting on the display, letting the user follow its value from any existing network (Fig. 6).

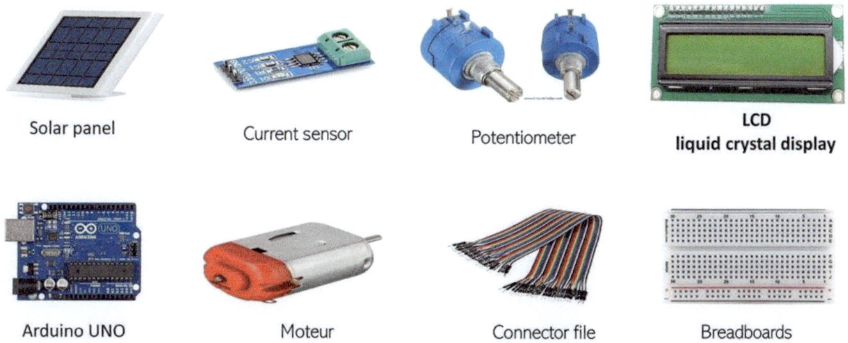

Fig. 5 Various components used for monitoring systems

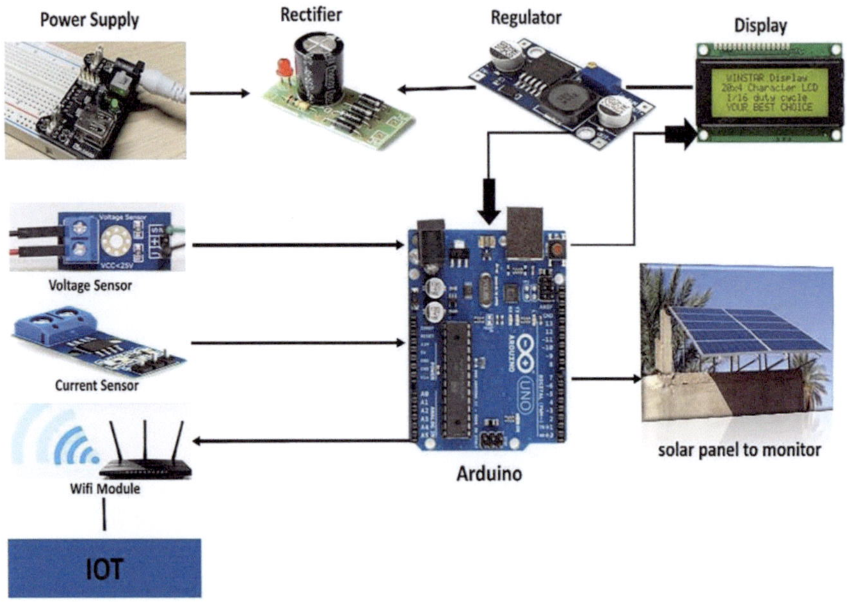

Fig. 6 Functional diagram of the IoT solar energy monitoring system

3 Results and Discussion

To achieve our objectives, we use the Proteus Professional program, which is principally used for the design and simulation of virtual boards. However, we use Proteus to import additional library files for specific system parts. The Arduino library UNO and the solar panel library have been imported.

3.1 Description of the Simulation

In Fig. 7 we present the simulation process used to simulate and examine the behavior of the system under study is explained in detail in this section. Via the potentiometer, the 24 V solar panel supplies current to the ACS712 collector, which detects the current value between 0 and 1024 (analog value). An additional voltage of 5 V DC is connected to the ACS712 current sensor. This current is then transferred to the DC motor via the ACS712 IP, and a second DC motor terminal is connected to the negative end of the solar panel.

Software part:

In our coding part, we must import Liquid-crystal.h because we used the LCD screen. Next, we create an instance of Liquid-crystal as:

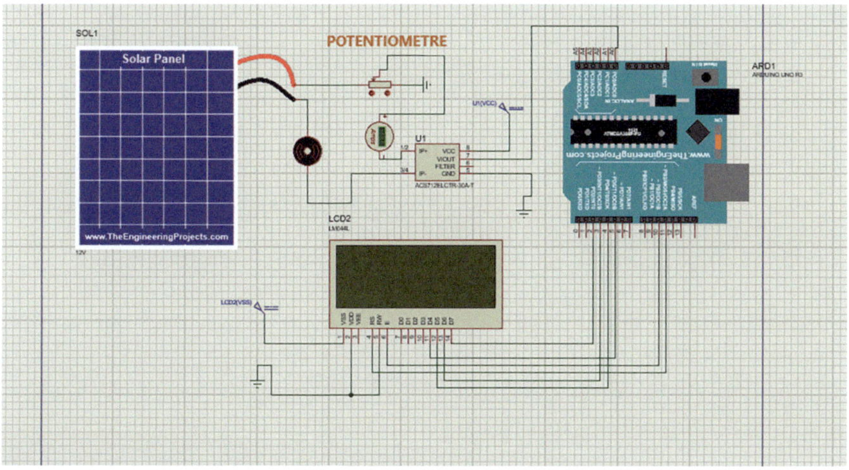

Fig. 7 Simulated circuit

Liquid-crystal LCD (11,10,5,4,3,2): the most important part is the calculation. We know that the ACS712 current sensor works with 5 V DC, so when there is no current in the IP+ and IP-terminals of the current sensor, the output voltage at VIOUT of the ACS712 is 2.5 V. This means we need to subtract 2.5 V from the voltage measured on the Arduino's A0 analog pin.

the "sensitivity" is in fact the internal resistance of the ACS712_30A current sensor. Equations:

$$P = V * I^2 \text{ with } V = I * R \tag{1}$$

The programs are saved on the programming computer as project data files with the extension. ino. In Arduino IDE versions prior to 1.0, they are stored with the. pde extension (Figs. 8 and 9). A minimal Arduino program, written in C/C++, consists of just two main functions:

```
// initialize the library with the numbers of the interface pins
LiquidCrystal lcd(11,10,5,4,3,2);

const int currentPin = A0; //  the input on analog pin A0
int sensitivity = 66; // use 100 for 20A Module and 66 for 30A Module
int adcValue= 0;
int offsetVoltage = 2500;//Sensor Callibration
double adcVoltage = 0;
double currentValue = 0;
```

Fig. 8 Parameters used

```
void loop()
{
  // Measure The Votage AND Current AND Power**********************************

  // read the input on analog pin A0:
  adcValue = analogRead(currentPin); // 0-1024
  // Convert the analog reading (which goes from 0 - 1023) to a voltage (0 - 5V):
  adcVoltage = (adcValue / 1024.0) * 5000; // mV
  currentValue = ( (adcVoltage - offsetVoltage) / sensitivity );

  int loadvoltage = currentValue * 12;

  int power = loadvoltage*currentValue*currentValue;
```

Fig. 9 Measuring current, voltage and power

Setup (): This function is executed only once when the device, called Tafroukht, is powered up or reset. It is used to configure variables, define input or output pin modes, and initialize the necessary libraries used in the sketch.

loop (): After executing the setup () function, the loop () function is called repeatedly to execute the main program.

It monitors the card until it's powered down or reset the programming code of the arduino uno card and sensors.

3.2 Simulated Circuit Design

Various phases of simulation

In this section, we present the various phases of simulation to ensure that the circuits really operate (Figs. 10 and 11).

We explained our system, how to install it and how to connect it to the Arduino Uno microcontroller.

We also looked at our program code and how the Arduino has a big role in extracting information from the solar panel and the method used to take and record this data.

The results of the simulation indicate that the ThingSpeak platform can be used to monitor settings such as current, voltage, power, humidity and temperature in a photovoltaic system. Data is collected and analyzed in real time, providing a detailed analysis of system performance (Figs. 12 and 13).

This study explores the use of IoT (Internet of Things) and embedded systems new technologies to optimize and monitor phophotovoltaic systems. It highlights the importance of data collection and analysis in improving the efficiency of solar installations.

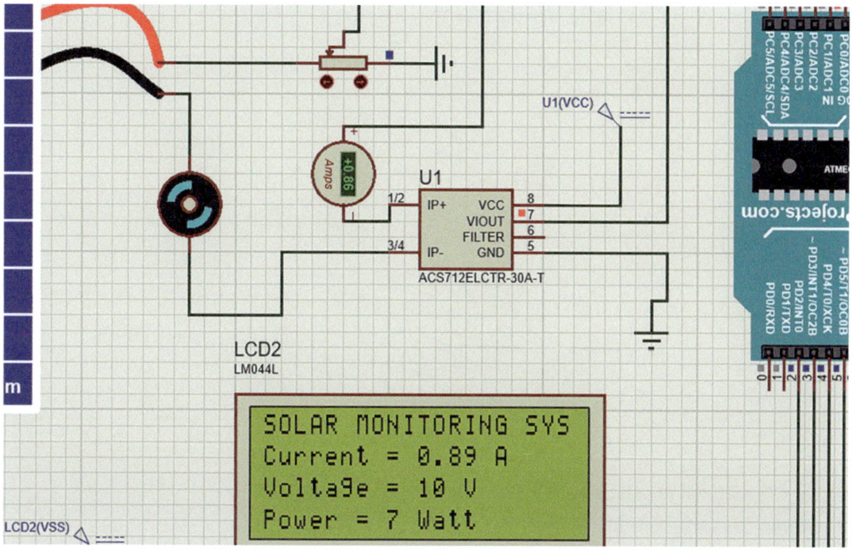

Fig. 10 The current value indicated by the current meter is the same as the value displayed by the LCD

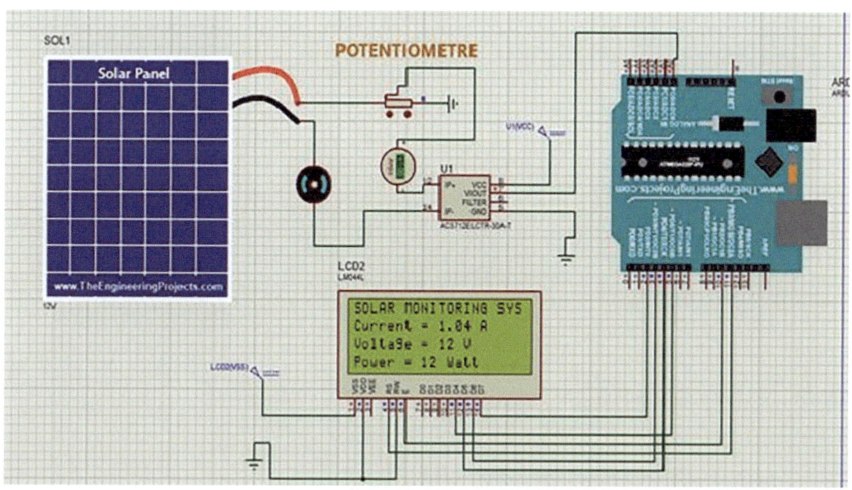

Fig. 11 The first simulation with a maximum potentiometer value

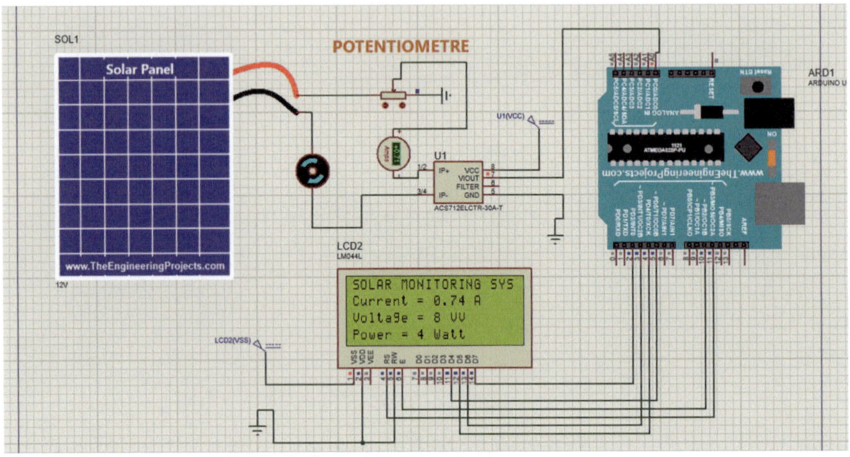

Fig. 12 The second simulation with an average potentiometer value

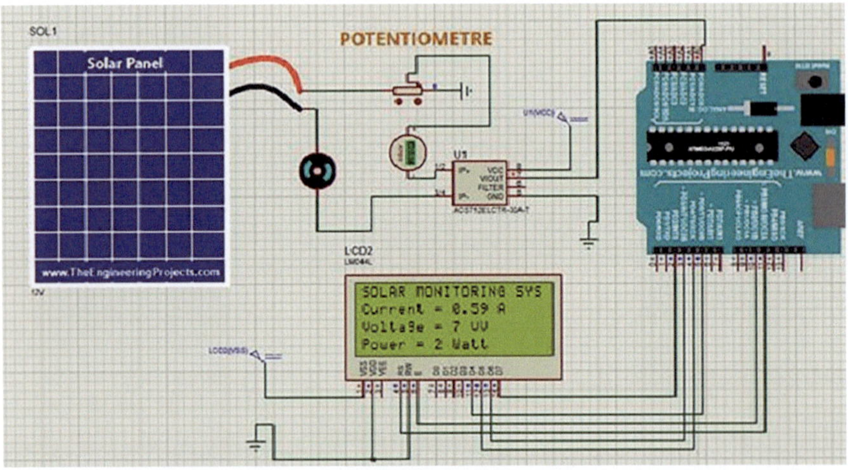

Fig. 13 The third simulation with a minimum potentiometer value

3.3 Data Collected by the Monitoring System

The Power

The photovoltaic systems work in isolation. An intelligent system provides remote monitoring for better detection of problems and more effective maintenance.

The data collected by SIM cards and transmitted via GSM can be used to monitor the voltage and power output of an inverter. The results then monitor power, voltage

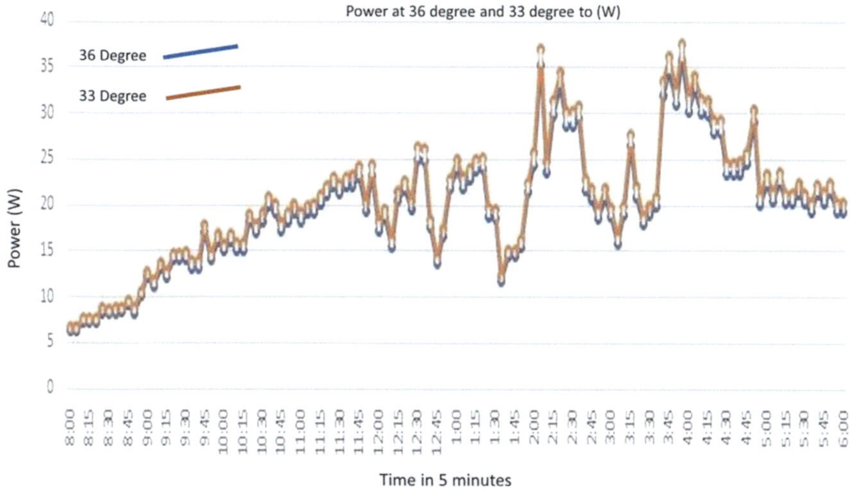

Fig. 14 The power connected through the day

and current performance as a function of time. However, a panel used in a non-optimal way is not at its maximum efficiency. So, the faster a problem is identified, the faster corrective action can be taken remotely or, in more serious cases, on-site, to limit the malfunction.

This Fig. 14 shows the power output of the solar panels in relation to the time (minutes). The value at the beginning of the experiment came to 6.35 watts., then began to increase progressively from 8 to 10 h. The power then falls again as a result of changes in current and voltage due to changes in solar radiation caused by natural factors. In this test, we can see that a system installed at an angle of 36° is more productive than one installed at 33°. This is due to the angle of inclination, which is optimal on this particular day.

The Currents

This graph shows the current (A) varying with time (in Minutes) in the solar panel.

As the current value is low in the first several minutes, the current value begins to increase from 9.10 AM until 5.02 PM, when it peaks at 2.66 A, and then gradually decreases until it stabilizes.

Voltage

This Fig. 15 shows a graph of the voltage (V) that varies with time (in minutes) in the solar panel.

We note that the voltage value starts to rise gradually from the beginning of the experiment until it reaches the value of 13.01 V at 11:30 a.m., then decreases and starts to rise another woman until it reaches its highest value, which is 14.21 V at 3:55 PM. It decreases until it progressively stabilizes due to natural solar panel changes.

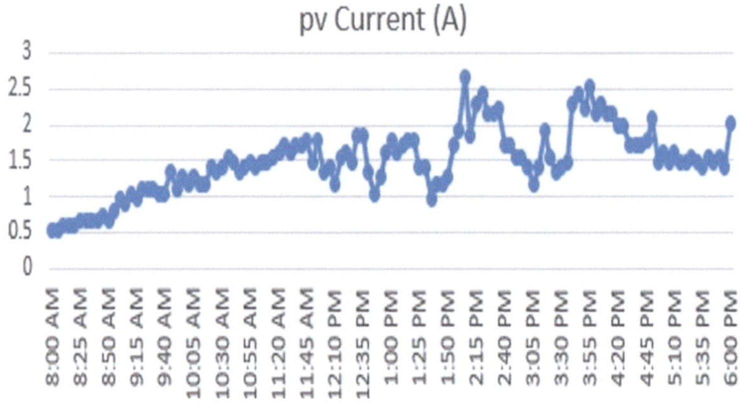

Fig. 15 PV current as a function of time

The aim is to evaluate the performance of this solar system. The results obtained are very encouraging, and it has been found that our solution offers many advantages over the classic solution, such as graphical data presentations for the time being (Fig. 16).

The proposed system reduces the need for human intervention, avoiding reading errors between multiple devices, reducing costs, and saving time. It also simplifies data collection and recording and generates customizable reports that help analyze and optimize the performance of the photovoltaic system.

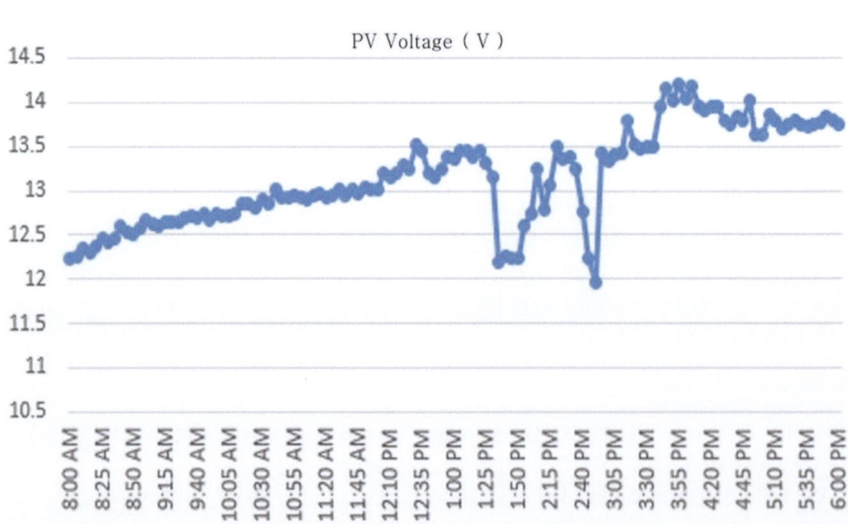

Fig. 16 Voltage of PV as function of time

We would like to propose improvements to this kit while always keeping the light side and the data acquisition system on board. The results express inclinitional with 36 degrees of welldetermined solar panel and good performance data than the other.

A reduction in energy efficiency of around 8 to 10% was observed during the hottest months. This reduction coincides with peak cell temperatures above 65 °C, which leads to losses due to the thermal effect (silicon temperature coefficient: –0.45%/°C).

3.4 *Influence of Temperature and Dust*

Dust accumulation reduces the useful irradiance received by the modules by 5–7% depending on the season. Photos of the modules reveal a gradual opacification of the glass, probably linked to the adsorption of fine particles and the degradation of encapsulation polymers by UV rays.

3.5 *Implications for Materials and Prospects*

The results indicate that conventional materials such as glass/EVA/polycrystalline silicon show moderate resistance in a semi-arid climate. To improve durability:

More UV-resistant sealing materials (e.g., POE) can be considered The use of self-cleaning coatings to limit fouling is recommended Passive or active cooling of the modules could limit heat loss At the same time, IoT monitoring enables predictive maintenance by identifying early signs of material-related failure.

4 Conclusion and Future Works

This study shows that intelligent monitoring of solar production not only optimizes energy management but also allows for the evaluation of photovoltaic material degradation under real-world conditions. The combination of materials science and IoT technologies paves the way for a new generation of more durable solar modules that are suited to harsh environments. Also, the present study highlights IoT technology integration in the management of photovoltaic systems. The use of IoT sensors and devices enables precise tracking of energy production, also monitor factors such as power, voltage, and current. By using tools such as LCD interfaces, Cloud platforms and applications like Thing Speak, we were able to track the evolution of these values in real time. To simulate this system, we used Proteus 8, where we created a model including a solar panel, an Arduino board for data processing and the calculation of power, voltage and current. The current sensor measures the amount of electricity generated, and mathematical relationships are used to calculate voltage and power. The results are displayed on an LCD screen for direct reading.

This system offers several advantages. It enables remote monitoring directly from a cell phone, offering simplified management and greater responsiveness in the event of a problem. It also contributes to the energy autonomy of schools, enabling teachers, for example, to power their computers using solar energy. The results of power, voltage and current monitoring are highly motivating

Finally, it ensures continuous power supply to the school, 24 hours a day, by guaranteeing optimized management of energy production and consumption.

If the quality of data collection is a decisive factor in the solar production chain generated by solar panels, so must be connection. In fact, it's the connection that not only ensures high-quality data collection but also enables real-time alerts to be set up to optimize the parks. It's not enough just to get the data, but to collect it continuously, without network interruptions. For this, efficient connectivity is essential.

In the future, we will work on a big photovoltaic park, and we will install sensors, such as: wind and precipitation temperature, and camera monitoring.

References

1. Ait Ali, A., et al.: The impact of desert regions on solar energy production with the evaluation of groundwater for maintenance: a case study in Morocco. Sustainability **16**(13), 5476 (2024). https://doi.org/10.3390/su16135476
2. Chahboun, M., et al.: Backstepping approach for the control of the double-fed asynchronous generator in a wind power system. Indones. J. Electr. Eng. Comput. Sci. **37**(1), 78–89 (2025). https://doi.org/10.11591/ijeecs.v37.i1.pp78-89
3. Koutroulis, E., Kalaitzakis, K.: Development of an integrated data-acquisition system for renewable energy sources systems monitoring. Renew. Energy **28**(1), 139–152 (2003)
4. Kalaitzakis, K., Koutroulis, E., Vlachos, V.: Development of a data acquisition system for remote monitoring of renewable energy systems. Measurement **34**(2), 75–83 (2003)
5. Forero, N., Hernández, J., Gordillo, G.: Development of a monitoring system for a PV solar plant. Energy Convers. Manage. **47**(15–16), 2329–2336 (2006)
6. Drews, A., De Keizer, A., Beyer, H.G., Lorenz, E., Betcke, J., Van Sark, W., Heydenreich, W.E. Wiemken, Stettler, S., Toggweiler, P., et al.: Monitoring and remote failure detection of grid-connected PV systems based on satellite observations. Sol. Energy **81**(4), 548–564 (2007)
7. Zahran, M., Atia, Y., Al-Hussain, A., El-Sayed, I.: Labview based monitoring system applied for PV power station. In: Proceedings of the 12th WSEAS International Conference on Automatic Control, Modelling and Simulation (ACMOS'10), pp. 65–70 (2010)
8. Mahjoubi, A., Fethi Mechlouch, R., Ben Brahim, A.: A low-cost wireless data acquisition system for a remote photovoltaic (PV) water pumping system. Energies **4**(1), 68–89 (2011)
9. Han, J., Choi, C.-S., Park, W.-K., Lee, I., Kim, S.-H.: Plc-based photovoltaic system management for smart home energy management system. IEEE Trans. Consum. Electron. **60**(2), 184–189 (2014)
10. Okhorzina, A., Bikbulatov, A., Yurchenko, A.V., Bernhard, N., Aldoshina, O.: The development of monitoring and control system of the low pv/t solar system. In: MATEC Web of Conferences. Vol. 79: Information-Measuring Equipment and Technologies (IME&T 2016), vol. 792016, p. 1010.Les Ulis [sn] (2016)

11. López-Vargas, A., Fuentes, M., García, M.V., Muñoz-Rodríguez, F.J.: Low-cost data- logger intended for remote monitoring of solar photovoltaic standalone systems based on arduino™. IEEE Sens. J. **19**(11), 4308–4320 (2019)
12. González, I., Calderón, A.J.: Integration of open-source hardware arduino platform in automation systems applied to smart grids/micro-grids. Sustain. Energy Technol. Assess. **36**, 100557 (2019)
13. Rosiek, S., Batlles, F.: A microcontroller-based data-acquisition system for meteorological station monitoring. Energy Convers. Manage. **49**(12), 3746–3754 (2008)
14. Zhang, B., Zuo, J., Mao, W.: SmartWAZ: Design and implementation of a smart WiFi access system assisted by Zigbee. IEEE Access **7**, 31002–31009 (2019)
15. Aghenta, L.O., Iqbal, M.T.: Low-cost, open source IoT-based SCADA system design using thinger. IO and ESP32 thing. Electronics **8**(8), 822 (2019)
16. Anwari, M., Dom, M.M., Rashid, M.: Small scale PV monitoring system software design. Energy Procedia **12**, 586–592 (2011)

Unveiling the Quantum World of Buckled Aluminene: DFT Insights with Van der Waals Corrections

Pape Sene, Allé Dioum, Sossé Ndiaye, Jean Paul Latyr Faye, Omar Faye, El Hadji Oumar Gueye, and Kharouna Talla

Abstract This present study focuses on the investigation of the structural, stability, and electronic properties of buckled aluminene, employing density functional theory (DFT) both with and without the inclusion of van der Waals (vdW) corrections. The equilibrium lattice parameter was calculated through the Birch-Murnaghan equation of state (EOS), resulting in 2.7332 Å without vdW corrections and 2.7211 Å when vdW interactions were considered, indicating a moderate influence of dispersion forces. Energetic and dynamical stability analyses confirm that buckled aluminene remains stable in both cases, with vdW corrections enhancing energetic stability by making the binding energy more negative (from − 3.5717 eV/atom to − 3.7110 eV/atom), indicating stronger interatomic cohesion. The dynamic stability and structural integrity are validated through the absence of imaginary frequencies in the phonon dispersion curves. Furthermore, the electronic structure analysis, comprising both band structure and density of states (DOS) calculations, indicates that buckled aluminene consistently exhibits a metallic character, regardless of the vdW corrections applied. Although minor shifts are observed in deeper electronic states due to vdW interactions, the overall electronic properties remain unchanged. These findings validate the reliability of DFT alone for describing buckled aluminene, with vdW corrections refining structural and stability parameters without altering its metallic character. This study aligns with previous research, highlighting the role of dispersion forces in accurately modeling 2D materials.

Keywords Density functional theory (DFT) · vdW corrections · Buckled aluminene · Stability · Electronic properties

P. Sene (✉) · A. Dioum
Material—Composite Systems and Applications (MASCA), Department of Physics, Cheikh Anta Diop University (UCAD), Dakar, Senegal
e-mail: pape4.sene@ucad.edu.sn

P. Sene · A. Dioum · S. Ndiaye · J. P. L. Faye · O. Faye · E. H. O. Gueye · K. Talla
Solid State Physics and Material Sciences Group, Department of Physics, Cheikh Anta Diop University of Dakar, Dakar, Senegal

O. Faye
Department of Mechanical Engineering, College of Engineering, University of Saskatchewan, Saskatoon, Canada

E. El Mehdi and M. Boutahir (eds.), *Advanced Materials for Sustainable Energy and Engineering*, Engineering Materials, https://doi.org/10.1007/978-3-032-10069-6_4

1 Introduction

Researchers have increasingly focused on two-dimensional (2D) materials due to their unique properties, which are highly relevant for applications in electronics, photonics, and spintronics [1]. Following the groundbreaking identification of graphene comprising a single atomic layer of carbon arranged in a honeycomb structure and known for its outstanding mechanical strength and high charge carrier mobility [2, 3], a multitude of other 2D materials have been explored, often inspired by theoretical models.

Besides graphene, other group IV materials, including silicene [4], germanene [5], stanene [6], and plumbene [7], exhibit diverse electronic characteristics. Additionally, group III-V compounds, including phosphorene, arsenene, and antimonene [8–13], as well as bismuthylene [14] and borophene [15–18], offer tunable band gaps and high transport anisotropy, making them attractive for optoelectronic and spintronic applications [19–21]. Other important members of the 2D material family include hexagonal boron nitride (h-BN) and boron–carbon-nitrogen (B-C-N) monolayers [22, 23], as well as transition metal dichalcogenides (TMDCs) like MoS_2, $NbSe_2$, and WS_2 [24–28]. More recently, 2D oxides such as MnO_2 and MoO_3 have been introduced [29–32]. Despite their potential, some of these materials face limitations such as low carrier mobility or insulating behavior restricting their electronic applications [33–35].

Among newly explored 2D materials, aluminene, a monolayer of aluminum, has drawn increasing interest due to its metallic properties and structural stability. Theoretical work by Kamal et al. [36] predicted a planar configuration featuring a honeycomb geometry and a Dirac cone distinct from that of graphene. Further studies conducted by Yuan et al. [37] and Nagarajan et al. [38] identified the presence of energetically favorable configurations, including the buckled and 8-Pmmn phases, in addition to a distorted hexagonal form. Aluminene has also been shown to remain stable under mechanical strain [39] and to exhibit potential for superconductivity [40] and gas detection [38].

In our previous Ab initio calculations (Sene et al. 2023), we confirmed the dynamic stability of various aluminene allotropes, particularly those resembling graphene and silicene, which exhibit a metallic nature [41]. Lukačević et al. [42] further proposed synthesis methods, highlighting the feasibility of producing stable puckered and buckled aluminene structures on appropriate substrates. Collectively, these findings underscore aluminene's potential as a versatile and structurally stable material for future technological innovations.

This study aims to examine the structural, stability, and electronic properties of buckled aluminene by employing density functional theory (DFT), both with and without Van der Waals (vdW) corrections. The analysis will first focus on aluminene's characteristics without corrections, followed by an evaluation considering vdW interactions to account for surface effects, and finally, a comparative study to understand the influence of these corrections on fundamental material properties.

2 Computational Method

The structural properties, energetic and dynamical stability, as well as the electronic characteristics of buckled aluminene, were explored using the Quantum ESPRESSO simulation suite [43–45], employing the framework of density functional theory (DFT). The exchange–correlation interactions were modeled with the PBEsol functional [46], known for its reliability in solid-state physics, in combination with scalar-relativistic projector augmented-wave (PAW) pseudopotentials [47]. To optimize computational performance while maintaining accuracy, a plane-wave energy cutoff of 50 Ry was employed.

For accurate electronic structure convergence, the Brillouin zone was sampled through a high-density Monkhorst–Pack grid of 60 × 60 × 1 k-points [48]. The Marzari–Vanderbilt cold smearing technique was applied to manage electronic occupations, with a smearing width set at 0.0147 Ry [49]. Structural optimization continued until the forces acting on each atom were reduced to below 10^{-4} eV/Å, while ensuring total energy convergence within 2×10^{-5} eV.

To account for long-range interactions, Van der Waals (vdW) forces were modeled using Grimme's DFT-D correction [50], as integrated within the Quantum ESPRESSO package, with a vdW radius $R_0 = 3.097$ Å and a dispersion coefficient $C_6 = 374.319$ eV·Å^6 for aluminum. Calculations were conducted both with and without this correction to analyze its influence on the material's properties, keeping the same computational parameters for consistency.

Vibrational properties were assessed through density functional perturbation theory (DFPT) as implemented in the PHonon module of Quantum ESPRESSO [51]. Phonon spectra were computed over a 5 × 5 × 1 q-point mesh to ensure adequate sampling of the Brillouin zone. Additionally, the energetic stability of the buckled aluminene structure, consisting of two aluminum atoms, was evaluated through the binding energy calculation as follows:

$$E_B = E_{Buckled\ Aluminene} - 2 \times E_{Al\ isolated} \tag{1}$$

Here: $E_{buckled\ aluminene}$ represents the total energy of the optimized buckled aluminene structure, while $E_{Al\ isolated}$ denotes the energy of a single isolated aluminum atom. This calculation is consistent with standard ab initio methods for determining binding energies.

This integrative methodology provides a reliable assessment of the impact of Van der Waals interactions on the fundamental properties of buckled aluminene.

3 Results and Discussion

3.1 *Structural Properties of Buckled Aluminene*

3.1.1 Determination of a_0 (Equilibrium Lattice Parameter)

Following the approach used by Shehu A. Yamusa et al. (2023) for MoS_2 [52], the equilibrium lattice parameter of buckled aluminene was determined by analyzing the total energy as a function of volume using the Birch-Murnaghan equation of state:

$$E(V) = E_0 + \frac{9V_0B_0}{16}\left[\left(\frac{V_0^{2/3}}{V} - 1\right)^2 B_0' + \left(\frac{V_0^{2/3}}{V} - 1\right)^2\left(6 - 4\frac{V_0}{V}\right)^{2/3}\right] \quad (2)$$

Here, E(V) is is the energy at volume V, E_0 is the equilibrium energy at V_0, B_0 is the bulk modulus, and B_0' its derivative. This method is illustrated in Fig. 1a, b.

Our energy-volume fit confirms that buckled aluminene is thermodynamically stable, as shown in Fig. 1a. To determine this, we used the *ev.x code* from Quantum Espresso to calculate the relevant quantities summerized in Table 1. The minimum total energy corresponds to an equilibrium lattice parameter of 2.93 Å (Fig. 1b).

The results obtained from the equation of state using the *ev.x* module indicate that buckled aluminene exhibits remarkable mechanical stability, with an equilibrium volume of approximately 1004.45 Å^3. The extracted bulk modulus ($k_0 = B_0 \approx 2.3$ GPa) reveals low mechanical stiffness, a characteristic feature of soft two-dimensional materials. This inherent flexibility, coupled with the observed energetic stability, positions buckled aluminene as a promising candidate for various technological applications. In particular, its mechanical integrity allowing it to withstand deformation without structural failure makes it a promising candidate for use in flexible electronic devices that require materials to endure repeated bending or strain.

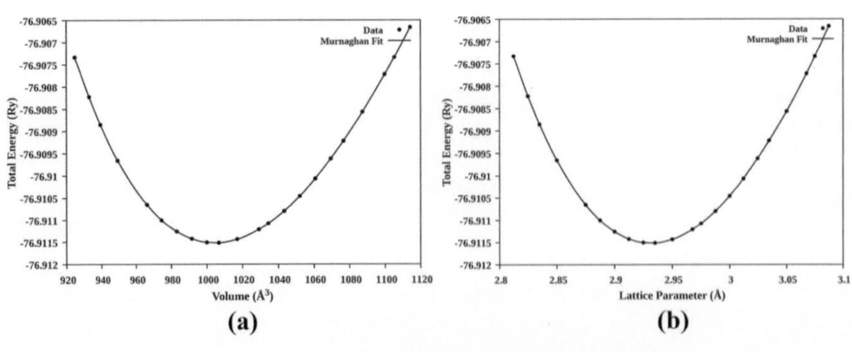

Fig. 1 **a** Total energy versus volume, **b** total energy versus lattice parameter

Table 1 Bulk modulus, volume, pressure derivative, and total energy of Buckled Aluminene calculated using the Birch-Murnaghan equation of state

# Equation of state: Murnaghan. chisq = 0.1995D-10					
# V0 = 6778.38 a.u.^{3}, k0 = 22 kbar, dk0 = 6.90 d2k0 = 0.000 emin = – 76.91151					
# V0 = 1004.45 Ang^3, k0 = 2.3 GPa					
Vol. (Å^3)	E_calc (Ry)	E_fit (Ry)	E_diff (Ry)	Pressure (GPa)	Enthalpy (Ry)
924.58	– 76.90733	– 76.90732	– 1e – 05	0.26	– 76.79846
932.81	– 76.90822	– 76.90822	0.0	0.22	– 76.81336
939.43	– 76.90885	– 76.90886	0.0	0.2	– 76.82469
949.39	– 76.90966	– 76.90966	1e – 05	0.16	– 76.84075
966.12	– 76.91065	– 76.91066	0.0	0.1	– 76.86524
974.54	– 76.911	– 76.91101	0.0	0.08	– 76.8765
983.0	– 76.91126	– 76.91126	– 0.0	0.05	– 76.88715
991.49	– 76.91142	– 76.91142	– 0.0	0.03	– 76.89723
1000.02	– 76.91151	– 76.9115	– 0.0	0.01	– 76.90678
1006.87	– 76.91152	– 76.91151	– 0.0	– 0.01	– 76.91404
1017.19	– 76.91144	– 76.91143	– 0.0	– 0.03	– 76.92437
1029.29	– 76.91121	– 76.91121	– 0.0	– 0.05	– 76.93559
1034.5	– 76.91108	– 76.91107	– 0.0	– 0.06	– 76.94015
1043.21	– 76.9108	– 76.9108	– 0.0	– 0.08	– 76.94742
1051.96	– 76.91046	– 76.91046	0.0	– 0.09	– 76.95431
1060.75	– 76.91007	– 76.91007	0.0	– 0.1	– 76.96085
1069.57	– 76.90962	– 76.90962	1e – 05	– 0.12	– 76.96705
1076.65	– 76.90922	– 76.90923	1e – 05	– 0.13	– 76.97178
1087.32	– 76.90857	– 76.90857	1e – 05	– 0.14	– 76.97851
1099.83	– 76.90772	– 76.90772	0.0	– 0.15	– 76.98585
1105.22	– 76.90734	– 76.90733	– 0.0	– 0.16	– 76.98884
1114.22	– 76.90666	– 76.90665	– 1e – 05	– 0.17	– 76.99362

3.1.2 Full Unit Cell Optimization

In this section, we perform a full optimization of the structure using DFT calculations on buckled aluminene without any corrections. Subsequently, Van der Waals (vdW) corrections are applied to assess their impact on the system.

Figure 2a and b illustrate the optimized structure of buckled aluminene. The resulting lattice parameters, along with specific interatomic and interlayer distances, are summarized in Table 2. Table 3 provides a comparative analysis with previously reported data to assess the consistency of our results.

Without vdW corrections, the lattice parameters are $a_1 = a_2 = 2.733$ Å and $a_3 =$ 20.881 Å. With vdW corrections, these values slightly change to $a_1 = a_2 = 2.721$ Å

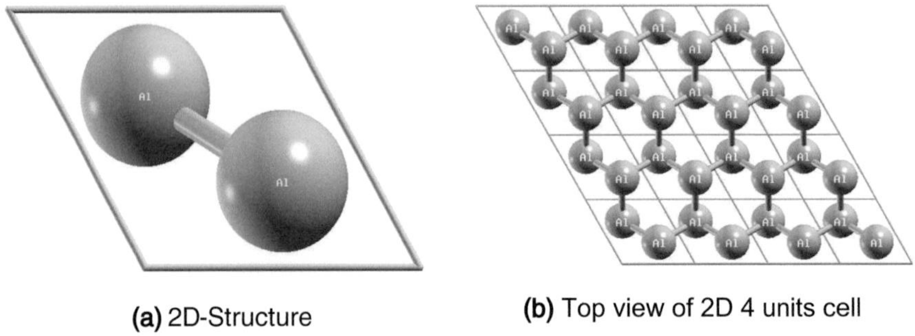

Fig. 2 Structure of buckled aluminene using xcrysden software [53]

Table 2 Lattice parameters of buckled aluminene after optimization

Lattice parameters (Å)	Aluminene without corrections	Aluminene with vdW corrections	Difference $\|\Delta a\|$
a_1 (Å)	2.7331 Å	2.7210 Å	0.0121 Å
Interatomic distance: $d_{(Al-Al)}$	2.8981 Å	2.9071 Å	0.009 Å

Table 3 Comparison of our results with reference literature

Authors	Lattice parameters (Å)	$\|$difference$\|$ $\|\Delta a\|$	Functional and pseudopotential
Our study	a = 2.7331 (without correction)	–	PBEsol, PAW
	a = 2.7210 (with correction)	0.0121 Å	PBEsol, PAW
Yeoh et al. (2018) [39]	a = 2.748	0.0148 Å (without correction) 0.0269 Å (with correction)	PW91, USPP
Yuan et al. (2017) [37]	a = 2.69, b = 2.72	0.0431 Å (without correction) 0.0310 Å (with correction)	PBE, PAW PP

and a_3 = 20.896 Å, This small in-plane contraction (Δa ≈ 0.012 Å) and out-of-plane expansion (Δc ≈ 0.015 Å) reflect the influence of dispersive forces, which tend to increase interlayer distances while slightly compressing in-plane bonds. This trend is expected, as vdW corrections account for dispersive forces that increase interlayer distances to compensate for interatomic compression. The interatomic distance increases slightly from 2.8981 to 2.9071 Å when vdW corrections are applied, indicating a marginal relaxation of the Al–Al bond lengths. This adjustment reflects the system's response to long-range interactions, allowing atoms to reposition slightly to achieve a lower total energy configuration.

Our optimized values show only minor deviations less than 0.05 Å compared to those reported by Yeoh et al. (2018) [39] and Yuan et al. (2017) [37], remaining well within the acceptable range for first-principles studies. These slight differences are mainly due to variations in computational parameters, particularly the exchange–correlation functionals and pseudopotentials. While we used the PBEsol functional with PAW pseudopotentials, Yeoh et al. employed PW91 with USPP, and Yuan et al. used PBE with PAW. Given the sensitivity of DFT results to these choices, such discrepancies are expected and do not compromise the reliability of our findings.

In conclusion, vdW corrections moderately affect the structural properties of buckled aluminene. The observed discrepancies with literature remain small, validating our approach and highlighting the influence of dispersive interactions on the lattice parameters.

3.2 *Study of Aluminene Stability*

The stability of buckled aluminene is examined through two complementary methods: assessing its dynamic behavior and evaluating its energetic favorability, considering cases both including and excluding Van der Waals (vdW) interactions.

3.2.1 Dynamical Stability

Phonon dispersion analysis is a crucial tool for assessing the dynamic stability of a material by examining atomic vibrations within its structure. We calculated the phonon spectrum of buckled aluminene using Quantum Espresso, both with and without Van der Waals (vdW) corrections. Figure 3a and b show the phonon spectra obtained without and with vdW corrections, respectively, while **Fig. 3c** provides a direct comparison between the two in a single plot.

The absence of negative (imaginary) phonon modes across the Brillouin zone indicates that buckled aluminene is dynamically stable under both conditions. This means that the structure resists spontaneous distortions when subjected to small perturbations, thereby confirming its robustness. Slight modifications in the phonon frequencies are observed when vdW forces are included.

Upon closer examination, the inclusion of vdW corrections causes a subtle downward shift in phonon frequencies, most notably within the optical phonon branches. Specifically, the highest optical phonon frequency decreases from 346.9368 cm^{-1} without vdW corrections to 346.0287 cm^{-1} when vdW interactions are taken into account, representing a reduction of approximately 0.9081 cm^{-1}. These phonon frequency values are somewhat lower than those reported by Sene et al. in 2023 [41], who found a maximum optical phonon frequency of 390.4325 cm^{-1} for a silicene-like structure.

A detailed inspection of the phonon branches highlights the following effects:

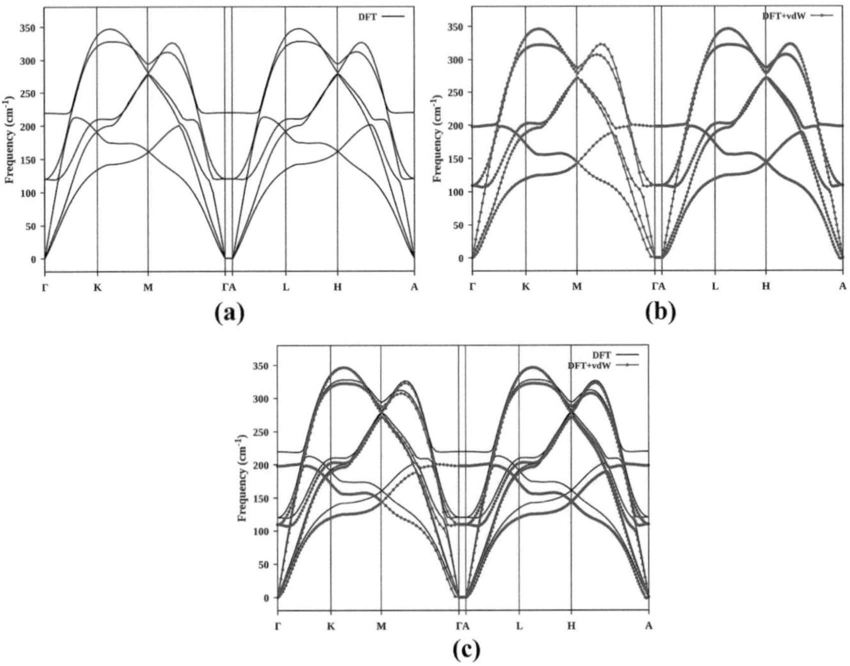

Fig. 3 **a** Non-corrected dispersion, **b** Van der Waals corrected dispersion, **c** comparison of non-corrected versus ndW-corrected phonon spectrum

- Acoustic modes (low frequencies): There is a moderate softening observed along the Γ–K and A–L directions, with a more pronounced decrease seen between the K–M and L–H paths. This suggests that vdW forces slightly suppress the long-wavelength collective vibrations of the lattice.
- Optical modes (high frequencies): Shifts are clearer near high-symmetry points such as Γ, K–M, L–H, and A, reflecting the influence of dispersion forces on short-range atomic interactions within the unit cell.
- In conclusion, while the vdW corrections induce a minor lowering of phonon frequencies, they do not compromise the overall dynamic stability of buckled aluminene. These findings further support its potential use in nanotechnological applications, particularly where vibrational properties play a key role, such as in nanoelectronic and phononic devices.

3.2.2 Energetic Stability

To evaluate the thermodynamic stability of buckled aluminene, the binding energy per atom was determined following the approach outlined in the "Computational Methodology" section, calculated as:

Table 4 Comparison of cohesive energy and structural parameters of buckled aluminene with and without Van der Waals (vdW) corrections

Methodology	E_{coh}/atom (eV)	Lattice constants « a» (Å)	Interlayer distance (Å)	Nearest neighbor distance 'd' (Å)
Our study (DFT)	– 3.5717	2.733	2.434	2.8981
Our study (DFT + vdW)	– 3.7110	2.721	2.447	2.9071
Lukačević et al. [42]	– 3.27	2.72	2.52	2.96

$$E_B = E_{Buckled\ Aluminene} - 2 \times E_{Al\ isolated} \tag{3}$$

Table 4 presents these findings alongside those reported by Lukačević et al. [42].

Our calculations reveal a more negative cohesive energy (– 3.57 eV/atom) than that of Lukačević et al. (– 3.27 eV/atom), indicating stronger bonds and greater stability in our model without vdW corrections. With vdW corrections, the cohesive energy becomes even more negative (– 3.71 eV/atom), further enhancing stability.

The lattice constants of buckled aluminene without vdW corrections (2.733 Å) are slightly larger than those reported by Lukačević et al. (2.72 Å), likely due to the absence of dispersion forces. Adding vdW corrections reduces this value to 2.721 Å, demonstrating a contraction of the crystal lattice and strengthening of atomic interactions. Similarly, the interlayer spacing shows a slight rise, moving from 2.434 angstroms without vdW corrections to 2.447 angstroms when these interactions are included, approaching Lukačević et al.'s 2.52 Å, indicating better consideration of dispersion forces. The nearest neighbor distance follows the same trend, increasing from 2.8981 Å to 2.9071 Å with vdW corrections, highlighting their stabilizing effect.

In conclusion, vdW corrections improve both energetic stability and geometric accuracy, emphasizing the importance of dispersion forces in predicting the structural and electronic properties of buckled aluminene.

3.3 *Electronic Band Structures of Buckled Aluminene*

This section presents an investigation of the electronic band structures of buckled aluminene, comparing results obtained with and without inclusion of Van der Waals (vdW) interactions to evaluate their influence on the material's electronic characteristics. The electronic band structures were computed along a high-symmetry path within the Brillouin zone, specifically following the sequence Γ–K–M–Γ–A–L–H–A, generated using the seeK-path tool from the Materials Cloud database [54–56]. For a direct comparison, Fig. 4 shows the electronic band structures derived from both calculations.

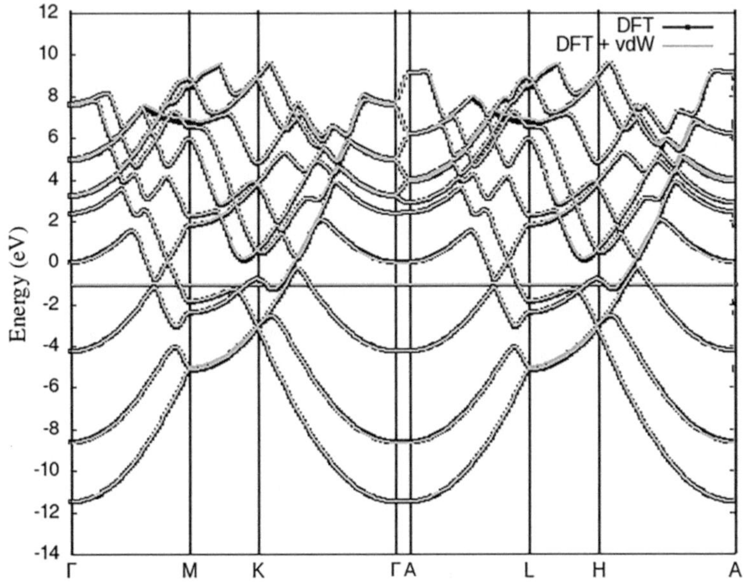

Fig. 4 Bands structure of buckled aluminene: no corrected versus vdw corrections

The results confirm that buckled aluminene remains metallic in both cases, as multiple bands cross the Fermi energy level (E_f). Without vdW corrections, E_f is located at − 1.1432 eV, while with vdW corrections, it shifts slightly to − 1.1025 eV, a minor difference of 0.0407 eV. However, the overall band dispersion remains essentially unchanged, with band crossings occurring at key symmetry locations within the Brillouin zone, reaffirming that vdW interactions do not modify the material's metallic nature.

A comparative study of Fig. 4 highlights:

- Around the Fermi level (E_f): the electronic bands exhibit almost no variation, confirming that the metallic character remains unaffected by the inclusion of vdW corrections.
- At lower energy levels: Small shifts in the band positions suggest that vdW interactions have a more noticeable effect on the deeper electronic states rather than those near E_f, which primarily determine conductivity.

These observations indicate that while vdW corrections slightly modify band energies, they do not significantly influence the metallic properties of buckled aluminene, confirming its stability as a conductive material.

3.4 Analysis of Total and Site-Projected Density of States

To gain a deeper understanding of the electronic properties of buckled aluminene, both the overall density of states (DOS) and the site-specific projected density of states (PDOS) were evaluated with and without accounting for Van der Waals (vdW) interactions. The outcomes, displayed in Fig. 5, clearly illustrate how long-range dispersion forces influence the electronic structure.

Regardless of the inclusion of vdW corrections, the material maintains its metallic nature, evidenced by a pronounced electronic density near the Fermi level, indicating the presence of freely mobile conduction electrons. Upon applying vdW corrections, subtle modifications appear in the DOS profile: some energy levels shift, peak structures become less pronounced, especially beyond 4 eV, and a general smoothing of the spectrum is observed. These changes reflect a damping of high-energy electronic excitations, which can be associated with enhanced structural stability.

Importantly, the p-orbital states remain the primary contributors near the Fermi level, reaffirming their role in defining the metallic nature of buckled aluminene.

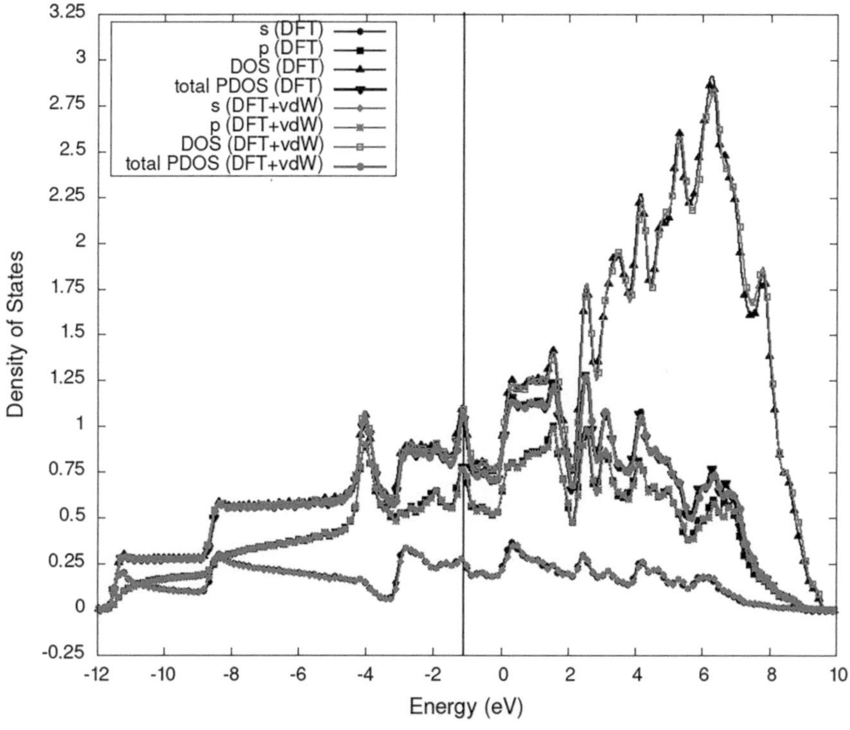

Fig. 5 No corrected versus vdW corrections DOS and PDOS of buckled aluminene

These findings agree with the earlier analysis of the band structure, demonstrating that incorporating vdW interactions causes only slight changes and does not substantially impact the electronic dispersion or the material's metallic behavior.

4 Conclusion

This work examined the structural stability and electronic properties of buckled aluminene through density functional theory (DFT) calculations employing the PBEsol-GGA functional, with particular attention to the role of Van der Waals (vdW) interactions. The results confirm that the material exhibits metallic behavior. The inclusion of vdW forces leads to a slight contraction of the lattice and enhances the overall structural stability, while only marginally influencing the electronic characteristics.

Our analysis of the electronic band structure and density of states indicates that buckled aluminene maintains its metallic nature, with minimal shifts in energy levels due to vdW effects. The density of states resolved by atomic orbitals reveals that p-orbitals dominate near the Fermi level, suggesting strong potential for electrical conductivity. These observations align with previous findings and highlight the importance of considering dispersion interactions for an accurate description of two-dimensional materials.

Given its inherent metallicity and mechanical robustness, buckled aluminene appears as a promising candidate for applications in nanoelectronics, flexible electronics, and energy-related technologies. Future investigations could explore how mechanical strain and atomic doping influence its properties, potentially enabling fine-tuning for advanced applications such as sensors, batteries, and high-performance electronic devices.

Acknowledgements The authors declare that this work did not receive any specific funding from public, commercial, or non-governmental agencies. The authors express their gratitude to the ***"African School on Electronic Structure Methods and Applications" (ASESMA 2023)*** for the excellent training provided on electronic structure techniques, which significantly contributed to this research.

This work is ***dedicated to the memory of Professor Aboubaker Chédikh BEYE***, Professor Emeritus of Physics and founder of the GPSSM Research Group at UCAD, who sadly passed away on November 24, 2024. His lifelong commitment to physics education and research continues to inspire us.

Data Availability Statement There are no datasets associated with this publication.

Disclosure of Potential Conflict of Interest The authors confirm that they have no financial or personal relationships that could have influenced the research presented in this article.

References

1. Geim, A.K., Novoselov, K.S.: The rise of graphene. Nature Mater. **6**(3), 183–191 (2007)
2. Novoselov, K.S., Geim, A.K., et al.: Electric field effect in atomically thin carbon films. Science **306**(5696), 666–669 (2004)
3. Jiang, Y., Miao, L., Jiang, G., Chen, Y., Qi, X., Jiang, X.-F., Zhang, H., Wen, S.: Broadband and enhanced nonlinear optical response of MoS_2/graphene nanocomposites for ultrafast photonics applications. Sci. Rep. **5**, 16372 (2015). https://doi.org/10.1038/srep16372
4. Jose, D., Datta, A.: Structures and chemical properties of silicene: unlike graphene. Acc. Chem. Res. **47**(2), 593–602 (2014). https://doi.org/10.1021/ar400180e
5. Dávila, M.E., Xian, L., Cahangirov, S., Rubio, A., Le Lay, G.: Germanene: a novel two-dimensional germanium allotrope akin to graphene and silicene. New J. Phys. **16**, 095002 (2014)
6. Zhu, F.F., Chen, W.J., Xu, Y., Gao, C.L., Guan, D.D., Liu, C.H., Qian, D., Zhang, S.C., Jia, J.F.: Epitaxial growth of two-dimensional stanene. Nat. Mater. **14**(10), 1020–1025 (2015)
7. Yuhara, J., Le Lay, G.: Beyond silicene: synthesis of germanene, stanene and plumbene. Jpn. J. Appl. Phys. **59**, SN0801 (2020)
8. Li, L., Yu, Y., Ye, G.J., Ge, Q., Ou, X., Wu, H., Feng, D., Chen, X.H., Zhang, Y.: Black phosphorus field-effect transistors. Nat. Nanotechnol. **9**(5), 372–377 (2014)
9. Kamal, C., Ezawa, M.: Arsenene: two-dimensional buckled and puckered honeycomb arsenic systems. Phys. Rev. B **91**, 085423 (2015)
10. Zhang, S.L., Yan, Z., Li, Y.F., Chen, Z.F., Zeng, H.B.: Atomically thin arsenene and antimonene: semimetal-semiconductor and indirect-direct band-gap transitions. Angew. Chem. Int. Ed. **54**, 3112–3115 (2015)
11. Tsai, H.-S., Wang, S.-W., Hsiao, C.-H., Chen, C.-W., Ouyang, H., Chueh, Y.-L., Kuo, H.-C., Liang, J.H.: Direct synthesis and practical bandgap estimation of multilayer arsenene nanoribbons. Chem. Mater. **28**, 2016 (2016)
12. Lei, T., Liu, C., Zhao, J.L., Li, J.M., Li, Y.P., Wang, J.Q., Wu, R., Qian, H.J., Wang, H.Q., Ibrahim, K.: Electronic structure of antimonene grown on Sb_2Te_3 (111) and Bi_2Te_3 substrates. J. Appl. Phys. **119**, 015302 (2016)
13. Ares, P., Aguilar-Galindo, F., Rodríguez-San-Miguel, D., Aldave, D.A., Díaz-Tendero, S., Alcamí, M., Martín, F., Gómez-Herrero, J., Zamora, F.: Mechanical isolation of highly stable antimonene under ambient conditions. Adv. Mater. **28**(30), 6332–6336 (2016)
14. Zhang, R.-W., Zhang, C.-W., Ji, W.-X., Yan, S.-S., Yao, Y.-G.: First-principles prediction on bismuthylene monolayer as a promising quantum spin Hall insulator. Nanoscale **9**(24), 8207–8212 (2017). https://doi.org/10.1039/c7nr01992k
15. Feng, B.J., Zhang, J., Zhong, Q., Li, W.B., Li, S., Li, H., Cheng, P., Meng, S., Chen, L., Wu, K.H.: Experimental realization of two-dimensional boron sheets. Nat. Chem. **8**, 563–568 (2016)
16. Peng, B., Zhang, H., Shao, H.Z., Ning, Z.Y., Xu, Y.F., Lu, H.L., Zhang, D.W., Zhu, H.Y.: Stability and strength of atomically thin borophene from first principles calculations. Mater. Res. Lett. **5**, 399–407 (2017)
17. Lopez-Bezanilla, A., Littlewood, P.B.: Electronic properties of 8-Pmmn borophene. Phys. Rev. B **93**, 241405 (2016)
18. Nagarajan, V., Chandiramouli, R.: Sensing properties of monolayer borophane nanosheet towards alcohol vapors: a first-principles study. J. Mol. Graph. Model. **73**, 208–216 (2017)
19. Wu, X., Shao, Y., Liu, H., Feng, Z., Wang, Y.L., Sun, J.T., Liu, C., Wang, J.O., Liu, Z.L., Zhu, S.Y., Wang, Y.Q., Du, S.X., Shi, Y.G., Ibrahim, K., Gao, H.J.: Epitaxial growth and air-stability of monolayer antimonene on $PdTe_2$. Adv. Mater. **29**(11) (2017). https://doi.org/10.1002/adma.201605407
20. Peng, B., Zhang, D., Zhang, H., Shao, H., Ni, G., Zhu, Y., Zhu, H.: The intrinsic properties of monolayer borophene and its potential applications. Nanoscale **9**, 7397–7407 (2017). https://doi.org/10.1039/c7nr01784d

21. Huang, Z., Liu, H., Hu, R., Qiao, H., Wang, H., Liu, Y., Qi, X., Zhang, H.: Structures, properties and application of 2D monoelemental materials (Xenes) as graphene analogues under defect engineering. Nano Today **35**, 100906 (2020). https://doi.org/10.1016/j.nantod.2020.100906
22. Nagashima, A., Tejima, N., Gamou, Y., Kawai, T., Oshima, C.: Electronic dispersion relations of monolayer hexagonal boron nitride formed on the Ni(111) surface. Phys. Rev. B **51**, 4606–4613 (1995)
23. Beniwal, S., Hooper, J., Miller, D.P., Costa, P.S., Chen, G., Liu, S.Y., Dowben, P.A., Sykes, E.C., Zurek, E., Enders, A.: Graphene-like boron-carbon-nitrogen monolayers. ACS Nano **11**, 2486–2493 (2017)
24. Radisavljevic, B., Radenovic, A., Brivio, J., Giacometti, V., Kis, A.: Single-layer MoS_2 transistors. Nat. Nanotechnol. **6**, 147–150 (2011)
25. Lopez-Sanchez, O., Lembke, D., Kayci, M., Radenovic, A., Kis, A.: Ultrasensitive photodetectors based on monolayer MoS_2. Nat. Nanotechnol. **8**, 497–501 (2013)
26. Yin, Z., Li, H., Li, H., Jiang, L., Shi, Y., Sun, Y., Lu, G., Zhang, Q., Chen, X., Zhang, H.: Single-layer MoS_2 phototransistors. ACS Nano **6**, 74–80 (2012)
27. Ataca, C., Sahin, H., Ciraci, S.: Stable, single-layer MX2 transition-metal oxides and dichalcogenides in a honeycomb-like structure. J. Phys. Chem. C **116**(16), 8983–8999 (2012)
28. Fang, H., Chuang, S., Chang, T.C., Takei, K., Takahashi, T., Javey, A.: High-performance single-layered WSe2 p-FETs with chemically doped contacts. Nano Lett. **12**, 3788–3792 (2012)
29. Azadmanjiri, J., Srivastava, V.K., Kumar, P., Wang, J., Yu, A.: Graphene-supported 2D transition metal oxide heterostructures. J. Mater. Chem. A **6**, 13509–13537 (2018)
30. Kumar, P., Liu, J., Ranjan, P., Hu, Y., Yamijala, S.S., Pati, S.K., Irudayaraj, J., Cheng, G.J.: Alpha lead oxide (α-PbO): a new 2D material with visible light sensitivity. Small **14**, 1703346 (2018)
31. Haque, F., Daeneke, T., Kalantar-Zadeh, K., Ou, J.Z.: Two-dimensional transition metal oxide and chalcogenide-based photocatalysts. Nano-Micro Lett. **10**, 23 (2018)
32. Xie, H., Li, Z., Cheng, L., Haidry, A.A., Tao, J., Xu, Y., Xu, K., Ou, J.Z.: Recent advances in the fabrication of 2D metal oxides. IScience **25**(1), 103598 (2021). https://doi.org/10.1016/j.isci.2021.103598
33. Kong, X., Liu, Q., Zhang, C., Peng, Z., Chen, Q.: Elemental two-dimensional nanosheets beyond graphene. Chem. Soc. Rev. **46**, 2127–2157 (2017). https://doi.org/10.1039/c6cs00937a
34. Qiu, M., Ren, W.X., Jeong, T., Won, M., Park, G.Y., Sang, D.K., Liu, L.-P., Zhang, H., Kim, J.S.: Recent progress in two-dimensional materials beyond graphene: synthesis, properties, and applications. Chem. Soc. Rev. **47**, 5588–5601 (2018). https://doi.org/10.1039/c8cs00254e
35. Zhang, S., Guo, S., Chen, Z., Wang, Y., Gao, H., Gómez-Herrero, J., Ares, P., Zamora, F., Zhu, Z., Zeng, H.: Recent progress in 2D group-VA semiconductors: from theory to experiment. Chem. Soc. Rev. **47**, 982–1021 (2018). https://doi.org/10.1039/c7cs00880j
36. Kamal, C., Chakrabarti, A., Ezawa, M.: Aluminene as highly hole-doped graphene. New J. Phys. **17**(8), 083014 (2015)
37. Yuan, J., Yu, N., Xue, K., Miao, X.: Stability, electronic and thermodynamic properties of aluminene from first-principles calculations. Appl. Surf. Sci. **409**, 85–90 (2017)
38. Nagarajan, V., Chandiramouli, R.: Investigation on adsorption properties of CO and NO gas molecules on aluminene nanosheet: a density functional application. Mater. Sci. Eng. B **229**, 193–200 (2018)
39. Yeoh, K.H., Yoon, T.L., Ong, D.S., Lim, T.L.: First-principles studies on the superconductivity of aluminene. Appl. Surf. Sci. (2018). https://doi.org/10.1016/j.apsusc.2018.03.133
40. Serifi, I., Kanga, N.B.J., Drissi, L.B., Saidi, E.H.: Electron-phonon contribution in aluminene: superconductive and transport properties. Superlattices Microstruct. **151**, 106822 (2021). https://doi.org/10.1016/j.spmi.2021.106822
41. Sene, P., Dioum, A., Ndiaye, S., Gueye, E.H.O., Talla, K., Beye, A.C.: Stability, electronic and structural properties of aluminene from first-principles calculations. MRS Adv. **8**, 666–680 (2023). https://doi.org/10.1557/s43580-023-00625-y
42. Lukačević, I., Varga Pajtler, M., Muževič, M., Gupta, S.K.: Prospects for experimental realization of two-dimensional aluminium allotropes. J. Mater. Chem. C (2019) **7**(9), 2666–2675. https://doi.org/10.1039/C8TC04718A

43. Giannozzi, P., Baroni, S., Bonini, N., Calandra, M., Car, R., Cavazzoni, C., Ceresoli, D., Chiarotti, G.L., Cococcioni, M., Dabo, I., et al.: QUANTUM ESPRESSO: a modular and open-source software project for quantum simulations of materials. J. Phys.: Condens. Matter **21**, 395502 (2009). https://doi.org/10.1088/0953-8984/21/39/395502
44. Giannozzi, P., Andreussi, O., Brumme, T., Bunau, O., Buongiorno Nardelli, M., Calandra, M., Car, R., Cavazzoni, C., Ceresoli, D., Cococcioni, M., et al.: Advanced capabilities for materials modelling with quantum ESPRESSO. J. Phys.: Condens. Matter **29**, 465901 (2017). https://doi.org/10.1088/1361-648X/aa8f79
45. Giannozzi, P., Baseggio, O., Bonfà, P., Brunato, D., Car, R., Carnimeo, I., Cavazzoni, C., de Gironcoli, S., Delugas, P., Ferrari Ruffino, F., Marrazzo, A., Timrov, I., Urru, A., Baroni, S.: Quantum ESPRESSO toward the exascale. J. Chem. Phys. **152**(15), 154105 (2020). https://doi.org/10.1063/5.0005082
46. Perdew, J.P., Ruzsinszky, A., Csonka, G.I., Vydrov, O.A., Scuseria, G.E., Constantin, L.A., Zhou, X., Burke, K.: Restoring the density-gradient expansion for exchange in solids and surfaces. Phys. Rev. Lett. **100**(13), 136406 (2008). https://doi.org/10.1103/PhysRevLett.100.136406. (Erratum in: Phys. Rev. Lett. **102**(3), 039902)
47. Blöchl, P.E.: Projector augmented-wave method. Phys. Rev. B **50**, 17953–17979 (1994). https://doi.org/10.1103/PhysRevB.50.17953
48. Monkhorst, H.J., Pack, J.D.: Special points for Brillouin-zone integrations. Phys. Rev. B **13**, 5188 (1976). https://doi.org/10.1103/PhysRevB.13.5188
49. Marzari, N., Vanderbilt, D., De Vita, A., Payne, M.C.: Thermal contraction and disordering of the Al(110) surface. Phys. Rev. Lett. **82**, 3296 (1999)
50. Grimme, S.: Semiempirical GGA-type density functional constructed with a long-range dispersion correction. J. Comput. Chem. **27**, 1787 (2006)
51. Quantum Espresso: http://www.quantum-espresso.org (2025). Last Accessed 09 May (2025
52. Yamusa, S.A., Shaari, A., Isah, I., Farzan, B., Ismail, M., Ahams, S.T., Itas, Y.S.: First-principles study on the structural, electronic, and elastic properties of transition metal dichalcogenides. Phys. Access **03**(01) (2023). https://doi.org/10.47514/phyaccess.2023.3.1.007
53. Kokalj, A.: XCrySDen-a new program for displaying crystalline structures and electron densities. J. Mol. Graph. Model. **17**, 176–179 (1999)
54. Materials Cloud: https://www.materialscloud.org/work/tools/seekpath (2025). Last Accessed 09 May 2025
55. Hinuma, Y., Pizzi, G., Kumagai, Y., Oba, F., Tanaka, I.: Band structure diagram paths based on crystallography. Comput. Mater. Sci. **128**, 140 (2017). https://doi.org/10.1016/j.commatsci.2016.10.015
56. Togo, A., Tanaka, I.: Spglib: a software library for crystal symmetry search. Sci. Technol. Adv. Mater.: Methods **4**(1), 2384822 (2024). https://doi.org/10.1080/27660400.2024.2384822

Heat Transfer Enhancement in Triangular Cavities Using Nanofluids: A Comparative Study of Heating Wall Orientation

Youness Ighris, Bilal El Hadoui, Mohsine Qaffou, Jamal Baliti, Youssef Elguennouni, Mohamed Hssikou, and Fouad Agoujil

Abstract This work presents a numerical study of natural convection in a right-angled triangular cavity filled with a Cu–water nanofluid, using the single-relaxation-time lattice Boltzmann method (SRT-LBM). The aim is to evaluate the influence of copper nanoparticles on thermal performance under different heating conditions and flow regimes. Two configurations are considered: Case 1, where the inclined wall is heated, and Case 2, where the vertical wall is heated. In both cases, the opposing wall is cold, and the third wall is adiabatic. The study explores a wide range of Rayleigh numbers ($10^3 \leq Ra \leq 10^6$) and nanoparticle volume fractions ($0\% \leq \varphi \leq 7\%$). The thermophysical properties of the nanofluid are calculated using Corcione's empirical model, taking into account temperature-dependent variations in thermal conductivity and viscosity. The thermal behavior is analyzed in terms of isotherms, streamlines, and the average Nusselt number. Results show that the addition of nanoparticles enhances heat transfer, particularly in the conduction-dominated regime (low Ra), where the effect of increased thermal conductivity outweighs the rise in viscosity. At higher Rayleigh numbers, however, an optimal concentration emerges—approximately 6% for $Ra = 10^5$ and 4% for $Ra = 10^6$, beyond which the negative impact of increased viscosity begins to limit convective transport. Comparative analysis between the two heating cases reveals that the orientation of the heated wall plays a crucial role in optimizing natural convection heat transfer. The numerical results are validated against benchmark data from the literature, confirming the robustness and accuracy of the SRT-LBM approach.

Y. Ighris (✉) · M. Qaffou · J. Baliti · M. Hssikou · F. Agoujil
Research Laboratory in Physics and Sciences for Engineers (LRPSI), Polydisciplinary Faculty, University of Sultan Moulay Slimane, Beni Mellal, Morocco
e-mail: ighrisyouness@gmail.com

B. E. Hadoui
Laboratory of Industrial Engineering and Surface Engineering, Faculty of Sciences and Technologies, University of Sultan Moulay Slimane, Beni Mellal, Morocco

Y. Elguennouni
Faculty of Science, Moulay Ismail University of Meknes, Meknes, Morocco

E. El Mehdi and M. Boutahir (eds.), *Advanced Materials for Sustainable Energy and Engineering*, Engineering Materials, https://doi.org/10.1007/978-3-032-10069-6_5

Keywords Heat transfer · Lattice Boltzmann method · Nanofluid · Triangular enclosure · Variable thermophysical properties

1 Introduction

Natural convection plays a fundamental role in a wide range of thermal engineering applications, including but not limited to electronic component cooling systems, solar thermal collectors, and thermal insulation technologies in buildings. Its efficiency and passive nature make it particularly attractive for systems where energy consumption and operational reliability are critical [1–5]. Among the various geometrical configurations studied in natural convection research, triangular cavities have attracted significant attention due to their geometric adaptability and ability to guide heat flow in compact environments [6–10]. These cavities often appear in applications with space constraints or non-rectangular enclosures, such as in roofing systems, solar panels, or microelectronic housings [11, 12].

In recent decades, researchers have increasingly turned to nanofluids as a means to improve the thermal performance of systems relying on natural convection. Nanofluids are engineered colloidal suspensions composed of a base fluid, such as water, ethylene glycol, or oil—and nanoparticles with high thermal conductivity, typically made of metals (e.g., Ag, Cu), metal oxides (e.g., TiO_2, SiO_2), or carbon-based materials (e.g., graphene, carbon nanotubes) [13–16]. The inclusion of nanoparticles in conventional fluids leads to a significant enhancement in effective thermal conductivity, which directly translates to improved heat transfer rates. This property makes nanofluids particularly attractive for use in compact enclosures, where space limitations demand efficient thermal management. Numerous studies have demonstrated that nanofluids can intensify buoyancy-driven flows, reduce thermal resistance, and improve temperature uniformity in enclosures. However, these benefits are influenced by several parameters, such as nanoparticle type, concentration, size, and the thermophysical models used to describe the nanofluid's behavior. Torki and Etesami [17] carried out experimental research focusing on SiO_2/water nanofluids within a rectangular enclosure, exploring the influence of both nanoparticle concentration and inclination angle. Their findings indicated that, at very low concentrations, the presence of nanoparticles had minimal effect on heat transfer. However, increasing the nanoparticle volume fraction beyond 0.005 led to a decline in the heat transfer coefficient. Interestingly, this adverse effect diminished when the cavity was inclined at higher angles or when the heated wall was oriented closer to vertical. Similarly, Scott et al. [18] investigated hybrid nanofluids composed of Al_2O_3 and multi-walled carbon nanotubes (MWCNTs) under natural convection conditions. They observed that the maximum enhancement in the Nusselt number occurred at a volume concentration of 0.10%. Beyond this optimal point, further increases in nanoparticle concentration resulted in a decline in thermal performance, regardless of the temperature gradient applied. In another combined experimental and numerical study, Kargarsharifabad [19] examined the behavior of Cu–water nanofluids in a cubic enclosure subjected to both steady and alternating magnetic fields. Their results revealed that, in the absence of any magnetic influence, the incorporation of copper nanoparticles led to improved heat transfer, with the Nusselt number increasing proportionally to the nanoparticle volume fraction. From a numerical standpoint, only a limited number of studies have captured this adverse behavior. It is widely believed that these discrepancies arise primarily from the choice of thermophysical models used to describe viscosity and thermal conductivity. For instance, El Hadoui and Kaddiri [20] pointed out that oversimplified assumptions in commonly used models may overlook key physical effects. In a recent study, Ighris et al. [21] optimized natural convection heat transfer in a complex cavity filled with Al_2O_3/water nanofluid, comparing two models—one theoretical and one experimental—that account for thermal conductivity and viscosity. The study highlights that the choice of nanofluid model strongly influences the results: the theoretical model improves thermal conductivity with little impact on viscosity, while the experimental model reveals greater viscous sensitivity. This research highlights the importance of geometric design and fluid model for effective thermal management. Selimefendigil et al. [22] conducted a numerical investigation of natural convection within a triangular annular cavity filled with an Al_2O_3/Cu–water hybrid nanofluid. The study explored the combined effects of the Rayleigh number, and nanoparticle volume fraction on flow behavior and thermal performance. The findings indicated that increasing the concentration of solid nanoparticles led to enhanced heat transfer. In a separate study, Rashad et al. [8] numerically examined natural convection in a triangular cavity filled with a hybrid nanofluid. The cavity was heated from the bottom by a constant heat flux element. The hypotenuse wall was maintained at a constant cold temperature, while the other two walls were thermally insulated. Their results demonstrated that increasing the nanoparticle volume fraction significantly improved heat transfer efficiency. Furthermore, Ghoben et al. [23] studied laminar natural convection heat transfer in a cavity with a triangular cross-section filled with a nanofluid. The configuration involved differential heating, with the two vertical walls maintained at different constant temperatures. The study considered two distinct nanoparticle volume fractions and compared the outcomes to those obtained using pure water, over a Rayleigh number range from 103 to 10^6. The results showed that higher nanoparticle concentrations led to a notable increase in the average Nusselt number, especially at higher Rayleigh numbers.

Numerous studies have investigated natural convection within triangular cavities, emphasizing the role of geometric configuration, boundary conditions, and the incorporation of nanofluids in enhancing thermal performance. Among these factors, the positioning of the heat source plays a critical role in shaping flow patterns and temperature distributions, offering opportunities for optimizing passive thermal management strategies. The addition of nanoparticles is known to enhance the thermal conductivity of the base fluid; however, this advantage may be offset by the concurrent rise in viscosity, which can hinder convective motion, particularly at higher concentrations. Although several theoretical models—such as those of Maxwell–Garnett and Brinkman—are commonly used to predict the effective thermophysical properties of nanofluids, most existing numerical studies rely on simplified formulations and often neglect the dynamic interplay between thermal conductivity and viscosity. Moreover, comparative analyses between different heating configurations in triangular enclosures remain scarce, especially in relation to the identification of optimal nanoparticle concentrations across various convection regimes. The present work aims to fill this gap by exploring two distinct heating scenarios and evaluating the thermal performance of Cu–water nanofluids under a wide range of Rayleigh numbers, while employing a more realistic thermophysical model to better capture the dual effect of nanoparticles.

2 Mathematical Formulation

2.1 Problem Statement

This study investigates natural convection heat transfer in a two-dimensional right-angled triangular cavity filled with a Cu–water nanofluid. The fluid is assumed to be Newtonian, incompressible, and thermally conductive, with its effective properties modeled using temperature-dependent correlations. The governing equations are solved using the single-relaxation-time lattice Boltzmann method (SRT-LBM), implemented on a uniform mesh. Natural convection is driven by thermal gradients and buoyancy forces due to gravity. Two heating configurations are considered to assess the influence of boundary conditions on thermal performance:

- Case 1: The inclined wall is heated at a constant high temperature (T_H), the vertical wall is cooled at T_C, and the horizontal wall is adiabatic.
- Case 2: The vertical wall is heated (T_H), the inclined wall is cooled (T_C) and the horizontal wall remains adiabatic.

All walls are assumed to be stationary with no-slip velocity conditions. The effects of varying Rayleigh numbers ($10^3 \leq Ra \leq 10^6$) and nanoparticle volume fractions ($0\% \leq \varphi \leq 7\%$) are examined to evaluate the heat transfer enhancement under different thermal boundary arrangements. This configuration enables a comparative analysis of flow structures and thermal behavior between inclined and vertical heating, providing insights into optimal thermal management in triangular geometries using nanofluids (Fig. 1).

Table 1 shows the thermophysical properties of the base fluid and nanoparticles.

The thermophysical properties of the nanofluid are defined by Eqs. (1)–(4), which describe its effective density ρ_{nf}, specific heat capacity c_{pnf}, thermal expansion coefficient β_{nf}, and thermal diffusivity α_{nf}, respectively, as outlined in reference [21]:

$$\rho_{nf} = (1 - \varphi)\rho_f + \varphi\rho_{Cu}, \tag{1}$$

$$(\rho c_p)_{nf} = (1 - \varphi)(\rho c_p)_f + \varphi(\rho c_p)_{Cu}, \tag{2}$$

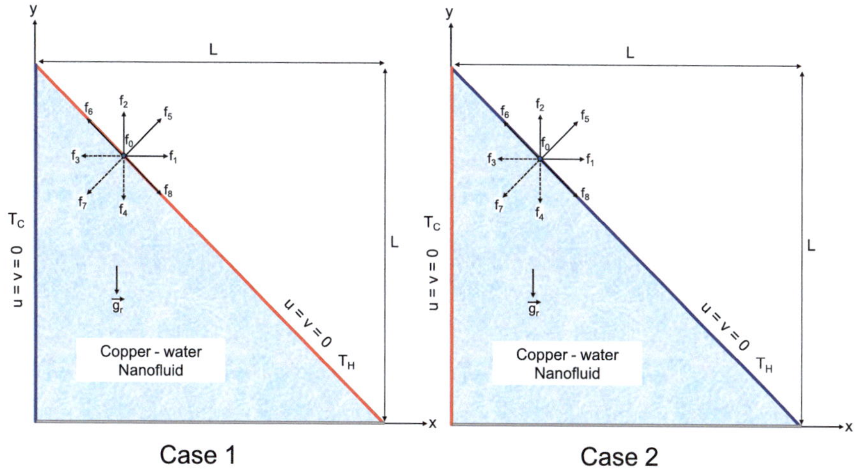

Fig. 1 Geometric configuration

Table 1 Physical and thermal properties of the water and copper

	$\rho(Kg/m^3)$	$c_p(J/Kg.K)$	$k(W/m.K)$	$\beta(1/K)$
H_2O	997.1	4179	0.613	21×10^{-5}
Cu	8933	385	401	1.167×10^{-5}

$$(\rho\beta)_{nf} = (1-\varphi)(\rho\beta)_f + \varphi(\rho\beta)_{Cu}, \tag{3}$$

where φ denotes the volume percentage of nanoparticles.

$$\alpha_{nf} = \frac{k_{nf}}{(\rho c_p)_{nf}}. \tag{4}$$

The thermal conductivity and dynamic viscosity were determined using the Corcione model [24], which is an empirical correlation derived from experimental data for Cu/water nanofluids.

$$k_{nf} = k_f\left(1 + Pr^{0.66}Re^{0.4}\left(\frac{T}{T_{fr}}\right)^{10}\left(\frac{k_p}{k_f}\right)^{0.03}\varphi^{0.66}\right), \tag{5}$$

$$\mu_{nf} = \mu_f\left(1 - 34.87\varphi^{1.03}\left(\frac{d_p}{d_f}\right)^{-0.3}\right)^{-1}. \tag{6}$$

In which Re represents the Reynolds number of the nanoparticles, Pr designates the Prandtl number of the base fluid, T is the temperature of the nanofluid, and T_{fr} its freezing point. φ refers to the volume percentage of nanoparticles, and d is the diameter.

2.2 Lattice Boltzmann Method

In this study, the single-relaxation-time lattice Boltzmann method (SRT-LBM) is employed to simulate two-dimensional natural convection of nanofluids within a triangular cavity. Two distribution functions are used: f_i for fluid flow and g_i for thermal transport. Their evolution is governed by the following discrete lattice Boltzmann equations [25]:

$$f_i(r + \boldsymbol{c}_i\Delta t, t + \Delta t) = f_i(r, t) - \tfrac{1}{\tau_f}\left(f_i - f_i^{eq}\right) + \Delta t\boldsymbol{F}_i, \tag{7}$$

$$g_i(r + \boldsymbol{c}_i\Delta t, t + \Delta t) = g_i(r, t) - \tfrac{1}{\tau_g}\left(g_i - g_i^{eq}\right). \tag{8}$$

Here, Δt is the time step, $\boldsymbol{c}_i$ represents the discrete lattice velocity in direction i, and τ_f, τ_g are the relaxation times related to fluid viscosity and thermal diffusivity, respectively. The terms f_i^{eq} and g_i^{eq} denote the equilibrium distributions, and $\boldsymbol{F}_i$ is the body force resulting from buoyancy, modeled using the Boussinesq approximation:

$$\boldsymbol{F}_i = 3\rho\beta\left(T - T_{ref}\right)\omega_{i,f}\boldsymbol{c}_i\boldsymbol{g}_r. \tag{9}$$

The reference temperature is defined as $T_{ref} = (T_H + T_C)/2$ while $\boldsymbol{g}_r$ and β denote the gravitational vector and the thermal expansion coefficient, respectively.

where $T_{ref} = (T_H + T_C)/2$ is the reference temperature, β is the thermal expansion coefficient, and $\boldsymbol{g}_r$ is the gravitational acceleration.

The equilibrium distribution functions used to compute macroscopic fields are defined as [26]:

$$f_i^{eq} = \rho\omega_{i,f}\left[1 + \tfrac{3\boldsymbol{c}_i.\boldsymbol{u}}{c^2} + \tfrac{9(\boldsymbol{c}_i.\boldsymbol{u})^2}{2c^4} - \tfrac{3\boldsymbol{u}.\boldsymbol{u}}{2c^2}\right], \tag{10}$$

$$g_i^{eq} = T\omega_{i,g}\left[1 + \tfrac{3\boldsymbol{c}_i.\boldsymbol{u}}{c^2}\right]. \tag{11}$$

In the aforementioned expressions, the symbols $\omega_{i,f}$ and $\omega_{i,g}$ denote the lattice weights associated with the momentum and thermal distribution functions, respectively. The parameter c refers to the magnitude of the lattice velocity vector which is directly related to the lattice spacing and time step via $c = \Delta x/\Delta t = \Delta x/\Delta t$, assuming uniform spacing in both x—and y—directions.

To discretize the two-dimensional computational domain, we adopt the widely used D2Q9 lattice model, illustrated in Fig. 2. This scheme involves nine discrete velocity vectors that enable the simulation of isotropic momentum and heat transfer in two dimensions. The velocities in the D2Q9 model are defined as:

- $c_0 = (0, 0)$: rest particle (center of the lattice),
- $c_{1,3} = (\pm 1, 0)$, $c_{2,4} = (0, \pm 1)$: motion along the primary axes,
- $c_{5,6} = (\pm 1, 1)$, $c_{7,8} = (\pm 1, -1)$: velocities in diagonal directions.

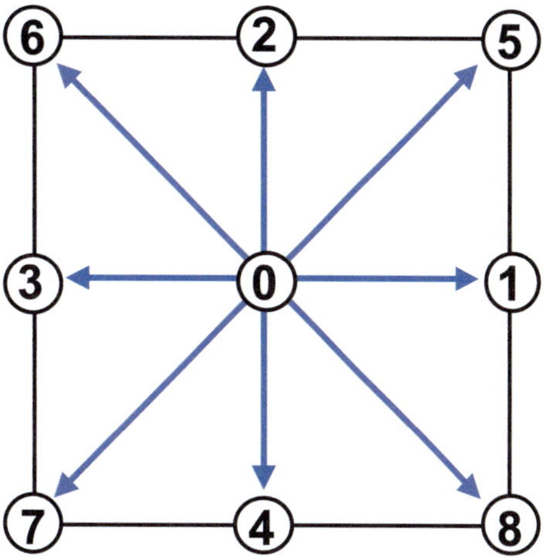

Fig. 2 D2Q9 model

The associated weight coefficients ω_k for each direction are:

- $\omega_0 = 4/9$: central direction,
- $\omega_1 = \omega_2 = \omega_3 = \omega_4 = 1/9$: axial directions,
- $\omega_5 = \omega_6 = \omega_7 = \omega_8 = 1/36$: diagonal directions.

This configuration provides the D2Q9 model with the necessary isotropy to accurately capture the dynamics of fluid flow and heat transfer. The combination of lattice velocities and weights ensures proper conservation of mass, momentum, and energy during the collision and streaming steps of the lattice Boltzmann method, thus enabling robust simulation of natural convection phenomena in two-dimensional cavities.

To characterize the thermal convection regime inside the triangular cavity, the Rayleigh number is calculated using the following expression:

$$Ra = \frac{g_r \beta L^3 (T_H - T_C)}{\nu\alpha}. \tag{12}$$

To assess the local heat transfer rate at the heated surface, the local Nusselt number is computed differently depending on the heating configuration. Two heating cases are considered:

- Case 1: inclined wall as the heated surface

$$Nu = \frac{\sqrt{2}}{2}\left(\frac{k_{nf}}{k_f}\right)\left(\frac{\partial\theta}{\partial X} + \frac{\partial\theta}{\partial Y}\right), \tag{13}$$

- Case 2: vertical wall as the heated surface

$$Nu = \left(\frac{k_{nf}}{k_f}\right)\left(\frac{\partial\theta}{\partial X}\right). \tag{14}$$

The macroscopic fields of temperature, density, and velocity are reconstructed from the distribution functions obtained through the lattice Boltzmann method using the following expressions:

$$\theta = \sum_{i=0}^{8} g_i,\ \rho = \sum_{i=0}^{8} f_i,\ u = \frac{1}{\rho} \sum_{i=0}^{8} f_i \boldsymbol{c}_i. \tag{15}$$

3 Numerical Simulation

In this study, a computational model based on the single-relaxation-time lattice Boltzmann method (SRT-LBM), implemented in FORTRAN, was developed to simulate natural convection in a right-angled triangular cavity filled with a Cu–water nanofluid. The aim is to optimize thermal performance by analyzing the influence of key parameters such as nanoparticle volume fraction ($0\% \leq \varphi \leq 7\%$) and Rayleigh number ($10^3 \leq Ra \leq 10^6$). A uniform grid resolution of 300×300 was employed to ensure numerical accuracy. Two distinct thermal boundary conditions were considered: in the first configuration, the inclined wall is heated while the vertical wall is cooled (Case 1), and in the second, the vertical wall is heated while the inclined wall is cooled (Case 2), with the horizontal wall remaining adiabatic in both cases. This section begins with the validation of the numerical model, followed by a detailed examination of the influence of heating configuration, nanoparticle concentration, and Rayleigh number on the overall heat transfer performance.

3.1 Validation Test

To ensure the reliability of the numerical model developed in this study, a validation procedure was carried out by comparing the present results with those reported by Ghasemi et al. [27], who investigated natural convection in a triangular cavity partially heated and filled with nanofluids. The comparison focused on the average Nusselt number along the heated surface under various Rayleigh numbers and nanoparticle concentrations. As summarized in Table 2, the numerical results obtained with our code show excellent agreement with the reference data, with only minor relative deviations. This close correspondence confirms the accuracy and robustness of the implemented model and demonstrates its ability to capture the essential physical mechanisms governing natural convection in nanofluid-filled enclosures.

Table 2 Comparison of the Nusselt number between our study and a literature study

		$\varphi = 0$	$\varphi = 0.02$	$\varphi = 0.04$
$Ra = 10^4$	Our study	5.802	6.034	6.281
	Ghasemi et al. [27]	5.823	6.069	6.331
	Relative error (%)	0.361	0.580	0.796
$Ra = 10^6$	Our study	19.422	19.937	20.440
	Ghasemi et al. [27]	19.171	19.646	20.114
	Relative error	1.309	1.481	1.620

3.2 Results and Discussion

Figure 3 presents a comparative analysis of the average Nusselt number (Nu_{avg}) as a function of the nanoparticle volume fraction (φ) for two distinct heating scenarios (Case 1 and Case 2) within a triangular cavity filled with nanofluid. In Case 1, the inclined wall is heated, the vertical wall is cooled, and the horizontal wall is adiabatic. Conversely, in Case 2, the vertical wall is heated, the inclined wall is cooled, while the horizontal wall remains adiabatic. The study spans a wide range of Rayleigh numbers, from 10^3 (conduction-dominated) to 10^6 (strong natural convection), providing valuable insight into the thermal response of the nanofluid system under varying flow regimes and configurations. At low Rayleigh numbers ($Ra = 10^3$ and 10^4), where thermal conduction is the dominant heat transfer mechanism, both configurations show a monotonic increase in the Nusselt number with increasing nanoparticle concentration. This behavior is primarily attributed to the enhancement of the nanofluid's effective thermal conductivity, as described by theoretical models such as Corcione's correlation. As more nanoparticles are introduced, the thermal transport capacity of the fluid increases, promoting better heat diffusion. Under these conduction-dominant conditions, Case 2 (vertical heated wall) consistently outperforms Case 1 (inclined heated wall). This indicates that positioning the heat source on the vertical wall facilitates a more favorable thermal gradient and streamlines the upward diffusion of heat, even in the absence of strong convective currents. The vertical orientation likely shortens the thermal path between hot and cold boundaries, enhancing the effectiveness of conduction. As the Rayleigh number increases to 10^5, the flow transitions from pure conduction to convection. The Nusselt number continues to rise with nanoparticle concentration up to an optimal point around $\varphi = 6\%$, beyond which it begins to decrease. This decline is linked to a rise in fluid viscosity, which tends to suppress buoyancy-driven motion and weakens convective transport. At this stage, the system exhibits a trade-off between the increased thermal conductivity (which favors heat transfer) and the reduced fluid mobility (which hampers convection). Although Case 2 maintains better thermal performance at most concentrations, the difference between the two configurations becomes less pronounced, especially near the optimal concentration. At high Rayleigh number $Ra = 10^6$, where natural convection dominates the heat transfer process, the influence of nanoparticle concentration becomes more complex. Initially, the Nusselt number increases with φ, peaking at approximately $\varphi = 4\%$, but decreases beyond this point. The reduction is again attributed to the adverse effects of increased viscosity, which inhibit the formation and circulation of convective rolls and lead to deteriorated heat transfer. Interestingly, at this high *Ra* regime, Case 1 shows a more competitive or even slightly better thermal performance than Case 2 at lower nanoparticle concentrations. This could be due to more stable and continuous convective cell development along the inclined hot surface, which may counteract the damping effects of viscosity. The inclined geometry might allow for more evenly distributed buoyancy-driven flow, preventing the stagnation zones that can occur along vertical surfaces when the fluid becomes too viscous. These findings underline the importance of selecting both the appropriate nanoparticle loading and heating configuration when designing thermally efficient triangular enclosures involving nanofluids under natural convection (Fig. 3).

Figure 4 illustrates the distributions of isotherms (a–d, i–l) and streamlines (e–h, m–p) for various Rayleigh numbers, with the nanoparticle volume fraction fixed at

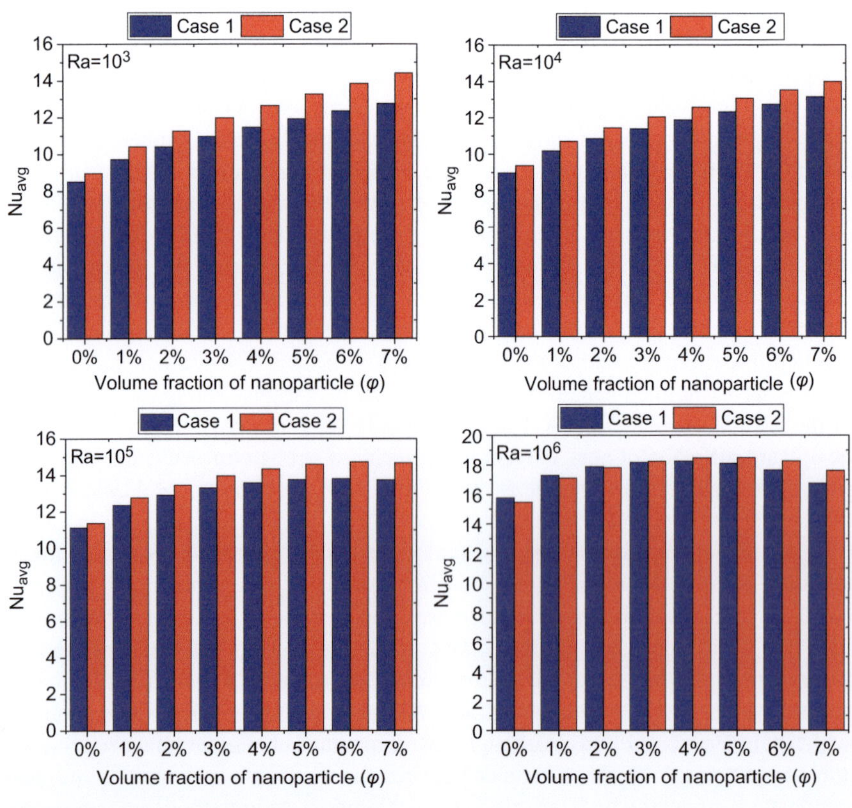

Fig. 3 Variation of the average Nusselt number as a function of nanoparticle volume fraction for two heating scenarios and different Rayleigh numbers

its optimal value ($\varphi \approx 4\%$), where natural convection is dominant. This analysis highlights the influence of both the Rayleigh number and the heating configuration on the thermal and flow behavior within the triangular cavity. In Case 1, where the inclined wall is the heated surface, the isotherms at $Ra = 10^3$ appear nearly parallel, indicating a regime dominated by thermal conduction. The flow is weak and characterized by a single, symmetric, low-intensity vortex at the center of the cavity ($\psi_{max} = 0.15$). As Ra increases to 10^4, the isotherms start to bend, signaling the onset of convective heat transfer. The vortex strengthens, with ψ_{max} rising to 1.79. At $Ra = 10^5$, the isotherms are significantly compressed near the hot and cold walls, indicating intensified convection. The flow becomes more dynamic, forming a structured vortex ($\psi_{max} = 7.48$) that enhances thermal exchange. At $Ra = 10^6$, the isotherms are strongly distorted, and sharp thermal gradients emerge, especially near the boundaries. The streamline patterns show a well-organized, powerful recirculating vortex ($\psi_{max} = 18.65$), demonstrating a regime where convection dominates heat transport. In Case 2, where the vertical wall is heated and the inclined wall is maintained at a cold temperature, similar trends are observed, yet with distinct differences in flow structure and thermal behavior. At $Ra = 10^3$, the isotherms remain nearly horizontal and evenly spaced, indicating dominant conduction, while the flow is very weak ($\psi_{min} = -0.15$). As Ra increases to 10^4, the isotherms begin to curve and tilt, and a clockwise vortex develops with greater strength ($\psi_{min} = -1.49$). At $Ra = 10^5$, a more vigorous circulation emerges ($\psi_{min} = -8.07$), accompanied by compressed isotherms near the heated and cooled boundaries, indicating enhanced convection. At $Ra = 10^6$, a strong, fully developed vortex forms ($\psi_{min} = -19.94$), and the isotherms exhibit sharp gradients, especially near the vertical heated wall, revealing effective thermal transport. Comparative analysis of both cases reveals that Case 2 generally produces stronger vortices and more concentrated thermal gradients near the active boundaries. These effects are more pronounced at higher Rayleigh numbers, suggesting that heating from the vertical wall leads to more efficient convective heat transfer than heating from the inclined wall. This is likely due to a more favorable alignment between buoyancy-driven flow and the orientation of the heat source. These findings underscore the significant influence of heating configuration on the performance of thermal management in triangular enclosures.

4 Conclusion

This study demonstrates the critical role of nanoparticle addition in enhancing natural convection heat transfer within a triangular cavity. At low Rayleigh numbers, the introduction of nanoparticles consistently improves heat transfer performance, primarily due to the dominance of enhanced thermal conductivity over the accompanying increase in viscosity. However, as the Rayleigh number increases, the interplay between these two effects becomes more complex. An optimal nanoparticle volume fraction emerges, estimated at around 6% for $Ra = 10^5$ and decreasing to approximately 4% for $Ra = 10^6$. This behavior suggests that beyond a certain

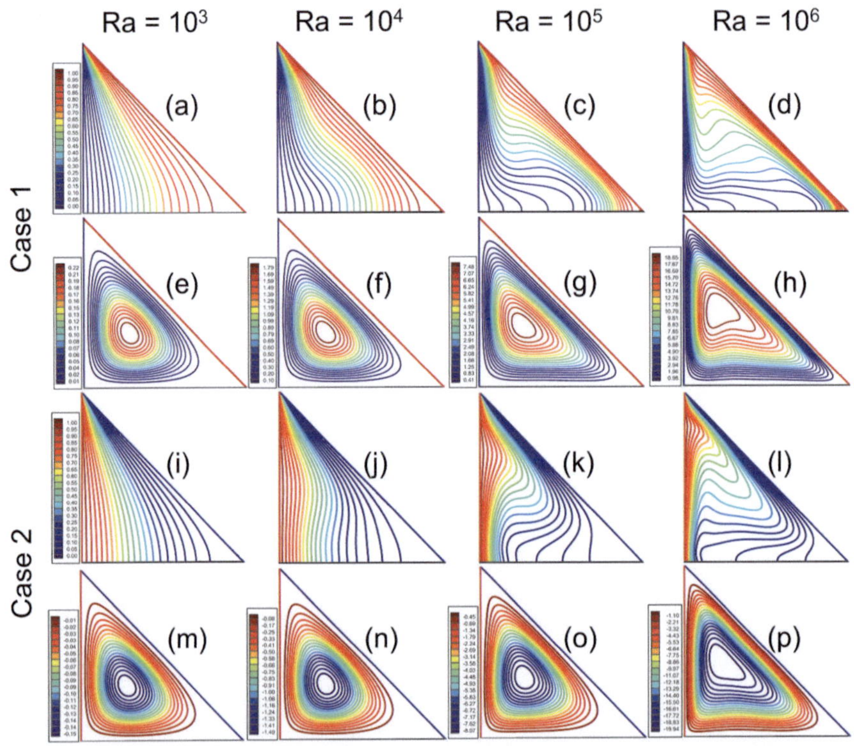

Fig. 4 Distributions of isotherms and streamlines for various Rayleigh numbers under two heating configurations at $\varphi = 4\%$

concentration, the rise in viscosity begins to hinder convective flow, partially offsetting the thermal conductivity advantage. Hence, careful tuning of the nanoparticle concentration is essential to maximize the thermal performance of nanofluids in convection-dominated regimes. Additionally, the comparative analysis of two heating scenarios—Case 1 (heating from the inclined wall) and Case 2 (heating from the vertical wall)—revealed that the orientation of the heated surface significantly influences the flow dynamics and heat transfer efficiency. Across all Rayleigh numbers studied, Case 2 consistently resulted in stronger vortex structures and more concentrated thermal gradients near the active boundaries. This configuration facilitated a more effective alignment between buoyancy-driven flow and the direction of heat input, thereby enhancing overall convection.

In summary, both the nanoparticle concentration and the geometric configuration of the heat source are key parameters in optimizing thermal management in triangular enclosures. Future work may explore the effects of other nanoparticle types, dynamic boundary conditions, or external forces such as magnetic fields to further improve system performance.

References

1. Florio, L.A., Harnoy, A.: Combination technique for improving natural convection cooling in electronics. Int. J. Therm. Sci. **46**(1), 76–92 (2007). https://doi.org/10.1016/j.ijthermalsci.2006.03.007
2. Baliti, J., Hssikou, M., Elguennouni, Y., Moussaoui, A., Alaoui, M.: Heat transfer and entropy generation for natural convection by adiabatic obstacles inside a cavity heated on the left side. FME Trans. **48**(4), 825–832 (2020). https://doi.org/10.5937/fme2004825B
3. Carreto-Hernandez, L.G., et al.: Numerical-experimental investigation of a wind tower-room sustainable system: A parametric analysis of the mixed convection with humidification. J. Build. Eng. **91**, 109624 (2024). https://doi.org/10.1016/j.jobe.2024.109624
4. Phiraphat, S., Prommas, R., Puangsombut, W.: Experimental study of natural convection in PV roof solar collector. Int. Commun. Heat Mass Transf. **89**, 31–38 (2017). https://doi.org/10.1016/j.icheatmasstransfer.2017.09.022
5. Agoujil, F., Elguennouni, Y., Ighris, Y., Baliti, J., Hssikou, M.: Modeling of natural convection by lattice Boltzmann in a partially heated cavity. In: E3S Web Conference, vol. 469 (2023). https://doi.org/10.1051/e3sconf/202346900071
6. Elguennouni, Y., Hssikou, M., Baliti, J., Alaoui, M.: Cavity inclination effect on convection flow in a triangular geometry. Heat Transf. **52**(4), 3253–3279 (2023). https://doi.org/10.1002/htj.22826
7. Sun, Q., Pop, I.: Free convection in a triangle cavity filled with a porous medium saturated with nanofluids with flush mounted heater on the wall. Int. J. Therm. Sci. **50**(11), 2141–2153 (2011). https://doi.org/10.1016/j.ijthermalsci.2011.06.005
8. Rashad, A.M., Chamkha, A.J., Ismael, M.A., Salah, T.: Magnetohydrodynamics natural convection in a triangular cavity filled with a Cu-Al_2O_3/water hybrid nanofluid with localized heating from below and internal heat generation. J. Heat Transfer **140**(7) (2018). https://doi.org/10.1115/1.4039213
9. Ighris, Y., Qaffou, M., Baliti, J., Elguennouni, Y., Hssikou, M.: Thermal management optimization of natural convection in a triangular chamber: role of heating positions and ternary hybrid nanofluid. Phys. Fluids **36**(9) (2024). https://doi.org/10.1063/5.0226427
10. Baliti, J., Elguennouni, Y., Hssikou, M., Alaoui, M.: Simulation of natural convection by multirelaxation time lattice Boltzmann method in a triangular enclosure. Fluids **7**(2) (2022). https://doi.org/10.3390/fluids7020074
11. Saleem, K.B., Alshara, A.K.: Natural convection in a triangular cavity filled with air under the effect of external air stream cooling. Heat Transf. Asian Res. **48**(7), 3186–3213 (2019). https://doi.org/10.1002/htj.21537
12. Tartibu, L.K.: Enhancing Natural Convection in an Isosceles Triangular Shaped Chamber Through Multi-objective Whale Optimization Algorithm, pp. 69–98 (2025). https://doi.org/10.1007/978-3-031-81237-8_3
13. B. El hadoui and M. Kaddiri, "Enhancement of carbon nanotubes/kerosene nanofluids on mixed convective heat transfer in rectangular enclosures. Int. J. Thermofluids 24, 100932 (2024). https://doi.org/10.1016/j.ijft.2024.100932
14. El Glili, I., Driouich, M.: Computational study of the pulsatile EMHD Sutterby nanofluid flow of blood through an inclined tapered porous artery having overlapping stenosis with body acceleration and slip effects. Arch. Appl. Mech. **94**(7), 1841–1869 (2024). https://doi.org/10.1007/s00419-024-02609-2
15. Foukhari, Y., Sammouda, M., Driouich, M.: Heat transfer enhancement by natural convection in a partially porous annular space between two coaxial cylinders saturated by Cu–water nanofluid. Eur. Phys. J. Plus **139**(3), 253 (2024). https://doi.org/10.1140/epjp/s13360-024-04988-5
16. Ighris, Y., Bouhouchi, Y., Elguennouni, Y., Baliti, J., Hssikou, M., Boumezzough, A.: Numerical study of natural convection in an inclined cavity filled with Al_2O_3/Cu-H_2O nanofluids. Numer. Heat Transf. Part A Appl. 1–25 (2024). https://doi.org/10.1080/10407782.2024.2314230

17. Torki, M., Etesami, N.: Experimental investigation of natural convection heat transfer of SiO_2/water nanofluid inside inclined enclosure. J. Therm. Anal. Calorim. **139**(2), 1565–1574 (2020). https://doi.org/10.1007/s10973-019-08445-9
18. Scott, T.O., Ewim, D.R.E., Eloka-Eboka, A.C.: Experimental investigation of natural convection Al_2O_3—MWCNT/water hybrid nanofluids inside a square cavity. Exp. Heat Transf., 1–19 (2022). https://doi.org/10.1080/08916152.2022.2136284
19. Kargarsharifabad, H.: Experimental and numerical study of natural convection of Cu-water nanofluid in a cubic enclosure under constant and alternating magnetic fields. Int. Commun. Heat Mass Transf. **119**, 104957 (2020). https://doi.org/10.1016/j.icheatmasstransfer.2020.104957
20. El hadoui, B., Kaddiri, M.: Enhancing the convective heat transfer in vertical and horizontal rectangular enclosures using nanofluids: The crucial role of aspect ratio. Phys. Fluids **36**(1) (2024). https://doi.org/10.1063/5.0186490
21. Ighris, Y., El hadoui, B., Baliti, J., Elguennouni, Y., Hssikou, M.: Optimizing thermal management of convective heat transfer in a complex nanofluid-filled cavity using the lattice Boltzmann method. Int. J. Numer. Methods Heat Fluid Flow **35**(5), 1845–1883 (2025). https://doi.org/10.1108/HFF-01-2025-0044
22. Selimefendigil, F., Chamhka, A.J.: Nanofluid-filled triangular annulus with an opening. Comput. Therm. Sci.: Int. J. **8**(6), 555–566 (2016)
23. Ghoben, Z.K., Hussein, A.K.: The natural convection inside a 3D triangular cross section cavity filled with nanofluid and included cylinder with different arrangements. Diagnostyka **23**(2), 1–13 (2022). https://doi.org/10.29354/diag/149734
24. Corcione, M.: Empirical correlating equations for predicting the effective thermal conductivity and dynamic viscosity of nanofluids. Energy Convers. Manage. **52**(1), 789–793 (2011). https://doi.org/10.1016/j.enconman.2010.06.072
25. Ighris, Y., et al.: Numerical simulation of fluid flows using the lattice Boltzmann method. In: Boutahir, M., El Mehdi, E. (eds.) Advanced Materials for Sustainable Energy and Engineering. Engineering Materials. Springer, Cham (2025). https://doi.org/10.1007/978-3-031-75032-8_28
26. Hssikou, M., Elguennouni, Y., Baliti, J., Alaoui, M.: Heat transfer of gas flow within a partially heated or cooled square cavity. Model. Simul. Eng. **2020**, 8886682 (2020). https://doi.org/10.1155/2020/8886682
27. Ghasemi, B., Aminossadati, S.M.: Brownian motion of nanoparticles in a triangular enclosure with natural convection. Int. J. Therm. Sci. **49**(6), 931–940 (2010). https://doi.org/10.1016/j.ijthermalsci.2009.12.017

Comparative Fatigue Analysis of Zirconia and Titanium Dental Implants with Varying Thread Designs

Hanane Elouirdini, Imane Touaiher, Chaïmaâ Hermama, Younes Moubachir, and Saoudi Taibi

Abstract This study presents a comprehensive fatigue analysis of dental implants, focusing on the combined effects of thread geometry and material type on mechanical performance and fatigue life. The implants were modeled using CAD software and analyzed through finite element analysis (FEA) with ANSYS, following ISO 14801 standards. Two widely used biomaterials titanium alloy (Ti-6Al-4 V) and yttria-stabilized tetragonal zirconia polycrystals (Y-TZP) were considered, along with different thread configurations: triangular and trapezoidal, and two thread angles (20° and 30°) for each material. A masticatory force of 118.2 N was applied at a 15° inclination with a fixed support to simulate physiological loading. The results of the static analysis indicate that triangular thread designs are prone to higher peak stresses (up to 290 MPa), especially at the thread edges, while trapezoidal threads significantly reduce stress concentrations (216 MPa) due to their larger contact surface and improved load distribution. Fatigue analysis reveals that trapezoidal threaded implants offer more then 4.7 times greater fatigue life than triangular ones. In terms of material specific performance, Y-TZP implants with a 30° thread angle exhibited lower maximum principal stress and better fatigue resistance than the 20° variant. Conversely, titanium implants with a 20° thread angle outperformed the 30° model, showing lower Von Mises stress and longer fatigue life.

Keywords Dental implant · Thread parameters · Fatigue analysis

1 Introduction

Dental implants have shown a great interest in dentistry due to their ability to provide a reliable and long-term solution for tooth replacement, offering both functional and aesthetic benefits. Their success depends on achieving both primary stability during placement and secondary stability through Osseointegration, where bone integrates directly the implant surface [1]. However, the long-term durability of implants is

H. Elouirdini (✉) · I. Touaiher · C. Hermama · Y. Moubachir · S. Taibi
Equipe de Recherche QSM, Université Mohammed V de Rabat, Rabat, Morocco
e-mail: elouirdinihanane@gmail.com

E. El Mehdi and M. Boutahir (eds.), *Advanced Materials for Sustainable Energy and Engineering*, Engineering Materials, https://doi.org/10.1007/978-3-032-10069-6_6

challenged by cyclic masticatory forces, which can lead to mechanical fatigue and eventual failure if not properly designed.

Hence, the design of dental implants plays a crucial role in their performance, particularly, thread geometry is one of the most critical factors [2]. In fact, it was demonstrated that threads enhance initial stability, increase bone-implant contact area, and improve load distribution [2]. Previous studies have shown that specific thread parameters including thread depth, width, pitch and angle, significantly influence stress distribution at the bone implant interface [3] while other studies showed that these same factors affect implant stability and fatigue performance [2].

For example, Advancements in numerical simulation techniques by finite element analysis (FEA), have allowed researchers to optimize thread designs by assessing their impact on stress distribution and bone adaptation. studies has also demonstrated that optimizing thread depth can reduce stress concentrations in cortical bone by up to 20% [3], while increasing implant diameter may lower stress at the implant neck by over 30% [1]. Identified TD, TW, and upper thread angle as the most critical parameters for reducing stress and improving implant performance. Vineet Khened et al. [5], have used the Taguchi method to confirm, that thread depth is the most influential factor in minimizing stress at the bone implant interface. Additionally, implant diameter has been shown as a key parameter that affect stress dissipation. Finite element studies indicate that increasing implant diameter can significantly reduce stress around the implant neck, with a 4.2 mm of implant diameter, a 31.5% of stress reduction was obtained compared to a standard diameter of 3.6 mm [2].

Despite these advances, with over 2000 implant designs available today [4], there remains a need for systematic evaluation of how thread geometry interacts with implant materials to affect fatigue resistance. Material selection is equally vital for implant success. Titanium alloy (Ti-6Al-4 V) has been the gold standard due to its excellent osseointegration and mechanical properties [5]. More recently, Yttria-stabilized zirconia (Y-TZP) has emerged as an alternative, offering superior aesthetics, biocompatibility, and high flexural strength (1000–1200 MPa) [1]. However, zirconia has brittle behavior and consequently low fracture toughness compared to that of titanium alloy which exhibits a high resistance to crack growth due to its ductile behavior [6]. While numerous studies have examined thread design or material type independently [1–6], few have investigated the interaction between these factors.

Given the importance of thread geometry in fatigue resistance, this study aims to compare two types of dental implants with significantly different thread designs and to analyze the fatigue performance of trapezoidal-threaded implants under standardized masticatory loads. Through fatigue testing and finite element analysis (FEA), we evaluate the impact of thread geometry on implant longevity under cyclic loading conditions. The analysis provides insights into how thread design influences stress distribution, mechanical stability, and overall fatigue behavior. Two thread angles (20° and 30°) are considered for both titanium alloy and Y-TZP implants.

2 Material and Methods

2.1 Material Properties

Ti-6Al-4 V and Y-TZP have different elastic characteristics given by Young's Modulus and Poisson's Ratio. As for the first, we are going to adopt 96 MPa [7] for Ti-6Al-4 V and 210 MPa for Y-TZP [6]. Later we will respectively work with 0.36 [7] and 0.32 [6]. Fatigue behavior was evaluated using the stress-life (S–N) approach, with testing conducted at a stress ratio (R) of 0.1 to simulate cyclic loading conditions. This stress ratio was selected in accordance with ISO 14801 standards to replicate physiological masticatory forces [8]. The resulting S–N curve for Ti-6Al-4 V is extracted from [9] while the curve of Y-TZP is extracted from [6].

2.2 Implant Design, Modeling, and Loading Conditions

The dental implant models were designed using CAD software to enable precise parametric control over all geometrical features, with two key variations investigated: thread profile (triangular, trapezoidal) and thread angle (20°, 30°). The triangular and trapezoidal thread profiles (Figs. 1 and 2) were analyzed using Ti-6Al-4 V titanium alloy while maintaining identical overall dimensions (length, diameter, pitch) to isolate the effect of thread design on fatigue performance. Additionally, thread angles of 20° and 30° were evaluated for both Ti-6Al-4 V and Y-TZP materials under the same dimensional constraints to assess the influence of thread angle and material properties. Structural parameters, including bone-implant contact length and prosthetic platform angulation, were kept consistent, while thread pitch, width, and angle variations were documented (Table 1, Fig. 3), ensuring that mechanical differences stemmed solely from thread geometry or material selection. This controlled approach allowed for a systematic comparison of stress distribution and fatigue resistance under finite element analysis.

The boundary conditions simulated physiological fixation by constraining all displacements at the bone-implant contact (BIC) interface, while masticatory forces were modeled using both axial and oblique loading components. A representative oblique force of 118.2 N was applied at 15° relative to the implant's longitudinal axis (Fig. 3) to replicate non-axial chewing loads, resulting in directional force components of $Fx = 17.1$ N (lingual), $Fy = 23.4$ N (distomesial), and $Fz = 114.6$ N (axial). Table 1 details the specific thread parameters for each implant variant (Zi and Ti for zirconia and titanium implants, respectively), highlighting the geometric implications of different thread angles under these loading conditions. This combined loading approach allowed for comprehensive evaluation of implant performance under clinically relevant masticatory forces.

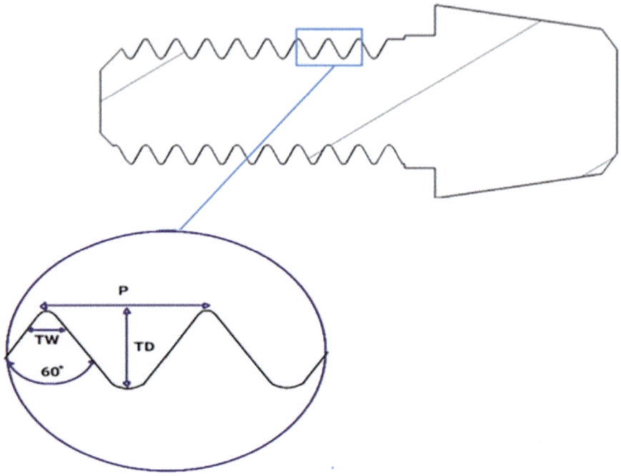

Fig. 1 Triangular thread profile

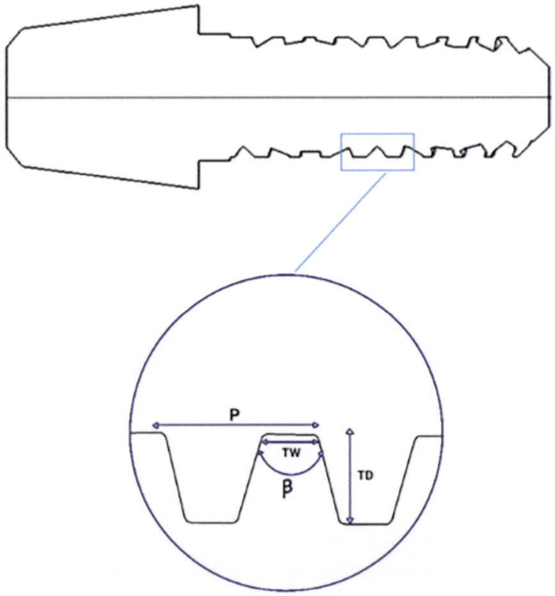

Fig. 2 Trapezoidal thread profile

Table 1 Dimensions of thread parameters

Parameters	Zi (20)	Ti (30)	Zi (20)	Ti (30)
Thread depth (TD) (mm)	0.866		0.866	
Thread width (TW) (mm)	0.144		0.144	
Thread pitch (TP) (mm)	1		1	
Thread angles (β) (deg)	20		30	

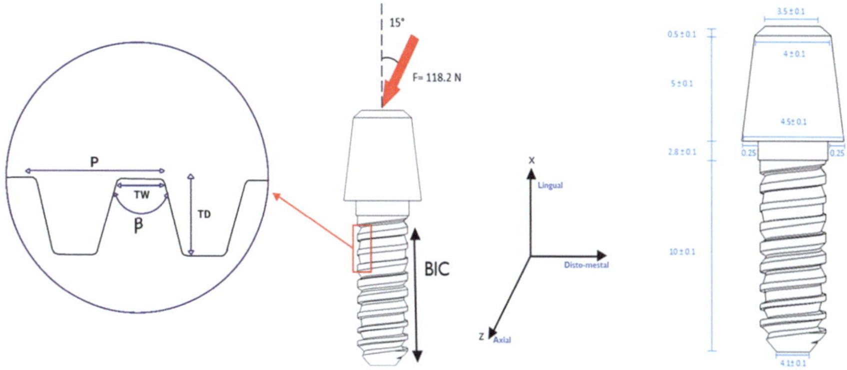

Fig. 3 Geometric dimensions of dental implant with applied boundary conditions

3 Results and Discussion

The Von Mises stress distribution shows peak stress concentrations at the superficial threads and at the implant-abutment connection. The triangular thread design exhibits a higher maximum stress (290 MPa) compared to the trapezoidal design, making it more susceptible to localized failure.

Fatigue life analysis indicates that the trapezoidal-threaded implant outperforms its triangular counterpart, achieving a minimum fatigue life of approximately 1.20 × 105 cycles about 4.7 times higher than that of the triangular design with 25,466 cycles. This enhanced fatigue resistance is attributed to the trapezoidal thread's more favorable stress distribution, resulting from its larger contact surface and optimized load transfer, which help reduce stress concentrations in critical regions. In contrast, the triangular design intensifies stress at the thread edges, accelerating fatigue failure and highlighting the mechanical advantages of the trapezoidal geometry for implant longevity.

Table 2 summarizes the results for various trapezoidal implant models with different thread angles (20° and 30°) and materials (Y-TZP and titanium alloy). The maximum stress corresponds to the first principal stress for Y-TZP and to the maximum equivalent Von Mises stress for the titanium alloy. This distinction is due to the differing mechanical behavior of the two materials: Y-TZP exhibits brittle behavior governed by the maximum principal stress, where the Rankine criterion is applicable, whereas the titanium alloy exhibits ductile behavior typically analyzed using the Von Mises criterion. Additionally, Fig. 2 presents the stress distribution in Y-TZP and Titanium alloy implants, illustrating the influence of thread angle on both stress distribution and fatigue behavior.

The analysis reveals two critical zones of stress concentration: the superficial threads especially the first few load-bearing threads and the implant-abutment connection, where mechanical forces and load transfer significantly influence performance. Among the configurations, Zi (30) shows optimal performance for Y-TZP,

Table 2 Maximum stress and fatigue life of Ti6Al4V and Y-TZP implants for 20° and 30° thread angles

Implant	Max stress (MPa)	Material yield strength (MPa)	Safety factor	Fatigue life (cycles)
Zi (30)	87	980 [6]	11	6.47×10^5
Zi (20)	134	980 [6]	7	3.58×10^5
Ti (30)	216	800 [7]	3.7	3.7×10^5
Ti (20)	85	800 [7]	9.4	10^6

with the highest safety factor (10), the lowest stress (87 MPa), and the best fatigue life (6.47 × 105 cycles). In contrast, Ti (30) presents concerning results, with the lowest safety factor (3.8), highest stress (216 MPa), and reduced fatigue life (3.7 × 105 cycles).

Comparative analysis reveals material dependent performance trends: Ti (20) offers the best outcome for titanium implants, with low stress (85 MPa) and extended fatigue life (10^6 cycles), while Zi (20) shows less favorable results (134 MPa, 3.58 × 105 cycles). These findings underscore the importance of material-specific thread geometry, with the 30° thread angle being optimal for Y-TZP and 20° for Ti6Al4V. Overall, Ti (20) emerges as the top-performing configuration, while Zi (20) ranks lowest, emphasizing the critical role of tailored design in dental implant engineering (Fig. 4).

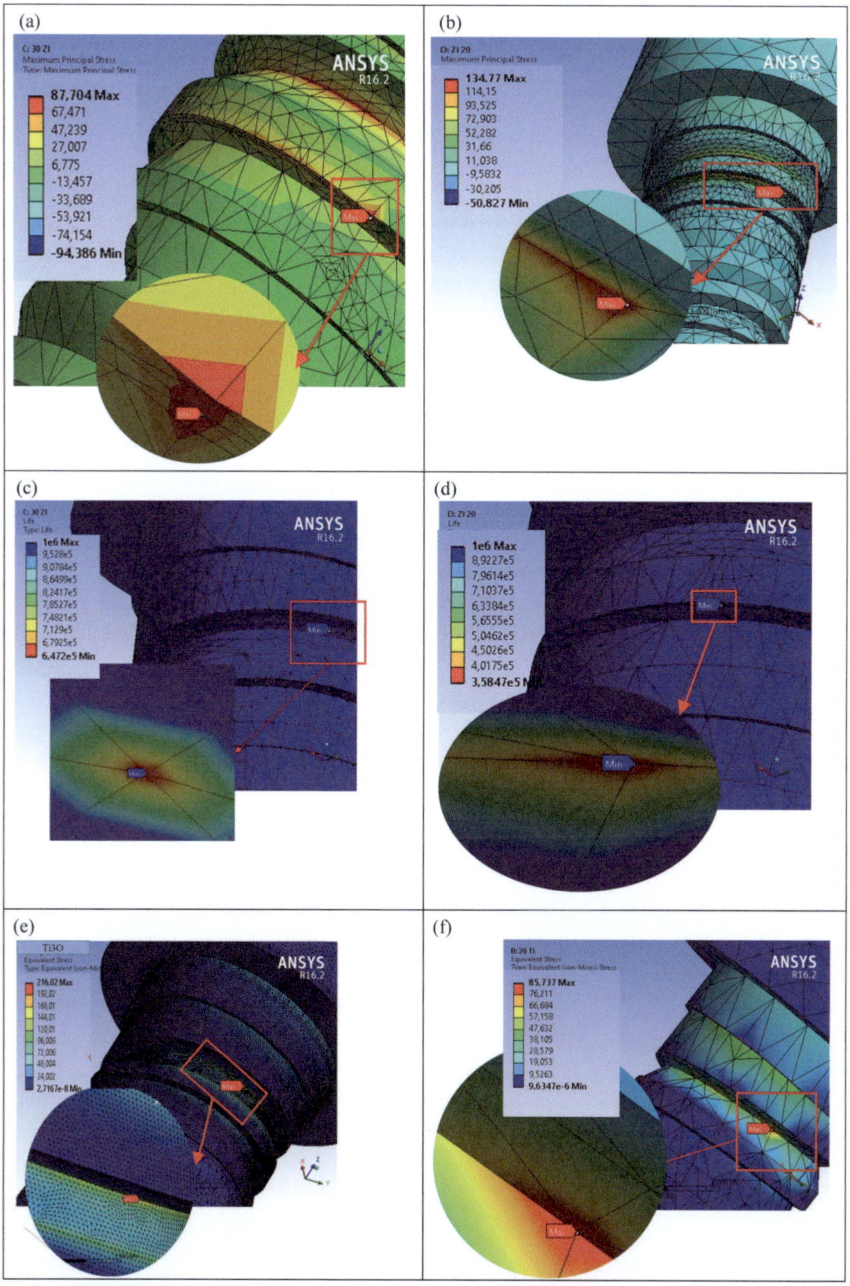

Fig. 4 Stress distribution and fatigue life in Y-TZP implants: 20° thread angle (**a**, **c**) versus 30° thread angle (**b**, **d**), and Von Mises stress distribution in Ti-6Al-4 V Implants (**e**, **f**: for respectively 30° and 20° thread angle)

4 Conclusion

This study highlights the influence of thread geometry, including shape and angle, as well as material type, on the fatigue performance and longevity of dental implants. Through Finite Element Analysis, it was found that the triangular thread design is prone to higher stress concentrations at the thread edges, leading to a greater risk of fatigue failure and reduced durability. In contrast, the trapezoidal thread design distributes stress more evenly, significantly enhancing fatigue resistance and achieving a lifespan approximately 4.7 times greater than that of the triangular design.

Additionally, the study demonstrates that the optimal thread angle for improving fatigue resistance varies depending on the implant material. For Y-TZP implants, a 30° thread angle provides better mechanical durability, while titanium implants benefit from a 20° thread angle for enhanced stress distribution and prolonged fatigue life. These findings underscore the importance of optimizing both thread geometry and material selection in implant design to maximize mechanical reliability and clinical success. By customizing implant designs based on material properties and geometric considerations, this research contributes to the development of more durable and high-performance prosthetic solutions.

References

1. Thesleff, M., Ortiz-Catalan, M., Brånemark, R.: The effect of cortical thickness and thread profile dimensions on stress and strain in bone-anchored implants for amputation prostheses. J. Mech. Behav. Biomed. Mater. **129**, 105148 (2022)
2. Armentia, M., Abasolo, M., Coria, I., Albizuri, J., Aguirrebeitia, J.: Fatigue performance of prosthetic screws used in dental implant restorations: rolled versus cut threads. J. Prosthet. Dent. **126**(3), 406.e1-406.e8 (2021)
3. Satpathy, M., Loeb, M., Jose, R.M., Sinclair, M.J., Duan, Y., Salazar Marocho, S.M., Roach, M.D., Griggs, J.A.: Screening dental implant design parameters for effect on the fatigue limit of reduced-diameter implants. Dent. Mater. **41**, 444–450 (2025)
4. Yang, B., Landa, A.I., Heuberger, P., Ploeg, H.-L.: Effects of dental implant diameter and tapered body design on stress distribution in rigid polyurethane foam during insertion. Med. Eng. Phys. **129**, 104181 (2024)
5. Ivanoff, C.-J., Gröndahl, K., Semerby, L., Bergström, C., Lekholm, U.: Influence of variations in implant diameters: a 3-to 5-year retrospective clinical report. Int. J. Oral Maxillofac. Implant. **14**(2), 173–180 (1999)
6. Bataineh, K., Al Janaideh, M.: Experimental and numerical fatigue investigations of CAD/CAM ceramic crowns under dynamic loads. Inform. Med. Unlocked **26**, 100730 (2021)
7. Chang, J.Z.-C., Hsu, J.-T., Li, M.-J., Lin, H.-Y., Sun, J., Tsou, N.-T., Sun, J.-S.: Optimizing dental implant design: structure, strength, and bone ingrowth. J. Dent. Sciences. Adv. Online Publ. **20**(2), 1016–1026 (2024)
8. ISO 14801:2016: Dentistry Implants Dynamic Loading Test for Endosseous Dental Implants. International Organization for Standardization (2016)
9. Kayabası, O., Yüzbasıoğlu, E., Erzincanlı, F.: Static, dynamic and fatigue behaviors of dental implant using finite element method. Adv. Eng. Softw. **37**(8), 649–658 (2006)

Influence of the Rayleigh Number on the Behavior of Carreau-Yasuda Non-Newtonian Fluids with Viscosity Varying with Temperature

Mohamed Rahmoun, Bilal El hadoui, Taoufik Makayssi, and Mohamed Lamsaadi

Abstract In this work, we carried out a numerical study of the thermal Rayleigh effect on natural double-diffusive convection in a non-Newtonian fluid obeying the Carreau-Yasuda rheological model, confined in a square cavity whose horizontal walls are adiabatic, while the vertical walls exhibit uniform temperatures and concentrations. The approach adopted is based on the finite volume method, enabling the complete conservation equations to be solved. The main parameters studied are the thermal Rayleigh number ($10 \le Ra_T \le 10^6$), the power index ($0.4 \le n \le 1$), the Pearson number ($0 \le m \le 3$), the time constant ($0 \le \lambda \le 10$) and the ratio between viscosities at the extreme shear limits ($0 \le s \le 1$). The results reveal that increasing the thermal Rayleigh number (Ra_T) and Pearson number (m), which models the thermo-dependent effect, leads to a significant improvement in heat transfer rate (*Nu*) and mass transfer rate (*Sh*). Furthermore, a decrease in the parameters n and s induces an intensification of exchanges, reflecting a pseudo-plastic behavior favorable to convection. Increasing the time constant λ also favors heat and mass transfer by reducing apparent viscosity. This study highlights the predominant influence of rheological and thermo-dependent parameters on double-diffusive convection mechanisms.

Keywords Natural convection · Double diffusion · Thermo-dependent · Heat and mass transfer · Carreau-Yasuda model

---

M. Rahmoun (✉) · T. Makayssi · M. Lamsaadi
Polydisciplinary Faculty, University of Sultan Moulay Slimane, Beni Mellal, Morocco
e-mail: mohamed.rahmoun@usms.ac.ma

B. El hadoui
Faculty of Sciences and Technologies, University of Sultan Moulay Slimane, Beni Mellal, Morocco

E. El Mehdi and M. Boutahir (eds.), *Advanced Materials for Sustainable Energy and Engineering*, Engineering Materials, https://doi.org/10.1007/978-3-032-10069-6_7

1 Introduction

Applications of natural convection and double diffusion are at the heart of many heat and mass transfer systems in fields ranging from geological engineering to biotechnology and advanced thermal processes. Javaheri et al. [1] have conducted a linear stability analysis on double-diffusive convection in a porous medium with application to geological storage of CO_2. Their results indicate that the stability of the system strongly depends on thermal and mass diffusion parameters, underlining the importance of understanding double-diffusion mechanisms in gas storage applications. The study by Chen and Johnson [2] highlights significant advances in the understanding of double diffusion phenomena, particularly in varied cavity and fluid configurations, and highlights the importance of control parameters such as Rayleigh number and buoyancy ratio. The study by Le et al. [3] examined double-diffusive convection and the Hall effect in viscous nanofluid creeping flow, with biotechnological applications. Their results demonstrate that the inclusion of the Hall effect significantly modifies velocity and temperature profiles, which is crucial for applications requiring precise control of flow conditions. Turner [4] offered a comprehensive review of double diffusion phenomena, highlighting the fundamental mechanisms governing these complex processes. This review serves as a theoretical basis for further studies of double-diffusive convection in various fluid systems. Jalili et al. [5] carried out a numerical investigation of natural double-diffusive convection in an inclined square cavity. Their results show that the angle of inclination of the cavity strongly influences heat and mass transfer, with a noticeable increase in the Nusselt number. In another study, Eshaghi et al. [6] examined optimal natural double-diffusive convection in a cavity.

Natural convection in Carreau fluids has attracted growing interest due to its relevance in many complex thermal applications involving nonlinear flows and particular rheological properties. The study by Bilal et al. [7] simulated a Carreau fluid flow in a square cavity, showing that decreasing the power number improves thermal stratification and that viscous stresses play a crucial role in energy dissipation. Shah et al. [8] included the effect of thermal radiation in their study of a boundary layer in Carreau fluid undergoing a chemical reaction; they demonstrated an increase in the rate of heat transfer. In addition, an increase in the power law index produces reduced shear rates, while in the case of shrinkage, skin friction increases. In the study by Zeb et al. [9], MHD double-diffusive Carreau fluid flow through a porous medium was analyzed. Their results reveal that varying thermal conductivity and suction/injection conditions significantly influence heat and mass transfer rates, with Nusselt number decreasing with λ and N, and Sherwood number decreasing with λ while increasing with Le. Dimitrienko and Li [10] presented a numerical simulation of heat transfer by natural MHD convection in a square cavity filled with Carreau fluids. They show that the heat transfer and convective flow of the pseudoplastic fluid are stronger than those of the Newtonian fluid, and that the heat transfer and convective flow of the dilatant fluid are weaker than those of the Newtonian fluid. In this study, Larbi et al. [11] examined the impact of the rheological behavior of non-Newtonian fluids on natural double-diffusive convection in a square enclosure. Their results show that rheological properties influence convection cell formation and heat and mass transfer rates, underlining the importance of fluid selection for specific applications. Makayssi et al. [12] studied natural double-diffusive convection for a shear-thinning Carreau fluid in a square cavity subjected to horizontal temperature and concentration gradients. The results demonstrate that Carreau model parameters influence flow structures and transfer rates, with convergence of results between Carreau and power-law models at high Rayleigh numbers. In the study by Lounis et al. [13], thermo-solutal convection of non-Newtonian Carreau-Yasuda fluids was explored in inclined square cavities under Dufour and Soret effects. The results indicate that increasing certain rheological parameters improves heat and mass transfer.

Natural convection in cavities filled with fluids with thermo-dependent viscosity represents a key area of research due to its importance in energy applications, geothermal energy, and passive cooling systems. The study by Ramamurthy and Krishnan [14] highlights the importance of non-Newtonian fluids with temperature-dependent viscosity in conventional thermal processes. Their results highlight that the variation of viscosity with temperature affects thermal performance, requiring particular attention in the design and optimization of thermal processes involving such fluids. El Hadoui and Kaddiri [15] explored the critical role of aspect ratio in enhancing natural convective heat transfer with nanofluids with temperature-dependent properties in rectangular enclosures. In another study, Shah et al. [16] analyzed convective heat transfer in a magnetodynamic Carreau fluid with temperature-dependent viscosity and thermal conductivity. The results reveal that the temperature dependence of these properties significantly influences temperature profiles and heat transfer rates, which is crucial for applications where these variations are important. Rahmoun et al. [17] demonstrated the effect of Lewis number on the double-diffusive convection of heat-dependent Carreau-Yasuda fluids in a square cavity; they concluded that high Lewis number values promote mass transport while reducing the intensity of heat transport. Another study by Astanina et al. [18] analyzes unsteady natural convection in a partially porous cavity containing a heat source, using a local thermal non-equilibrium model. The results show that varying the viscosity parameter leads to better cooling of the heat transfer element thanks to more intense circulation, as liquid viscosity decreases with increasing temperature. Finally, the study by Daghab et al. [19] used finite volume simulation to study a temperature-dependent viscosity power law fluid in a square cavity with a localized source, demonstrating that the heated zone induces intense local recirculation and a significant increase in mean Nusselt.

A review of the specialized literature reveals a lack of studies focusing specifically on the effect of thermal Rayleigh number (Ra_T) on the behavior of non-Newtonian Carreau-Yasuda type fluids, whose viscosity varies with temperature, in a cavity subjected to thermal gradients and concentration imposed on the vertical walls. As part of this research, we examined the influence of Ra_T on several determining parameters, including the flow behavior index (n), Pearson number (m), as well as the characteristic parameters of the Carreau-Yasuda rheological model (s and λ). Taken together, this work highlights the richness and complexity of natural double-diffusive convection in thermo-viscosity-dependent fluids, where the interplay between thermophysical, geometrical, and rheological parameters decisively determines the efficiency of heat and mass transfer.

2 Mathematical Formulation

The domain of study is a closed square cavity in two dimensions, with equal sides of length H', completely filled with a non-Newtonian fluid whose rheological behavior is governed by the Carreau model. In this configuration, the two vertical walls of the cavity are considered to be thermally and solutally active, i.e., they are subject to constant and uniform temperature and concentration conditions. In contrast, the horizontal walls are treated as perfectly insulated and adiabatic. The geometric setup of the enclosure and the boundary conditions applied are shown schematically in Fig. 1.

Fluid flow is characterized using the Carreau-Yasuda rheological model, taking into account the non-Newtonian nature of the fluid. In this framework, the density of the fluid, noted ρ, is assumed to depend linearly on temperature T' and concentration S'. This dependence is translated by the following relationship:

$$\rho = \rho_0\left[1 - \beta_T\left(T' - T_r'\right) + \beta_S\left(S' - S_r'\right)\right] \tag{1}$$

where ρ_0 denotes the density of the fluid at reference conditions defined by a temperature T_r' and a concentration S_r'. The coefficients β_T and β_S correspond to the coefficients of thermal and solutal expansion, respectively.

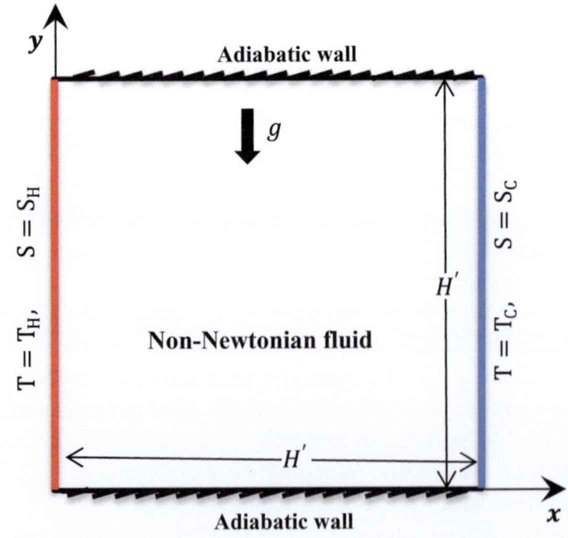

Fig. 1 Schematic representation of the problem studied

Furthermore, the dynamic viscosity μ of the fluid varies as a function of the shear rate according to the Carreau-Yasuda model law [20], expressed as follows:

$$\frac{\mu - \mu_\infty}{\mu_0 - \mu_\infty} = \left[1 + \left(\lambda' \dot{\gamma}'\right)^2\right]^{\frac{n-1}{2}} \times k_T \tag{2}$$

In this expression, μ_0 and μ_∞ represent the limiting viscosities of the fluid at very low and very high shear rates, respectively. The parameter λ' defines a characteristic time related to fluid behavior, while $\dot{\gamma}'$ designates the second invariant of the shear rate tensor. The index *n*, known as the power law index, reflects the degree of non-linearity of viscosity with respect to shear rate. Typically, for fluids with pseudoplastic behavior (or thinning shear fluids), n varies between 0.4 and 1, and λ' takes values between 0.2 and 10. Note that when $n = 1$ or $\lambda' = 0$, the behavior becomes that of a Newtonian fluid. On the other hand, if λ' tends towards very high values, the model approaches that of the power law.

The thermal consistency index K_T is a temperature-dependent empirical function. Its variation with temperature is governed by an exponential law inspired by the Arrhenius relationship, known as the Frank-Kamenetski law [21], formulated as follows:

$$k_T = ke^{-b\left(T' - T'_r\right)} \tag{3}$$

where k represents the consistency index evaluated at the reference temperature T'_r, and b is a coefficient describing the thermal sensitivity of consistency. This relationship expresses the exponential decrease in fluid consistency as temperature rises, reflecting the thermo-dependent behavior characteristic of the non-Newtonian fluids.

The flow behavior inside the cavity is described using fundamental conservation equations expressed in a two-dimensional framework and Cartesian coordinates. These equations include the conservation of mass (continuity equation), momentum in the horizontal and vertical directions (Navier–Stokes equations), as well as the energy (heat) and concentration equations. In their dimensionless form, they can be written as follows:

Continuity equation:

$$\frac{\partial u}{\partial x} + \frac{\partial v}{\partial y} = 0 \tag{4}$$

Equations of motion:

$$u\frac{\partial u}{\partial x} + v\frac{\partial u}{\partial y} = \Pr\left[\mu_a\left(\frac{\partial^2 u}{\partial x^2} + \frac{\partial^2 u}{\partial y^2}\right) + 2\frac{\partial \mu_a}{\partial x}\frac{\partial u}{\partial x} + \frac{\partial \mu_a}{\partial y}\left(\frac{\partial u}{\partial y} + \frac{\partial v}{\partial x}\right)\right] - \frac{\partial p}{\partial x} \tag{5}$$

$$u\frac{\partial v}{\partial x}+v\frac{\partial v}{\partial y}=\Pr\left[\mu_a\left(\frac{\partial^2 v}{\partial x^2}+\frac{\partial^2 v}{\partial y^2}\right)+2\frac{\partial\mu_a}{\partial y}\frac{\partial v}{\partial y}+\frac{\partial\mu_a}{\partial x}\left(\frac{\partial u}{\partial y}+\frac{\partial v}{\partial x}\right)+Ra_T(T+NS)\right]-\frac{\partial p}{\partial y} \tag{6}$$

Energy and concentration equations:

$$u\frac{\partial T}{\partial x}+v\frac{\partial T}{\partial y}=\nabla^2 T \tag{7}$$

$$u\frac{\partial S}{\partial x}+v\frac{\partial S}{\partial y}=\frac{1}{Le}\nabla^2 S \tag{8}$$

The dimensionless apparent viscosity μ_a is given by:

$$\mu_a=s+(1-s)\left[1+(\lambda\dot{\gamma})\right]^{2\frac{n-1}{2}}\times K_T \tag{9}$$

With the shear rate defined by:

$$\dot{\gamma}=\left(2\left(\left(\frac{\partial u}{\partial x}\right)^2+\left(\frac{\partial v}{\partial y}\right)^2\right)+\left(\frac{\partial u}{\partial y}+\frac{\partial v}{\partial x}\right)^2\right)^{\frac{1}{2}},\ K_T=e^{-mT} \tag{10}$$

where s represents the ratio between viscosities at the extreme shear limits (zero and infinite), λ is a dimensionless time constant, n is the power law index characterizing rheofluidification, and m designates the Pearson number quantifying the thermal influence on viscosity.

In addition, several key dimensionless parameters are involved in modeling the problem, namely the thermal Rayleigh number, the Prandtl number, the Lewis number, and the buoyancy ratio, which were presented respectively as:

$$Ra_T=\frac{g\beta_T H'^3\Delta T}{\left(\frac{\mu_0'}{\rho}\right)\alpha},\ Pr=\frac{\left(\frac{\mu_0'}{\rho}\right)}{\alpha},\ Le=\frac{\alpha}{D},\ N=\frac{\beta_S\Delta S}{\beta_T\Delta T} \tag{11}$$

Parameter m can also be interpreted as:

$$m=-\frac{1}{K_T}\frac{dK_T}{dT}=-\frac{d\ln\left(\frac{K_T}{K}\right)}{dT} \tag{12}$$

The boundary conditions, in dimensionless formulation, are as follows:

On horizontal walls: $y = 0$ and $y = 1$, for all x, velocities are zero ($u = v = 0$) and gradients along x for T and S are also zero.

On the left wall: $x = 0$, temperature and concentration are imposed at $T = 1$, $S = 1$, with zero velocities.

On the right wall: $x = 1$, temperature and concentration are fixed at $T = 0$, $S = 0$, and velocities are also zero.

Finally, heat and mass transfers within the cavity are quantified by the Nusselt and Sherwood numbers, defined respectively as:

$$Nu = -\int_0^1 (\partial T/\partial x)_{x=0} dy \quad \text{and} \quad Sh = -\int_0^1 (\partial S/\partial x)_{x=0}\, dy \tag{13}$$

3 Numerical Method

This study rigorously analyzes the thermo-solutal behavior of a non-Newtonian Carreau-Yasuda type fluid in the context of double-diffusive convection. It is based on the numerical resolution of the system of dimensionless equations (Eqs. (4)–(8)). These equations, coupled with the corresponding boundary conditions, are discretized using the finite-volume method, widely recognized for its robustness and ability to deal efficiently with transport problems in fluid media. In this context, the discretization of convection and diffusion terms is a crucial step, as it directly influences the accuracy of the results. For this purpose, the power law scheme is adopted. It is particularly well suited to cases where gradients are pronounced, as it ensures greater stability and accuracy in modeling the combined transport phenomena.

In order to solve the discretized equations, the well-known SIMPLE algorithm (Semi-Implicit Method for Pressure-Linked Equations), originally proposed by Patankar [22, 23], was implemented. This algorithm is a benchmark in numerical fluid mechanics for pressure–velocity coupling, particularly in the context of the Navier–Stokes equations. It offers a structured and efficient iterative approach, guaranteeing convergence of solutions even in complex configurations. The numerical implementation was carried out using a code developed in FORTRAN.

Careful adjustment of the sub-relaxation parameters was carried out to promote algorithm stability and accelerate convergence, although there is no universal formula for systematically determining these coefficients. Finally, all numerical simulations were carried out considering a high value of the Prandtl number (Pr) to reduce computation time while maintaining the fidelity of the results. This strategy improves the speed of convergence of the scheme while preserving the precision required in the analysis of heat and mass transfers.

Table 1 Comparison of maximum stream function values ($|\psi_{max}|$) and Nusselt number (Nu), between the present results and data from Vahl Davis et al. [24] and Bennacer and Gobin [25] for different Ra

Ra	Vahl Davis et al. [24]		Bennacer and Gobin [25]		Present study	
	$\|\Psi_{max}\|$	Nu	$\|\Psi_{max}\|$	Nu	$\|\Psi_{max}\|$	Nu
10^5	9.644	4.514	9.590	4.524	9.632	4.538
10^6	16.960	8.798	16.976	8.815	16.920	8.930

Table 2 Comparison of maximum stream function values ($|\psi_{max}|$) Nusselt number (Nu), and Sherwood number (Sh) between the present results and data from Makayssi et al. [12] for different n

Ra_T	n	Makayssi et al. [12]			Present study		
		$\|\Psi_{max}\|$	Nu	Sh	$\|\Psi_{max}\|$	Nu	Sh
10^3	0.4	4.014	2.101	6.333	3.954	2.086	6.270
	0.6	2.524	1.527	4.587	2.525	1.537	4.609
	0.8	1.730	1.267	3.681	1.732	1.276	3.701

3.1 *Numerical Code Validation*

In order to verify the accuracy and reliability of our numerical code, we carried out a series of validations by comparing it with results available in the literature. First, the reference data obtained by de Vahl Davis et al. [24] were used for the case of natural convection. Then, the work of Bennacer and Gobin [25] was used as a basis for validating our model in a similar context; these comparisons are summarized in Table 1. For the double-diffusive convection of non-Newtonian fluids of the Carreau type, the results reported by Makayssi et al. [12], considering the specific parameters $s = 0.01$ and $\lambda = 0.2$, enabled us to assess the performance of our approach. The comparison, summarized in Table 2, validated our numerical methodology by ensuring good agreement with published data, thus confirming the robustness of our governing equation solver scheme.

3.2 *Mesh Sensitivity Analysis*

As part of this study, an assessment of the influence of mesh density on the accuracy of numerical results was carried out in order to guarantee the reliability of simulations while optimizing computation time. Table 3 illustrates the effects of uniform mesh refinement on the characteristic quantities of a square cavity subjected to double-diffusive convection, for parameters set at $Ra_T = 10^4$, $m = 2$, $N = 1$, $Le = 5$, $n = 0.6$, and $A = 1$.

Table 3 Analysis of the mesh effect for $Ra_T = 10^4, m = 2, N = 1, Le = 5, n = 0.6$ and $A = 1$

	40 × 40	60 × 60	80 × 80	100 × 100	120 × 120	140 × 140
$\vert\Psi_{max}\vert$	13.160	13.036	13.015	13.015	13.006	13.003
Nu	6.649	6.513	6.464	6.443	6.432	6.426
Sh	13.534	13.106	12.953	12.886	12.851	12.831

Simulations were carried out using different uniform meshes of increasing dimensions: 40 × 40, 60 × 60, 80 × 80, 100 × 100, 120 × 120, and 140 × 140. For each mesh, the values of the maximum stream function (ψ_{max}), as well as the Nusselt (Nu) and Sherwood (Sh) numbers, were determined and recorded in Table 3.

The comparative analysis shows that increasing the mesh resolution improves the accuracy of the results, although this improvement tends to level off after a certain threshold. More specifically, it was observed that the 100 × 100 mesh represents an optimum compromise between computational cost and numerical accuracy, offering a faithful representation of transfer phenomena without unnecessarily lengthening computation time. This mesh was therefore selected for the main simulations in the study.

4 Results and Discussion

Figure 2 illustrates the combined impact of the thermal Rayleigh number Ra_T, the Pearson number m, and the characteristic time parameter λ on flow as well as heat and mass transfer in a cavity. From Fig. 2a, representing the variation of the maximum stream function ψ_{max}, it can be seen that flow intensity increases significantly with increasing Ra_T, reflecting an intensification of natural convection. This intensification is even more marked as m increases, due to the decrease in the apparent viscosity of non-Newtonian fluids, making the flow more vigorous. What's more, for a high value of λ ($\lambda = 10$), the effect of m is amplified, further promoting the development of convection cells. From Fig. 2b, which shows the variation of the Nusselt number Nu, an indicator of convective heat transfer, we can see that Nu increases with Ra_T, confirming the increase in heat transfer efficiency as convection intensifies. This effect is further enhanced for high values of m, particularly when λ is large, as the reduction in viscosity facilitates fluid movement and hence heat transfer. Finally, Fig. 2c shows the variation in the Sherwood number Sh, representative of mass transfer. The trends are similar to those observed for Nu, with an increase in Sh as Ra_T increases, and an increasing influence of m, highlighting the joint enhancement of heat and mass transfer in the double-diffusive convection regime. Importantly, at low values of Ra_T, the curves for ψ_{max}, Nu, and Sh remain almost constant, irrespective of m, reflecting the dominance of conductive effects.

Figure 3 shows the combined effects of the thermal Rayleigh number Ra_T, the flow parameter s, and the Pearson number m on flow intensity, heat and mass transfer in

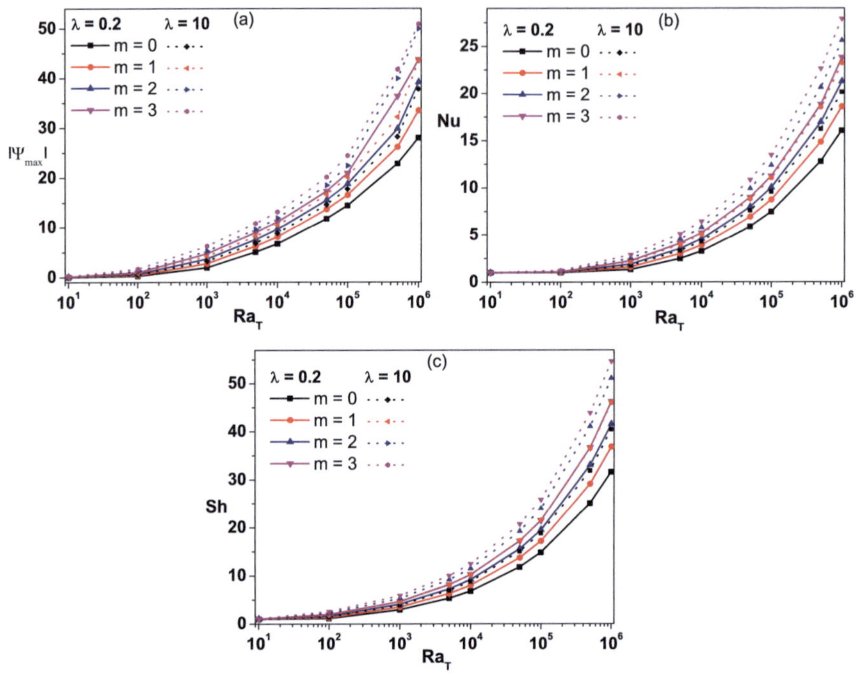

Fig. 2 Rayleigh number effect on flow intensity (Ψ_{max}) and Nusselt (*Nu*) and Sherwood (*Sh*) numbers for $n = 0.8$, $Le = 5$, $N = 1$, $s = 0.01$, ($\lambda = 0.2$, $\lambda = 10$), and ($m = 0, 1, 2, 3$)

a cavity containing a non-Newtonian fluid. In Fig. 3a, the variation of the maximum current function ψ_{max} clearly shows that by comparing the two values of s (0.01 and 0.5), we observe that low values of s amplify the effect of Ra_T, due to a greater sensitivity of viscosity to shear stress, thus enhancing convective dynamics. Figure 3b shows the variation of the Nusselt number *Nu* as a function of Ra_T, reflecting the efficiency of heat transfer. It can be seen that *Nu* increases with Ra_T, and that this growth is more pronounced when m is high and s is low. This combination favors more pseudoplastic fluid behavior and improved fluidity, facilitating thermal convection. Figure 3c shows the variation in Sherwood number *Sh*, an indicator of mass transfer. The trend is similar to that of *Nu*: *Sh* increases with Ra_T, and this increase is amplified for high values of m and low s, indicating that the rheological characteristics of the fluid (low viscosity at high shear) also favor mass transport. In all three sub-figures, a common behavior is observed: for low values of Ra_T, the curves are almost horizontal, showing a regime dominated by conduction. This behavior persists up to a certain critical threshold of Ra_T, beyond which the convective regime becomes dominant.

Figure 4 illustrates the influence of relaxation time constant λ, Pearson number m, and power law index n on flow intensity as well as heat and mass transfer in a non-Newtonian fluid. From Fig. 4a, representing the variation of the maximum stream function ψ_{max}, it can be seen that flow intensity decreases with increasing index n, reflecting less pseudoplastic behavior and therefore higher apparent viscosity,

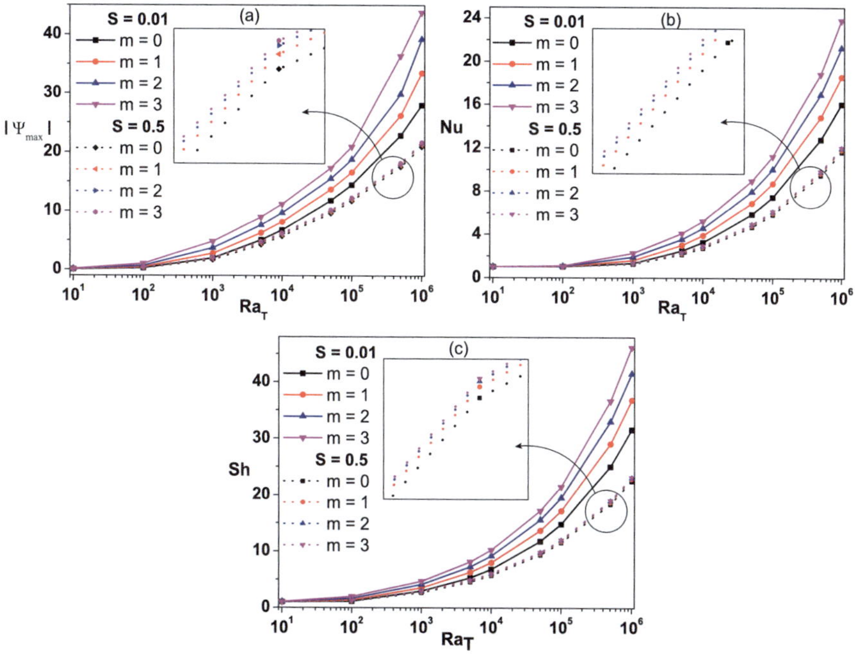

Fig. 3 Rayleigh number effect on flow intensity (Ψ_{max}) and Nusselt (*Nu*) and Sherwood (*Sh*) numbers for $n = 0.8$, $Le = 5$, $N = 1$, $\lambda = 0.2$, ($S = 0.01$, $S = 0.5$), and ($m = 0, 1, 2, 3$)

reducing fluid mobility. On the other hand, an increase in λ leads to a significant rise in ψ_{max}, especially for highly pseudoplastic fluids (low n). This effect is accentuated when m is high: for example, for $\lambda = 0$, ψ_{max} increases from 5.066 to 8.571 when m increases from 0 to 3, demonstrating the thermo-dependent effect on flow dynamics. Figure 4b shows the variation of the Nusselt number *Nu*. The results reveal an increase in *Nu* as n decreases (enhanced pseudo-plasticity), and a significant improvement in heat transfer with increasing λ, particularly for high m: for $\lambda = 0$, *Nu* increases from 2.467 to 3.895 between $m = 0$ and $m = 3$. This improvement is attributed to lower viscosity in high-shear zones, which intensifies thermal convection. Figure 4c shows the variation in the Sherwood number *Sh*. The trends are similar to those observed for *Nu*: *Sh* decreases with increasing n, and increases with λ, while an increase in m further enhances mass transfer. For instance, for $\lambda = 0$, *Sh* increases from 5.156 to 7.640 when m increases from 0 to 3. This behavior can be explained by the same rheological mechanism: the reduction in apparent viscosity in the pseudoplastic regime favors mass transport by intensifying convective motions.

Figure 5 highlights the joint effects of the parameter s, representing the ratio of viscosities at shear rate, and the Pearson number m, expressing the thermo-dependence of viscosity, on flow intensity and on heat and mass transfer in a non-Newtonian fluid. In Fig. 5a, we observe that the maximum current function ψ_{max} increases markedly with m, but only for small values of s (such as $s = 0$ or $s =$

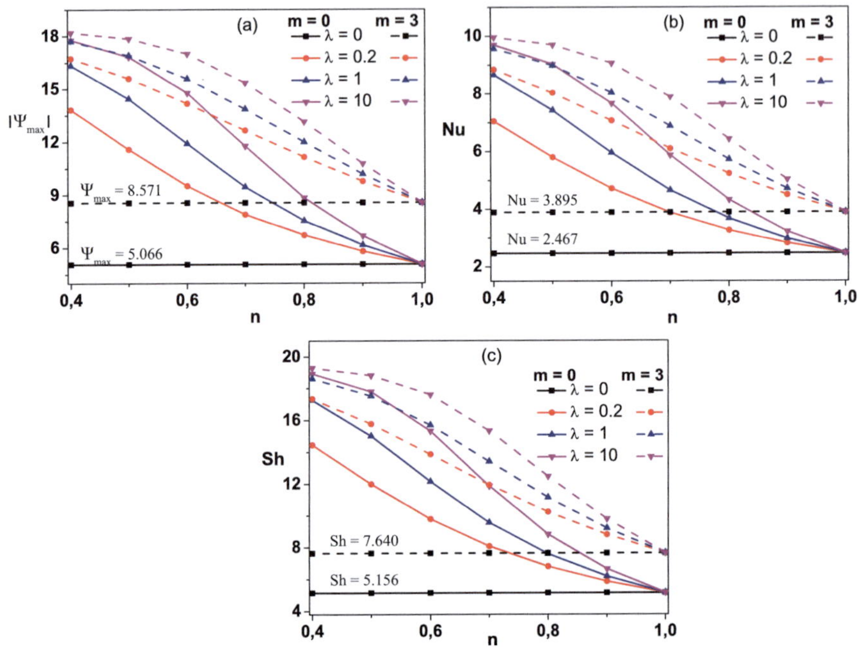

Fig. 4 Effect of λ, m and the power index n on **a** flow intensity (Ψ_{max}), **b** Nusselt number Nu, **c** Sherwood number Sh, for $n = 0.6$, $Le = 5$, $N = 1$, $S = 0.01$, and ($m = 0, 3$)

0.01). This behavior reflects a flow intensification linked to a significant reduction in apparent viscosity in high-shear zones, thus favoring convection. On the other hand, for high values of s (such as s = 0.5 or $s = 1$), the ψ_{max} curve becomes almost horizontal whatever the value of m, indicating that the beneficial effect of thermo-dependence on flow intensity is neutralized by a viscosity less sensitive to shear variations. This implies that, in high-viscosity regimes, flow remains weakly influenced by temperature variations, thus limiting convective development. This is confirmed in Fig. 5b and c, which show the evolution of the Nusselt Nu (heat transfer) and Sherwood Sh (mass transfer) numbers, respectively. For small values of s, increasing m leads to a significant improvement in Nu and Sh, showing that lower apparent viscosity enables better heat and mass transfer. Conversely, for high values of s, both numbers become almost constant, irrespective of m, reflecting a limitation of convective transfer in fluids whose rheology is highly resistant to shear effects.

Figure 5 shows the distributions of the vertical (at $y = 0.5$) and horizontal (at $x = 0.5$) velocity components in the central plane of the cavity, as a function of different values of the time constant λ and Pearson number m, for a non-Newtonian fluid with $n = 0.6$. Results are reported for two contrasting cases of $m = 0$ (without thermo-dependence) and $m = 3$ (with strong thermo-dependence).

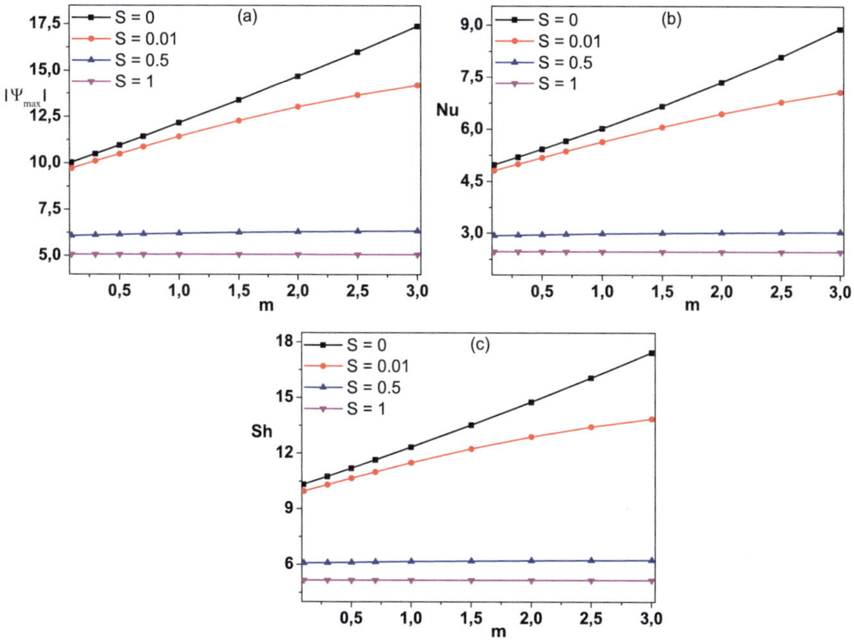

Fig. 5 Effect of parameter s, and Pearson number m on **a** flow intensity, (Ψ_{max}), **b** Nusselt number Nu, **c** Sherwood number Sh, for $n = 0.6$, $Le = 5$, $N = 1$, and $\lambda = 0.2$

We first observe that, whatever the velocity component, increasing the parameter λ leads to an increase in the amplitude of the velocity profiles, both for $m = 0$ and for $m = 3$. This indicates that the time constant λ, associated with the retarded or memory effect in shear-dependent viscosity models, tends to decrease the apparent viscosity of the fluid, thus promoting flow acceleration.

Furthermore, by comparing the profiles for $m = 0$ and $m = 3$, we can see that the effect of m is particularly marked when λ is large. Indeed, increasing m, which models temperature-dependent viscosity sensitivity, accentuates peak velocities in both vertical and horizontal directions. For $m = 3$, the V and U peaks are significantly more pronounced, indicating a significant intensification of natural convection (Fig. 6).

These observations confirm that the combined effect of a high time constant λ and a large Pearson number m favors the development of a more intense flow inside the cavity. This behavior can be attributed to the combined reduction in apparent viscosity induced by both the delayed effect (via λ) and the thermo-dependent effect (via m), enabling the fluid to respond more rapidly to thermal and mass gradients.

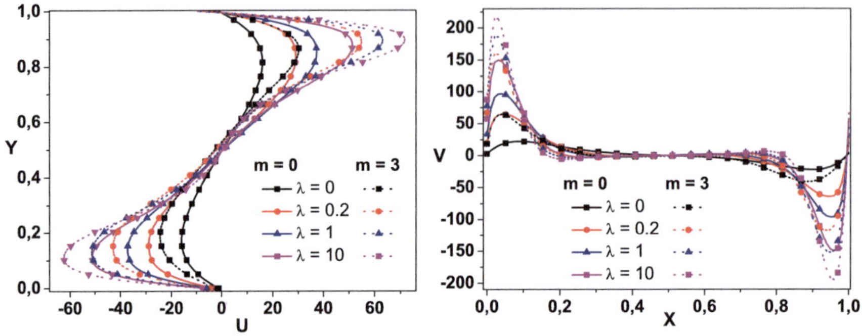

Fig. 6 Vertical (V) and horizontal (U) velocity profiles, in cavity medium ($x = 1/2, y$), ($x, y = 1/2$) respectively, for different m and λ

5 Conclusion

In this study, we investigated the effect of thermal Rayleigh on natural double-diffusive convection in a non-Newtonian fluid obeying the Carreau-Yasuda rheological model. The computational domain is a square cavity whose horizontal walls are considered adiabatic and impermeable, while the vertical walls maintain constant values of temperature and concentration. The analysis focuses on the combined effect of the thermal Rayleigh number (Ra_T), the Pearson number (m), and the characteristic parameters of the Carreau-Yasuda model: the behavior index n, the time constant λ, and the parameter s. The numerical results obtained allow us to draw several important conclusions:

- The increase in the thermal Rayleigh number Ra_T leads to a significant improvement in heat and mass transfer rates.
- The increase in Pearson number m, representing the thermal dependence of viscosity, significantly intensifies heat and mass transfer rates.
- A decrease in the behavior index n and the parameter s leads to an improvement in heat and mass exchanges, reflecting a pseudoplastic behavior favorable to convection.
- An increase in the time constant λ heat and mass transfers, reducing the apparent viscosity of the fluid in zones of high shear, which favors the development of more intense convective structures.

References

1. M Javaheri J Abedi H Hassanzadeh 2010 Linear stability analysis of double-diffusive convection in porous media, with application to geological storage of CO_2 Transp. Porous Media 84 441 456 https://doi.org/10.1007/s11242-009-9513-z

2. CF Chen DH Johnson 1984 Double-diffusive convection: a report on an engineering foundation conference J. Fluid Mech. 138 405 416 https://doi.org/10.1017/S0022112084000173
3. H Ge-JiLe K Javid SU Khan M Raza MI Khan S Qayyum 2021 Double diffusive convection and Hall effect in creeping flow of viscous nanofluid through a convergent microchannel: a biotechnological applications Comput. Methods Biomech. Biomed. Eng. 24 1326 1343 https://doi.org/10.1080/10255842.2021.1888373
4. JS Turner 1974 Double-diffusive phenomena Annu. Rev. Fluid Mech. 6 37 54 https://doi.org/10.1146/annurev.fl.06.010174.000345
5. B Jalili M Emad EH Malekshah P Jalili A Akgül MK Hassani 2024 Investigating double-diffusive natural convection in a sloped dual-layered homogenous porous-fluid square cavity Sci. Rep. 14 7193 https://doi.org/10.1038/s41598-024-57395-2
6. S Eshaghi F Izadpanah AS Dogonchi AJ Chamkha MB Ben Hamida H Alhumade 2021 The optimum double diffusive natural convection heat transfer in H-Shaped cavity with a baffle inside and a corrugated wall Case Stud. Therm. Eng. 28 101541https://doi.org/10.1016/j.csite.2021.101541
7. Muhammad Bilal, S., Zeb Khan, N., Nisar, R.: FEM simulations to analyze flow and thermal characteristics of Carreau non-Newtonian fluid in a square cavity. Int. J. Emerg. Multidiscip.: Math. **2** (2023). https://doi.org/10.54938/ijemdm.2023.02.1.146
8. SAGA Shah A Hassan H Karamti A Alhushaybari SM Eldin AM Galal 2023 Effect of thermal radiation on convective heat transfer in MHD boundary layer Carreau fluid with chemical reaction Sci. Rep. 13 4117 https://doi.org/10.1038/s41598-023-31151-4
9. S Zeb S Ahmad M Ibrahim T Saeed 2022 MHD double-diffusive Carreau fluid flow through a porous medium with variable thermal conductivity and suction/injection Entropy 24 377 https://doi.org/10.3390/e24030377
10. YuI Dimitrienko S Li 2020 Numerical simulation of MHD natural convection heat transfer in a square cavity filled with Carreau fluids under magnetic fields in different directions Comput. Appl. Math. 39 252 https://doi.org/10.1007/s40314-020-01300-w
11. Larbi, A.O., Rebhi, R., Rahal, S., Lorenzini, G., Maamar, L., Menni, Y., Ahmad, H.: Impact of non-Newtonian fluids' rheological behavior on double- diffusive natural convection in an inclined square porous layer. J. Adv. Res. Fluid Mech. Therm. Sci. **99**, 17–47 (2022). https://doi.org/10.37934/arfmts.99.2.1747
12. T Makayssi M Lamsaadi M Kaddiri 2021 Natural double-diffusive convection for the Carreau shear-thinning fluid in a square cavity submitted to horizontal temperature and concentration gradients J. Nonnewton Fluid Mech. 297 104649https://doi.org/10.1016/j.jnnfm.2021.104649
13. Lounis, S., Rebhi, R., Hadidi, N., Lorenzini, G., Menni, Y., Ameur, H. and Sidik, N.A.C: Thermo-solutal convection of Carreau-Yasuda non-Newtonian fluids in inclined square cavities under Dufour and Soret Impacts. CFD Lett. **14**, 96–118 (2022). https://doi.org/10.37934/cfdl.14.3.96118
14. Ramamurthy, T., Krishnan, S.: Importance of temperature-dependent non-Newtonian fluid in conventional thermal processing. J. Therm. Sci. Eng. Appl. **15** (2023). https://doi.org/10.1115/1.4056468
15. El hadoui, B., Kaddiri, M.: Double diffusive natural convection with variable properties of nanofluid using Lattice Boltzmann method. In: Ali-Toudert, F., Draoui, A., Halouani, K., Hasnaoui, M., Jemni A., Tadrist L. (eds.) Advances in Thermal Science and Energy, pp. 22–32. Cham, Springer Nature Switzerland (2024)
16. SAGA Shah A Hassan N Alsubaie A Alhushaybari FM Alharbi AM Galal D-P Burduhos-Nergis C Bejinariu 2022 Convective heat transfer in magneto-hydrodynamic Carreau fluid with temperature dependent viscosity and thermal conductivity Nanomaterials 12 4084 https://doi.org/10.3390/nano12224084
17. Rahmoun, M., Makayssi, T., Lamsaadi, M.: Lewis number's effect on natural double diffusive convection of thermo-dependent non-Newtonian Carreau-Yasuda liquids in a square enclosure. In: ICTEA: International Conference on Thermal Engineering, vol. 1 (2024)
18. MS Astanina M Sheremet CJ Umavathi 2019 Unsteady natural convection in a partially porous cavity having a heat-generating source using local thermal non-equilibrium model Int. J. Numer. Methods Heat Fluid Flow 29 1902 1919 https://doi.org/10.1108/HFF-06-2018-0338

19. Daghab, H., Kaddiri, M., Raghay, S., Arroub, I., Lamsaadi, M., Rayhane, H.: Finite volume simulation of natural convection for power-law fluids with temperature-dependent viscosity in a square cavity with a localized heat source. Int. J. Heat Technol. **39**, 1405–1416 (2021). https://doi.org/10.18280/ijht.390502
20. K Yasuda RC Armstrong RE Cohen 1981 Shear flow properties of concentrated solutions of linear and star branched polystyrenes Rheol. Acta 20 163 178 https://doi.org/10.1007/BF0151 3059
21. Balmforth, N.J., Craster, R.V.: Geophysical Aspects of Non-Newtonian Fluid Mechanics (2001). https://doi.org/10.1007/3-540-45670-8_2
22. Patankar, S.V.: Numerical Heat Transfer and Fluid Flow. CRC Press (2018). https://doi.org/10.1201/9781482234213
23. El Hadoui, B., Kaddiri, M.: Comparing two numerical methods in the case of aiding and opposing natural double diffusion in a square enclosure. In: 2024 4th International Conference on Innovative Research in Applied Science, Engineering and Technology (IRASET), pp. 1–5. IEEE (2024). https://doi.org/10.1109/IRASET60544.2024.10548769
24. G Vahl Davis De IP Jones 1983 Natural convection in a square cavity: a comparison exercise Int. J. Numer. Methods Fluids 3 227 248 https://doi.org/10.1002/fld.1650030304
25. Bennacer, R., Gobin, D.: Cooperating thermosolutal convection in enclosures—I. Scale analysis and mass transfer. Int. J. Heat Mass Transf. **39**, 2671–2681 (1996). https://doi.org/10.1016/0017-9310(95)00350-9

Theoretical Investigation of the Structural, Optical, and Thermoelectric Properties of the Hybrid Organic-Inorganic Perovskite $[NH_3-(CH_2)_4-NH_3]CdCl_4$ Compound: A First-Principles Approach

Hafida Ziouani, Jean-Pierre Tchapet Njafa, Sanaa Mazouar, Taoufik Abdelillah, El Mostafa Khechoubi, and Mahmoud Ettakni

Abstract Hybrid organic-inorganic halide perovskites like $[NH_3-(CH_2)_4-NH_3]CdCl_4$ are promising for optoelectronic applications, yet their temperature-dependent properties are not fully characterized. This work employs density functional theory (DFT) with the GGA-PBE functional and plane-wave pseudopotentials, as implemented in ABINIT, to explore the structural, optical, and thermoelectric behavior of $[NH_3-(CH_2)_4-NH_3]CdCl_4$. Simulations used a 35 Ha energy cutoff and $11\times8\times8$ k-point mesh, with experimental crystal data as input. The structure features inorganic ($[CdCl_4]^{2-}$) layers, organic ($[NH_3-(CH_2)_4-NH_3]^{2+}$) cations, and N–H–Cl hydrogen bonds linking both sublattices. Thermal cycling induces notable bandgap narrowing and enhanced optical absorption, which in turn improve thermoelectric performance and carrier transport. These findings highlight $[NH_3-(CH_2)_4-NH_3]CdCl_4$'s potential for temperature-tunable optoelectronic and thermoelectric applications.

Keywords DFT-GGA · Electronic properties · Bandgap · Optical properties · Solar cells

H. Ziouani (✉) · S. Mazouar · T. Abdelillah · E. M. Khechoubi · M. Ettakni
Department of Physics, Faculty of Science, Materials and Renewable Energy Team, LP2MS Laboratory, Moulay Ismail University, Meknes, Morocco
e-mail: ha.ziouani@edu.umi.ac.ma

J.-P. Tchapet Njafa
Department of Physics, Faculty of Science, University of Yaoundé 1, Yaoundé, Cameroon
e-mail: jean-pierre.tchapet@facsciences-uy1.cm

E. El Mehdi and M. Boutahir (eds.), *Advanced Materials for Sustainable Energy and Engineering*, Engineering Materials, https://doi.org/10.1007/978-3-032-10069-6_8

1 Introduction

The historical development of perovskite materials represents a remarkable scientific journey spanning nearly two centuries. It began in 1839 with German mineralogist Gustav Rose's discovery of a calcium titanate mineral, $CaTiO_3$, in the Ural Mountains, which he named "perovskite" in honor of Lev Aleksevich von Perovski [1–3]. This established the nomenclature for an entire family of structurally related materials. Half a century later, in 1893, Horace L. Wells achieved the first successful synthesis of halide perovskite materials, preparing crystals of $CsPbX_3$ (X = Cl, Br, I) and Cs_4PbX_6 (X = Cl, Br) from aqueous solutions [4]. This pioneering synthetic work transitioned perovskites from purely mineralogical curiosities to scientifically engineered materials. The crystal structure of perovskite was systematically described by Victor Goldschmidt in 1926 through his seminal work on tolerance factors, providing the theoretical framework for understanding their structural stability and enabling the prediction and design of new perovskite materials [5].

The early twentieth century marked the emergence of perovskites as materials of significant technological importance. During the 1940s, military and industrial demands for ferroelectric materials with high dielectric constants led to the groundbreaking discovery of barium titanate ($BaTiO_3$) ceramics [6, 7]. $BaTiO_3$ distinguished itself as the first discovered ferroelectric perovskite, exhibiting properties intimately connected with structural phase transitions [8]. This facilitated the development of the phenomenological theory of ferroelectricity by A.F. Devonshire in the 1950s [9, 10]. This period transformed perovskites into cornerstones of modern electronic and optoelectronic technologies [11], with ferroelectric perovskites like $BaTiO_3$ and lead zirconate titanate $Pb(Zr,Ti)O_3$ finding applications in piezoelectric transducers, multilayer ceramic capacitors, and electro-optical devices [12–14].

The photoconductive properties of cesium lead halides were first recognized in the 1950s by Møller, laying the groundwork for their modern renaissance in photovoltaic applications [15]. Further expanding their versatility, metallic antiperovskites with the formula M_3AB were discovered, exhibiting extraordinary properties such as superconductivity and giant magnetoresistance [16–18]. The modern era of perovskite research is largely defined by the emergence of organic-inorganic hybrid perovskites for photovoltaic applications, culminating in breakthrough work demonstrating their exceptional potential for high-efficiency solar cells [19, 20]. This pivotal development has led to an unprecedented surge in perovskite research, with applications extending from photovoltaics to light-emitting devices, lasers, and photodetectors [21, 22]. The historical trajectory from Rose's initial mineralogical discovery to today's sophisticated hybrid organic-inorganic perovskites exemplifies the profound impact of systematic materials research and the continuous evolution of scientific understanding, positioning these materials at the forefront of next-generation technologies due to their unique combination of structural flexibility, exceptional optoelectronic properties, and solution processability.

Given their abundance in nature, low cost, and ease of deposition into thin films, organic- inorganic halide hybrid perovskites offer considerable potential for usage in commercial applications [23].

Thermoelectric materials are of universal interest due to their capacity to transform heat into electric energy, rendering them a significant source of energy [24, 25]. Currently, the world requires thermoelectric supplies to reduce its reliance on fossil fuels. Perovskite compounds exhibit exceptional thermoelectric properties, which have prompted the development of thermoelectric device appliances. In the field of high-temperature thermoelectric research, germanium-based chloro perovskites have garnered significant attention in recent years [26]. Lead is a toxic substance for the environment, whereas hybrid perovskite is a free material and therefore environmentally benign. Consequently, it can be employed in photovoltaic applications with minimal environmental impact.

The crystal structure of $[NH_3-(CH_2)_4-NH_3]CdCl_4$ is composed of an infinite number of inorganic layers. Each layer is formed by corner-sharing CdCl6 octahedra, arranged in a two- dimensional perovskite-like structure. The alkyl ammonium (or alkylenedi ammonium) chains are oriented almost perpendicular to the planes of the layer. The NH_3 groups at the end of the organic chain engage in N–H–Cl hydrogen bonds with the octahedra.

Previous studies have investigated thermal and crystallographic analyses [27]. Evidence of five crystalline forms, designated V, IV, III, II, and I, has been identified. It has been established that the various phase transitions are either irreversible or reversible. The various phase transitions of this complex can be associated with the conformational evolutions of the organic part $[NH_3-(CH_2)_4-NH_3]^{2+}$, thanks to structural determinations of the "limit" Phases five (**V**: the initial room-temperature form) and one (**I**: the new room-temperature form) [28]. A review of the literature reveals that an increase in temperature at **Phase V** results in an irreversible change from the "gauche" conformation of the organic chain to an "all-trans" conformation, with the final conformation retained during subsequent phase transitions. In addition to these crystallographic studies, recent work has investigated the electronic and optical properties of similar hybrid perovskites, such as $[NH_3–(CH_2)_4–NH_3]MX_4$ (M = Cu, Mn), using Density Functional Theory (DFT) [29]. These studies have revealed that the substitution of Cu and Mn significantly alters the electronic and optical behaviors, making these compounds promising candidates for optoelectronic applications.

In this work, we will study the effect of Phase transitions, especially the **Phase V** and **I**, on the structural, electronic, optical, and thermoelectric properties of the compound $[NH_3-(CH_2)_4-NH_3]CdCl_4$.

2 Computational Details

The comprehensive theoretical investigation of the structural, electronic, optical, and thermoelectric properties of $[NH_3-(CH_2)_4-NH_3]CdCl_4$ was conducted using the advanced Density Functional Theory+U (DFT+U) methodology as implemented within the highly sophisticated ABINIT computational package [30, 31]. This state-of-the-art approach represents a significant enhancement over conventional DFT methods, incorporating a carefully calibrated Hubbard-like correction term specifically designed to address the inherent limitations of standard DFT in treating strongly correlated electron systems. The DFT+U formalism provides explicit and accurate treatment of on-site electron-electron Coulomb repulsion interactions within spatially localized d and f orbitals, which are particularly crucial for transition metal compounds and systems exhibiting strong electron correlation effects [32, 33].

The computational framework employed the rigorously validated double counting correction scheme known as the Full Localized Limit (FLL) approach [34], which ensures proper treatment of the electron-electron interaction energy while avoiding spurious double counting contributions. The exchange-correlation energy functional was described using the Generalized Gradient Approximation (GGA) within the Perdew-Burke-Ernzerhof (PBE) parametrization [35], which provides an excellent balance between computational efficiency and accuracy for describing the electronic structure of hybrid organic-inorganic perovskite materials. The PBE functional has been extensively benchmarked and demonstrated to yield reliable structural parameters and electronic properties for similar perovskite systems.

The sophisticated treatment of electron-ion interactions was accomplished through the implementation of the Projector Augmented-Wave (PAW) method [36, 37], a highly accurate all-electron approach that maintains the computational efficiency of pseudopotential methods while preserving the complete nodal structure of the electronic wavefunctions in the core regions. This methodology ensures accurate representation of both valence and semi-core electronic states, which is essential for obtaining reliable electronic band structures and optical properties. The PAW approach eliminates the transferability issues commonly associated with norm-conserving pseudopotentials and provides superior accuracy for calculations involving transition metal atoms such as cadmium.

The electronic wavefunctions throughout the computational domain were systematically expanded using a comprehensive plane-wave basis set, with a carefully optimized kinetic energy cutoff of 35 Ha (approximately 952 eV) applied to all calculations. This cutoff value ensures that the total energy, atomic forces, and stress tensor components were converged to within acceptable numerical precision [38]. The chosen cutoff represents an optimal compromise between computational accuracy and efficiency, ensuring that all relevant electronic states are adequately described while maintaining reasonable computational costs for the large supercell calculations required for this hybrid perovskite system.

The ground-state electronic structure determination involved a systematic two-step computational protocol. Initially, self-consistent field (SCF) calculations were performed with stringent convergence criteria to obtain the equilibrium charge density distribution and the corresponding ground-state total energy. Subsequently, non-self-consistent field (NSCF) calculations were carried out using the converged charge density as a fixed input, enabling the precise determination of electronic band structures and density of states over extended energy ranges without the computational overhead of charge density optimization.

The numerical integration over the irreducible Brillouin zone was performed using the well-established Monkhorst-Pack (MP) k-point sampling scheme [39], with a carefully optimized grid density of $11 \times 8 \times 8$ k-points. This k-point mesh was selected based on systematic convergence tests to ensure that all calculated properties, including total energies, band gaps, and density of states, were adequately converged with respect to k-point sampling density. The asymmetric grid dimensions reflect the different periodicities along the crystallographic axes of the layered perovskite structure, with denser sampling along directions of higher electronic dispersion.

The initial structural parameters and atomic coordinates for both crystalline phases of $[NH_3-(CH_2)_4-NH_3]CdCl_4$ were obtained from the comprehensive experimental crystallographic data reported in the literature [27], which served as the fundamental reference point for all subsequent computational investigations. These experimental structures provided essential starting geometries for the theoretical optimization procedures and enabled direct validation of the computational methodology through comparison with established structural parameters. The computational approach was further validated by reproducing key experimental observations, including lattice parameters, bond lengths, and phase transition characteristics, thereby establishing confidence in the theoretical framework employed throughout this investigation.

3 Results and Discussion

3.1 Structural Properties

Following comprehensive geometry optimization procedures, the systematic fitting of the Birch-Murnaghan equation of state [40, 41] was employed to determine the equilibrium structural parameters. This rigorous approach resulted in fully optimized lattice parameters for both crystalline phases of $[NH_3-(CH_2)_4-NH_3]CdCl_4$ using the GGA-PBE exchange-correlation functional approximation [42]. The calculated structural data are comprehensively presented in Fig. 1 and Table 1, demonstrating remarkable agreement with previously reported experimental measurements [27], thereby validating the computational methodology's accuracy and reliability. The structural optimization protocol involved iterative energy minimization until convergence criteria were satisfied, ensuring thermodynamically stable configurations [43].

Furthermore, it is particularly noteworthy that all subsequent material properties investigated in this comprehensive study—including electronic band structures,

Table 1 Parameter of the two Phases of $[NH_3-(CH_2)_4-NH_3]CdCl_4$

	Crystal lattice structure	
	Monoclinic	Monoclinic
Crystaline system	**Phase V**: initial room-temperature form	**Phase I**: new room-temperature form
a	7.657 Å	7.344 Å
b	7.585 Å	7.485 Å
c	9.541 Å	10.775 Å
β	101.560°	91.060$_\circ$
Z	2	2
Space group	$P2_1/a$	$P2_1/a$

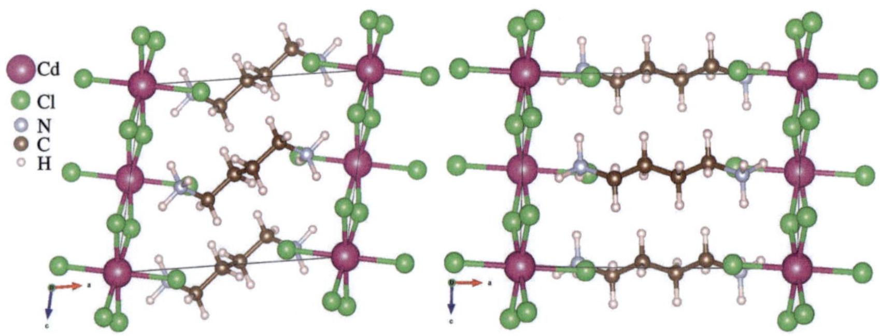

(a) **Phase V** (b) **Phase I**

Fig. 1 Crystal structure of $[NH_3-(CH_2)_4-NH_3]CdCl_4$

optical absorption characteristics, and thermoelectric transport coefficients—were systematically calculated based on these fully relaxed structural configurations. This meticulous approach involved complete optimization of lattice constants, precise determination of atomic positions within the unit cell, and minimization of total free energies of the supercell structures [44]. Such thorough structural relaxation ensures that the calculated properties accurately reflect the ground-state configurations of the material phases under investigation, providing a solid foundation for reliable property predictions and comparative analysis between the different polymorphic forms.

3.2 *Electronic Properties*

The electronic structure of a material plays a pivotal role in determining its physical characteristics, including electrical conductivity, optical absorption, and potential

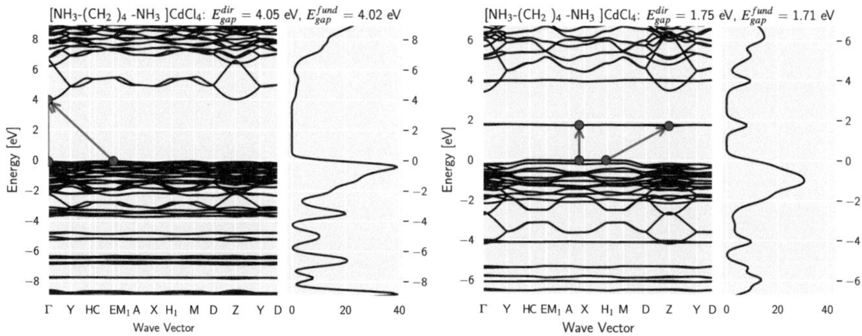

(a) **Phase V** (b) **Phase I**

Fig. 2 Electronic band structures of $[NH_3{-}(CH_2)_4{-}NH_3]CdCl_4$. The Fermi level is set at 0 eV in both plots

Table 2 Gap energy values obtained by DFT+U method for the two structural phases of $[NH_3{-}(CH_2)_4{-}NH_3]CdCl_4$

Crystal structure	E_g (eV)	Fermi level (eV)	Gap type
Phase V	4.020	− 0.463	Indirect
Phase I	1.710	0.074	Indirect

applicability in electronic and optoelectronic devices. Among these properties, the density of states (DOS) serves as a fundamental descriptor, offering insight into the available electronic states at each energy level and elucidating the nature of chemical bonding and atomic orbital hybridization [45]. In this study, both total (TDOS) and partial (PDOS) densities of states were computed at equilibrium geometries using the Generalized Gradient Approximation (GGA) within the PerdewâŁ"BurkeâŁ"Ernzerhof (PBE) exchange-correlation functional framework [42]. The energy spectrum was examined in the range of −12 to 12 eV, taking the Fermi level as the reference point at 0 eV, as shown in Fig. 2.

To determine the bandgap with improved accuracy, the Density Functional Theory plus Hubbard U (DFT+U) method was employed in conjunction with GGA-PBE. This approach is particularly effective for accounting for localized electronic states, especially in systems with transition metals, where standard DFT often underestimates bandgap values [?] [46]. The obtained energy gaps are summarized in Table 2.

Electronic Structure of the Phase V A detailed analysis of the DOS and PDOS plots for the **Phase V** polymorph of $[NH_3{-}(CH_2)_4{-}NH_3]CdCl_4$ (see Figs. 2 and 3) reveals the contributions from atomic orbitals of each constituent element. These insights are essential for understanding orbital hybridization, charge distribution, and electronic transitions [47].

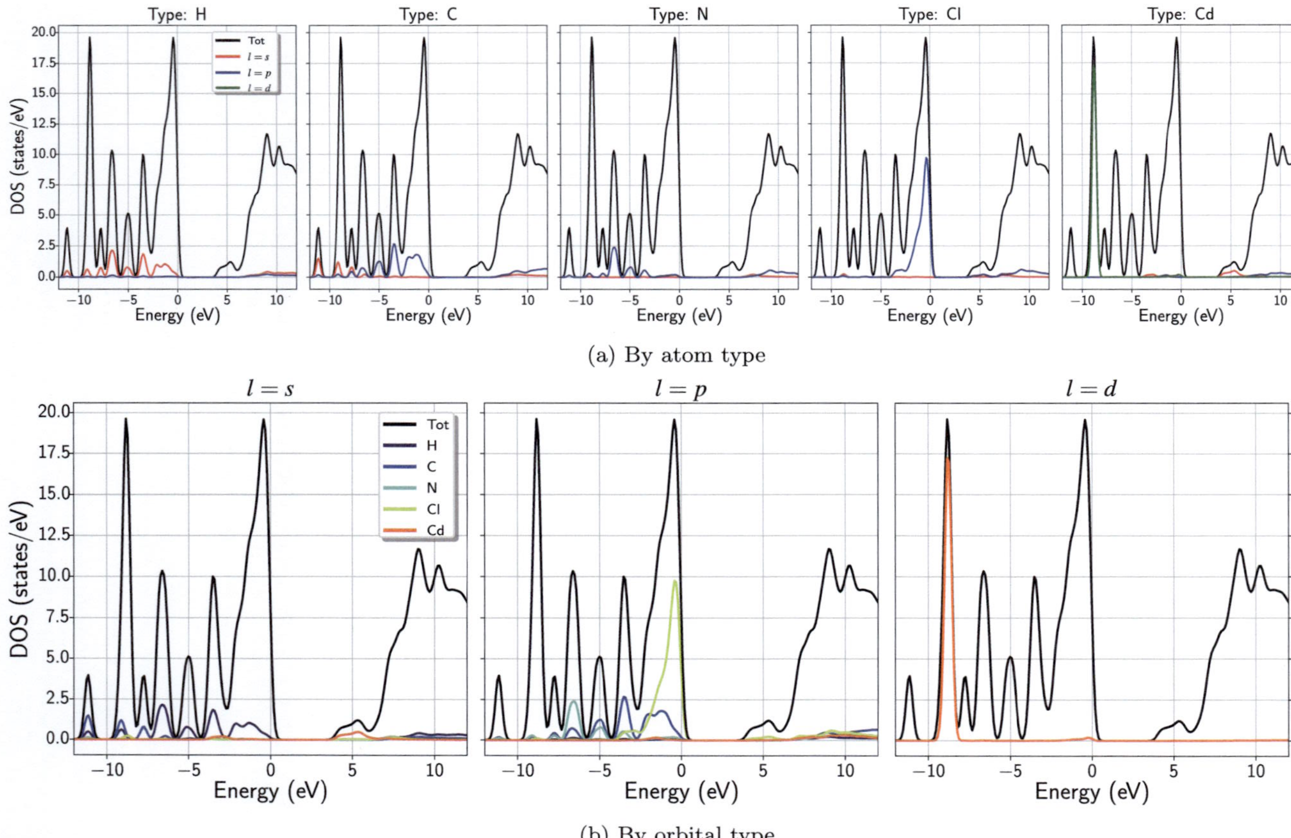

Fig. 3 Total and partial density of states (TDOS and PDOS) for $[NH_3-(CH_2)_4-NH_3]CdCl_4$ in **Phase V**. The graphs illustrate orbital contributions from each atomic species

- **Hydrogen (H)**: The PDOS indicates that hydrogen's contribution is almost exclusively through its $1s$-orbital, which is expected due to its single valence electron. This orbital exhibits non-negligible intensity throughout the energy window, especially in the valence band region.
- **Carbon (C)**: Carbon atoms contribute predominantly through their $2p$-orbitals, which display multiple sharp peaks across the valence and conduction bands. The $2s$-orbital contribution is minimal, reflecting the hybridization state of carbon in the molecular structure, possibly sp^2 or sp^3-like character depending on local geometry.
- **Nitrogen (N)**: Similar to carbon, nitrogen shows a dominant $2p$-orbital contribution. The presence of lone pair electrons and higher electronegativity of nitrogen may result in more localized states in the valence band, as reflected in the PDOS peaks.
- **Chlorine (Cl)**: The $3p$-orbitals of chlorine contribute significantly across the energy spectrum, particularly in the valence region. The $3s$-orbital contribution is marginal, consistent with the electronic configuration of chlorine and its role as an electronegative, bonding atom.
- **Cadmium (Cd)**: Notably, cadmium exhibits substantial $4d$-orbital activity. Peaks in the valence band are strongly associated with these orbitals, while minor $5p$ contributions are discernible in the conduction band. This behavior is typical of group 12 elements and aligns with earlier electronic structure studies.

The calculated bandgap for this phase, with $E_g^{dir} = 4.050\,\text{eV}$ and $E_g^{fund} = 4.200\,\text{eV}$, is relatively wide, clearly indicating a semiconducting behavior. The indirect nature of the bandgap, as seen in the band structure, implies reduced optical absorption efficiency compared to direct-bandgap materials. However, the magnitude of the gap situates this material as a promising candidate for ultraviolet optoelectronic devices, including UV photodetectors and transparent electronics [48].

In conclusion, the electronic properties of the **Phase V** phase of $[NH_3-(CH_2)_4-NH_3]CdCl_4$ underscore the important role of p-orbitals in light elements and d-orbitals in cadmium. The observed bandgap supports potential applications in high-bandgap semiconductors, where thermal excitation of carriers is minimized, thus enhancing device stability at elevated temperatures.

Electronic Structure of the Phase I Following a similar analytical approach, the electronic structure of the **Phase I** polymorph of $[NH_3-(CH_2)_4-NH_3]CdCl_4$ is examined through PDOS and TDOS, as illustrated in Figs. 2 and 4. This analysis provides a deeper understanding of orbital interactions, the electronic environment of each atom, and the nature of the states surrounding the Fermi level, which are critical in determining the charge transport and optical characteristics of the material [49].

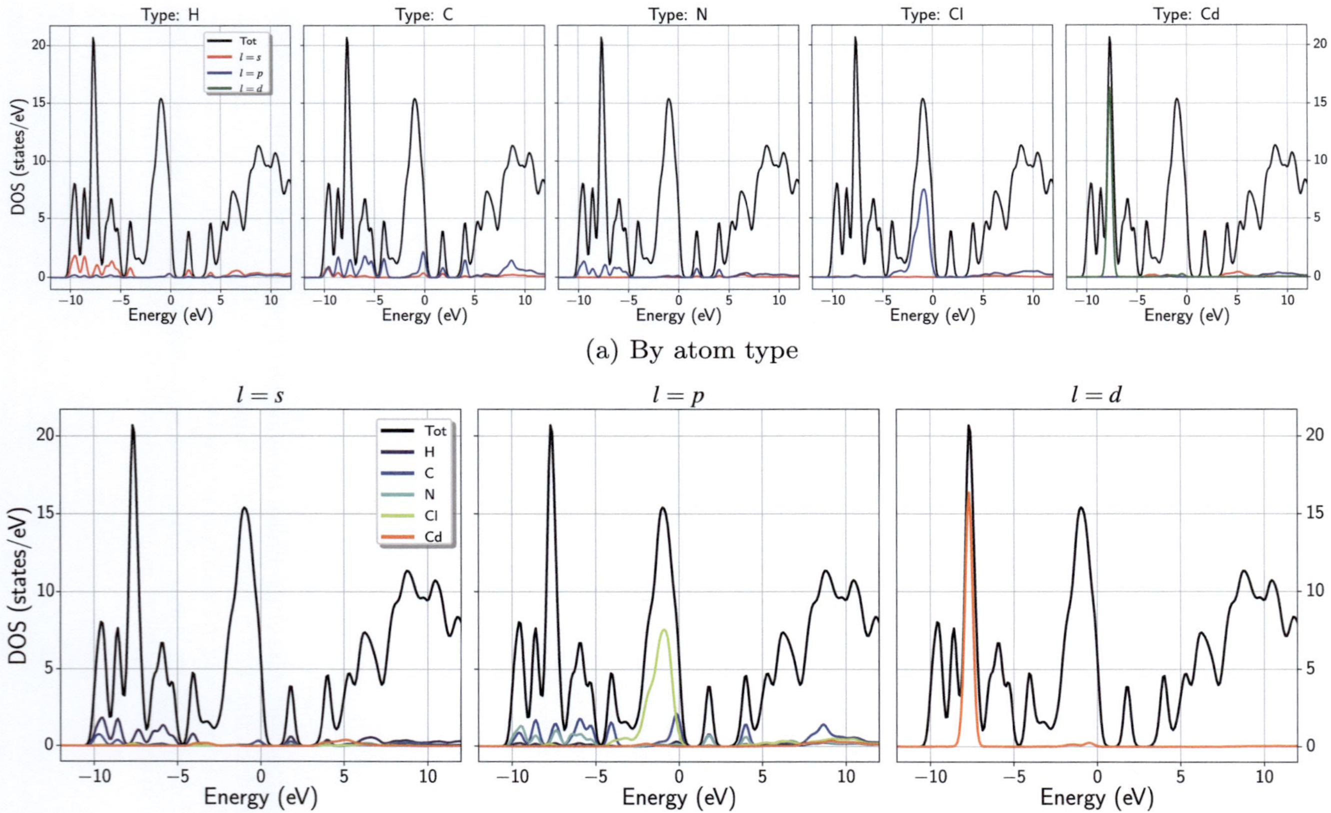

(b) By orbital type

Fig. 4 Total and partial density of states (TDOS and PDOS) for $[NH_3-(CH_2)_4-NH_3]CdCl_4$ in **Phase I**. The graphs illustrate orbital contributions from each atomic species

- **Hydrogen (H)**: The PDOS indicates that hydrogen mainly contributes through its $1s$-orbital in the energy range below the Fermi level, with a modest and expected decay beyond it. The negligible presence of p-orbital character reaffirms hydrogen's electron configuration and limited bonding complexity in this matrix.
- **Carbon (C)**: Carbon atoms exhibit a significant presence of $2p$-orbital states both below and just above the Fermi level. This suggests possible delocalized π-bonding interactions or conjugation effects, depending on the molecular geometry and bonding topology. The $2s$-orbital also contributes, although to a lesser extent, indicating a degree of orbital hybridization.
- **Nitrogen (N)**: The nitrogen PDOS reveals dominant $2p$-orbital peaks around the Fermi level, with minor $2s$ contributions. These sharp features are indicative of localized states, potentially associated with lone pair electrons or bonding within electron-rich functional groups. Such a configuration can influence the dielectric properties and reactivity of the structure [50].
- **Chlorine (Cl)**: Similar to the **Phase V** phase, chlorine's $3p$-orbitals dominate its contribution to the DOS. The peaks are distributed symmetrically around the Fermi level, suggesting active participation in both bonding and anti-bonding states. The $3s$-orbital remains nearly inactive in the energy window of interest, consistent with chlorine's electronic configuration.
- **Cadmium (Cd)**: The cadmium atoms again exhibit a strong $4d$-orbital presence, with multiple peaks below the Fermi level, and a moderate $5p$-orbital contribution in the conduction band. The $5s$-orbitals are nearly absent, which aligns with the filled nature of the $4d^{10}$ configuration and weak hybridization of s-states in such systems [47].

The band structure reveals a direct bandgap of $E_g^{dir} = 1.750\,\text{eV}$ and a fundamental bandgap of $E_g^{fund} = 1.710\,\text{eV}$, confirming the semiconducting nature of the **Phase I** phase. This lower bandgap compared to the **Phase V** structure suggests enhanced absorption in the visible spectrum, making this phase suitable for applications such as photovoltaic devices or visible-light photodetectors [51].

An important observation is the presence of electronic states near the Fermi level, as indicated by the PDOS. This implies potential for increased electrical conductivity due to thermally activated charge carriers. The proximity of occupied and unoccupied states also suggests that excitonic effects could be non-negligible, which may influence the optical response and carrier dynamics. Such features are often exploited in organic-inorganic hybrid systems and layered semiconductors for tunable performance characteristics.

In summary, the PDOS analysis of the **Phase I** phase reveals that the p-orbitals of H, C, N, and Cl are the principal contributors to the valence and conduction bands, with cadmium's d-orbitals playing a vital role in shaping the deeper valence states. The presence of a moderate, direct bandgap enhances the suitability of this phase for visible-light optoelectronic applications. Furthermore, the electronic configuration suggests a more conductive and responsive behavior under illumination or external bias, which can be harnessed in devices requiring efficient charge separation and transport.

3.3 Optical Properties

The study of the optical properties of solids has proven to be a powerful tool in understanding the electronic properties of materials. In this section, the study of optical properties is carried out using the GGA-PBE approximation, which has been shown to be successful in determining the optics bandgap with appreciable precision. The optical properties of any material can be described by the complex dielectric function (see Eq. (1)) [40]:

$$\varepsilon(\omega) = \varepsilon_1(\omega) + i\varepsilon_2(\omega). \tag{1}$$

It describes the optical response of the medium at all photon energies $E = \hbar\omega$. The real part of the dielectric function follows from the KramersâŁ"Kronig relation (2) [40]:

$$\varepsilon_1(\omega) = 1 + \frac{2}{\pi} P \int_0^\infty \frac{\varepsilon_2(\alpha)}{\alpha^2 - \omega^2} \alpha \mathrm{d}\alpha, \tag{2}$$

where P implies the principal value of the integral.

The imaginary part, $\varepsilon_2(\omega)$, in the long wavelength limit, has been obtained directly from the electronic structural calculation (see Eq. (3)) [52]:

$$\varepsilon_2(\omega) = \frac{e^2\hbar}{\pi m^{2\omega^2}} \sum_{v,c} \int_{BZ} |M_{cv}(k)|^2 \, \delta\left[\omega_{cv}(k) - \omega\right] d^3k, \tag{3}$$

with $M_{cv}(k)$ the transition moment elements.

The following formula (4) is employed to calculate the absorption coefficient [40, 52, 53]:

$$\alpha(\omega) = \frac{\sqrt{2}}{c} \omega \sqrt{-\varepsilon_1(\omega) + \sqrt{\left(\varepsilon_1^2(\omega) + \varepsilon_2^2(\omega)\right)}}. \tag{4}$$

The optical conductivity is given by the Eq. (5) [35]:

$$\mathtt{Re}\left(\sigma_{\alpha\beta}(\omega)\right) = \frac{\omega}{4\pi} \mathtt{Im}\left(\varepsilon_{\alpha\beta}(\omega)\right). \tag{5}$$

The calculation of reflectivity is very important to evaluate the optical properties of materials; it makes it easy to understand the status and behavior of transparency, based on the reflectivity that we can draw from the transmittance. The reflectivity $R(\omega)$ (see Eq. (6)) follows directly from the Fresnel's formula [52]:

$$R(\omega) = \left| \frac{\sqrt{\varepsilon(\omega)} - 1}{\sqrt{\varepsilon(\omega)} + 1} \right|^2 . \tag{6}$$

It is also recommended to calculate the energy versus the wavelengths by varying the refractive index. This allows for an understanding of how the material can behave and how it can be canalized. To confirm the results found by the optical absorption and the transmittance, the refractive index $n(\omega)$ is given as follows in Eq. (7) [40]:

$$n(\omega) = \sqrt{\frac{\varepsilon_1(\omega) + \sqrt{\varepsilon_1^2(\omega) + \varepsilon_2^2(\omega)}}{2}} . \tag{7}$$

This section presents a study of the effect of temperature on the optical properties of $[NH_3-(CH_2)_4-NH_3]CdCl_4$, including reflectivity and optical conductivity. Due to symmetry, the optical properties differ for the three directions, xx, yy and zz. The subsequent analysis will focus on the xx direction which is the most favorable for light production (the same consideration is also made for Sect. 3.4).

It is well established that the imaginary part of the dielectric function $\varepsilon_2(\omega)$ is closely related to the electronic band structure and exhibits light absorption in the material [54]. Conversely, the real part of the dielectric function $\varepsilon_1(\omega)$ refers to the dispersion of the incident photons by crystals [55] and provides information regarding polarization [56]. The calculated static dielectric constants $\varepsilon_1(\omega)$ are equal to 121.046 0, and 55.535 9 for **Phase V** and **Phase I** of the compound, respectively (see Fig. 5g, h). As shown by the results, $\varepsilon_1(\omega)$ reaches a maximum at 0.0027 eV for **Phase V**. For **Phase I**, $\varepsilon_1(\omega)$ reaches a maximum at 0.1170 eV, then decreases and becomes negative reaching a minimum at 0.1905 eV. Figure 5i, j show a good correspondence between the imaginary part $\varepsilon_2(\omega)$ and the absorption peaks. For **Phase I**, the dielectric function increases sharply after 7 eV and 8 eV, giving the absorption edge. This threshold value comes mainly from the electronic transitions between the maxima of the valence band and the minima of the forbidden conduction band. Usually, it refers to the energy of the forbidden band, which agrees with the total density of this Phase. This result indicates that there is a reduction in the bandgap, which is in agreement with the results found in electronic properties (Table 2).

Among the essential properties for photovoltaic applications is the reflectivity and the capacity of a sample surface to reflect incident radiation flux. Figure 5c, d illustrate that reflectivity $R(\omega)$ decreases with an increase in wavelength, with the highest value of reflectivity which is 12 for **Phase V** and 8 for **Phase I**.

The extinction coefficient for the xx direction is described in Fig. 5e, f. They demonstrate that the extinction coefficient is high at low photon energies and decreases as the energy increases. The values of the peaks for **Phase V** are 4.701 3, 2.214 8, 1.893 7 and 1.721 5 at an energy range of 0 to 1.7 eV, indicating specific

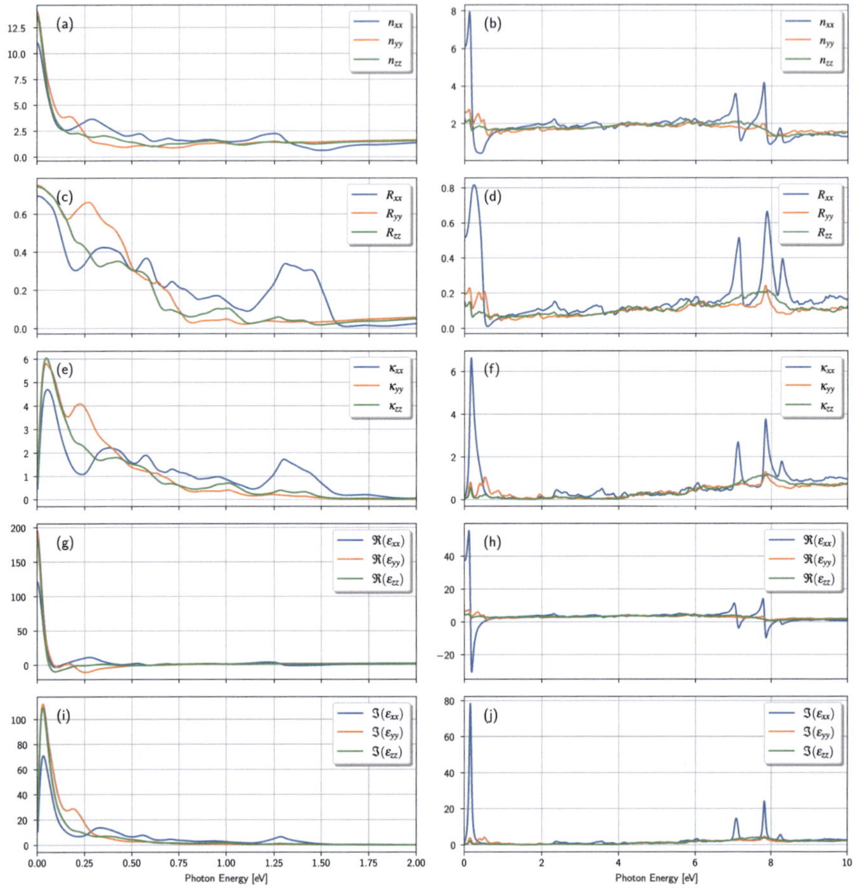

Fig. 5 Optical properties of $[NH_3{-}(CH_2)_4{-}NH_3]CdCl_4$. The left panel displays the optical properties of **Phase V**, while the right panel presents the optical properties of **Phase I**. Subplots **a** and **b** illustrate the reflective indexes, **c** and **d** depict the refractivities, **e** and **f** demonstrate the extinction coefficient k, **g** and **h** present the real part of the dielectric function, and **i** and **j** illustrate the imaginary part of the dielectric function

absorption bands. While the peaks for **Phase I** are 6.651 4, 2.688 9, 3.761 1 and 1.784 6 at an energy range of 0 to 8.5 eV. A high extinction coefficient at low photon energies indicates significant absorption, which corresponds to electronic transitions or interband absorption. Additionally, these peaks represent resonant absorption where the material absorbs light more effectively at specific energies.

These results indicate that $[NH_3{-}(CH_2)_4{-}NH_3]CdCl_4$ exhibits strong interaction with light at low photon energies, with high refractive indices, high reflectivity, and significant absorption. As the photon energy increases, these optical properties indicate reduced interaction, with lower refractive indices, lower reflectivity, and reduced absorption. These properties are consistent with typical dielectric materials, showing strong responses at lower energies and stabilizing at higher energies, making them potentially useful in applications where strong light interaction at low energies is desired, such as in optical coatings, sensors, or devices that require high refractive indices and strong absorption. The decrease in these properties at higher energies suggests good transparency and reduced reflection, which is beneficial for optoelectronic applications.

3.4 Thermoelectric Properties

We have employed BoltzTrap2 software [57] to obtain thermo-electric properties of the two phases (**Phase V** and **Phase I**) of $[NH_3-(CH_2)_4-NH_3]CdCl_4$ with a constant relaxation time τ_0. In order to ascertain the thermoelectric (TE) behaviour of both **Phase V** and **Phase I** in detail, the following characteristics have been expressed as a function of chemical potential: the Seebeck coefficient (S), electrical conductivity (σ), power factor (PF), electronic specific heat (Cv), electronic thermal conductivity (κ), number of concentration carriers (n) and density of states (DOS). The calculated characteristics have been plotted for a range of temperatures between 200 and 600 K.

The figure of merit, ZT, defines thermoelectric efficiency:

$$ZT = \frac{S^2 \sigma T}{\kappa}, \tag{8}$$

where S is the Seebeck coefficient, σ the electrical conductivity, κ the total thermal conductivity (electronic and lattice contributions), and T the absolute temperature. Materials with $ZT \geq 1$ are considered promising for thermoelectric applications [58, 59]. Strategies like doping, band engineering, and nanostructuring are often used to enhance ZT by decoupling electrical and thermal transport [60].

In Fig. 6b, the electric conductivity as a function of chemical potential at room temperature first increases for both phases. It reaches its global maximum value. Then it decreases and reaches its global minimum after the null value of the chemical potential at $\mu - \varepsilon_F = 0.0193$ Ha for **Phase V** and $\mu - \varepsilon_F = 0.0206$ Ha for **Phase I**.

Figure 6c displays the electronic thermal conductivity as a function of chemical potential. The behavior of the electronic thermal conductivity curves is nonlinear increasing for both compounds. The maximal electronic thermal conductivity of **Phase V** at RT is equal to 1.0416×10^{14} W m^{-1} K^{-1}, while it is 0.3264×10^{14} W m^{-1} K^{-1} for **Phase I**.

Figure 6d shows how the power factor (PF) curves changed when the chemical potential increased for both compounds. They reach their maximum values of 1.2868×10^{11} W m^{-1} K^{-1} s^{-1} for **Phase V** and 0.3249×10^{11} W m^{-1} K^{-1} s^{-1} for **Phase I**. The findings indicate that the compound in **Phase V** exhibits superior performance compared to the compound in **Phase I**, contingent upon the utilization

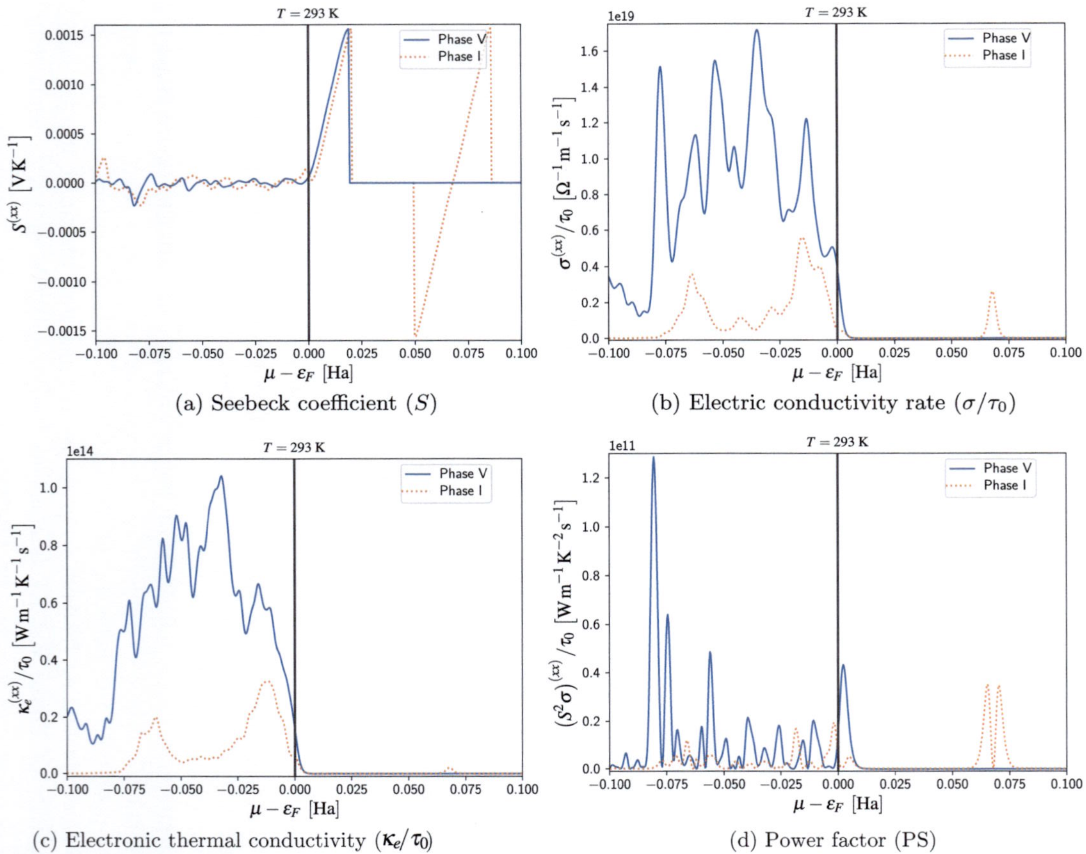

Fig. 6 Transition coefficients as a function of different ground state chemical potential at room temperature for the **Phase V** (blue line) and **Phase I** (dotted orange line) compounds

of an appropriate element dopant at specified concentrations and/or through band engineering or transition to nanostructures. It is feasible to synthesize perovskite at the nanoscale, as evidenced by the reduction in lattice thermal conductivity. This will enhance the augmentation of thermoelectric efficiency.

Figure 7 illustrates the electronic-specific heat capacity as a function of chemical potential at different temperatures. The calculated maximum value of the specific heat capacity of **Phase V** in this study is 16.884 2 J mol^{-1} K^{-1} and 11.964 3 J mol^{-1} K^{-1} for **Phase I** at RT. The Cv values for **Phase V** are slightly greater than those of **Phase I**.

Figure 8 illustrates the variation in the density of the state around the Fermi level as a function of temperature. For the two alloys, **Phase V** and **Phase I**, the bandgap persists over a wide range of temperatures.

Figure 9 presents the electrical conductivity in **Phase V** (left panel) and **Phase I** (right panel). In **Phase V**, the electrical conductivity of $[NH_3-(CH_2)_4-NH_3]CdCl_4$ exhibits multiple peaks at different temperatures. These peaks suggest changes in conductivity associated with Phase transitions or structural modifications. The complex behavior likely arises from interactions between the $[CdCl_4]^{2-}$ anions and the organic cations $[NH_3-(CH_2)_4-NH_3]^{2+}$. In **Phase I**, there is a single prominent peak in the electrical conductivity which means that the overall conductivity is significantly lower compared to **Phase V**. This finding indicates that **Phase I** has different conductive properties, due to a distinct arrangement of ions and altered crystal structure caused by the effect of temperature. Structural Considerations like the structural dynamics of $[NH_3-(CH_2)_4-NH_3]CdCl_4$ likely play a crucial role in determining its electrical behavior. And Changes in hydrogen bonding (such as N–H–Cl interactions) and the arrangement of $[CdCl_4]^{2-}$ anions contribute to the observed conductivity variations. This is consistent with the sensitivity of thermoelectric properties to hydrogen bonding and lattice dynamics [61].

Figure 10 illustrates the carrier concentration in the two phases. The carrier concentration for both phases appears to be relatively constant across the distance measured. The carrier concentration for **Phase I** is slightly higher than that of **Phase V**. As is well known, the carrier concentration is typically used to refer to the number of charge carriers per unit volume in a semiconductor. In this context, a charge carrier could be an electron or a hole. The results demonstrate that the temperature at which the measurement is taken affects the carrier concentration.

4 Conclusion

Density Functional Theory calculations have elucidated the structural, electronic, optical, and thermoelectric properties of $[NH_3-(CH_2)_4-NH_3]CdCl_4$ in its **V** and **I** phases. Phase transitions markedly influence these properties. The GGA-PBE optimized lattice parameters align well with experiment, confirming computational accuracy. The bandgap reduction from 4.020 eV (**Phase V**) to 1.710 eV (**Phase I**) reflects enhanced electronic conductivity. Optical analyses revealed stronger absorption in

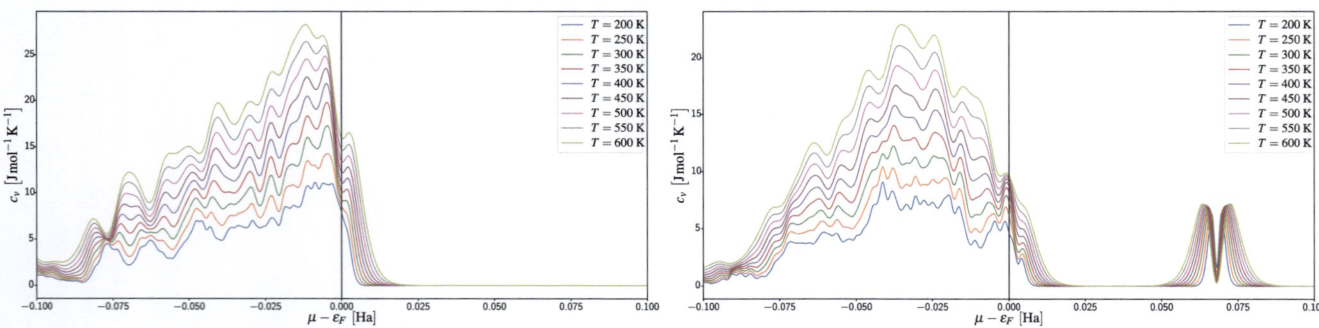

(a) **Phase V**

(b) **Phase I**

Fig. 7 The electronic specific heat as a function of chemical potential at different temperatures

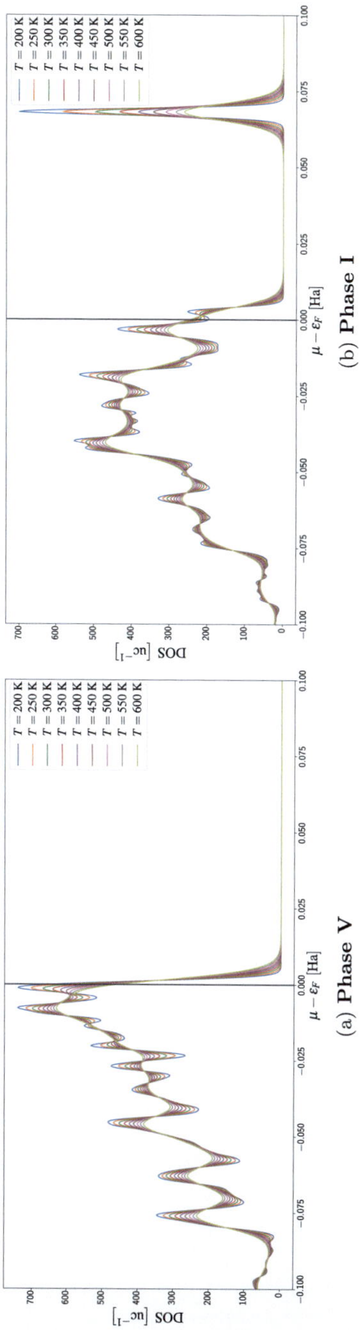

Fig. 8 The density of state at Fermi level DOS (uc^{-1}) as a function of chemical potential at different temperatures

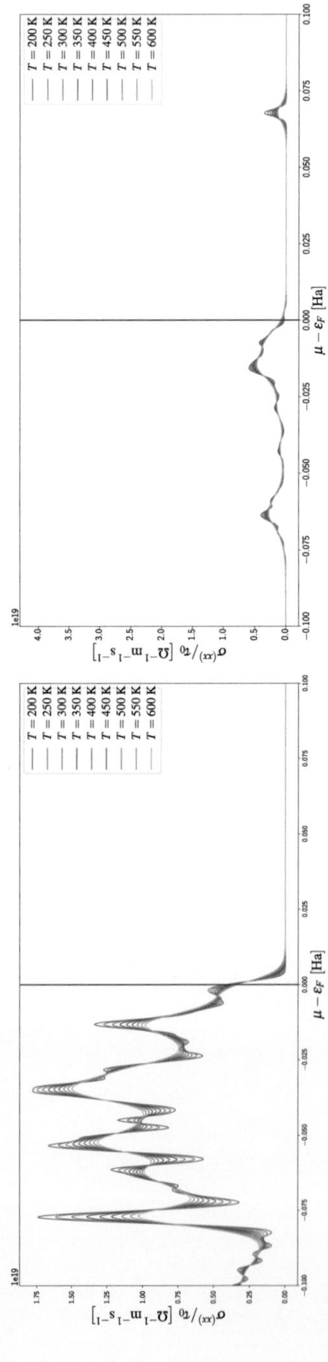

(a) **Phase V**

(b) **Phase I**

Fig. 9 Electrical conductivity rate (σ/τ) as a function of chemical potential at different temperatures

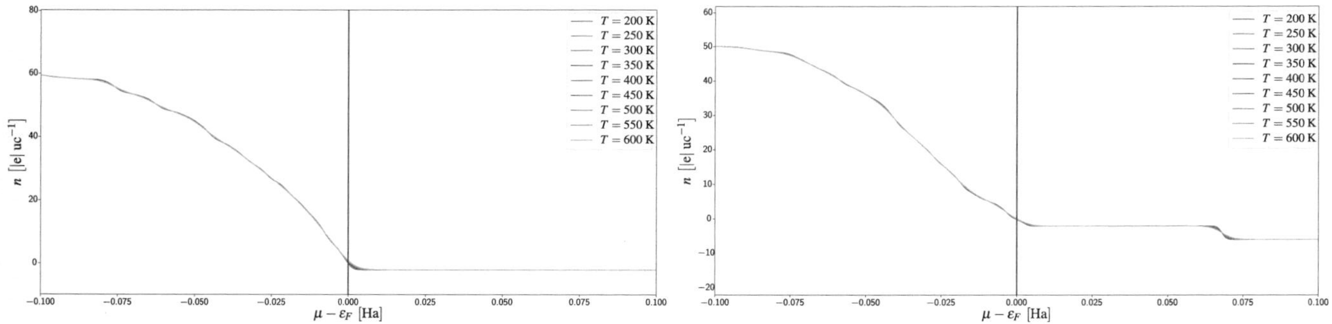

(a) **Phase V** (b) **Phase I**

Fig. 10 Number of carrier concentration as a function of chemical potential at different temperatures. The efficiency of a thermoelectric device (n) is not determined by maximum ZT at a single temperature but depends on the average ZT over a wide temperature range following

Phase I, suggesting photovoltaic potential, while thermoelectric evaluations identified **Phase V** as the superior phase for energy conversion. These results underscore the role of structural dynamics in tuning optoelectronic and thermoelectric performance. Further studies on doping and nanoscale engineering are warranted to optimize these functionalities.

Acknowledgements This research was supported through computational resources of HPC-MARWAN (hpc.marwan.ma) provided by the National Center for Scientific and Technical Research (CNRST), Rabat, Morocco.

References

1. Rose, G.: Beschreibung einiger neuen mineralien des urals. Annalen der Physik und Chemie **124**(12), 551–572 (Jan 1839) (Original description of perovskite mineral)
2. Katz, E.A.: Perovskite: name puzzle and German-Russian odyssey of discovery. Helv. Chim. Acta **103**(6), e2000061 (2020)
3. Tilley, R.J.D.: Perovskites: Structure-Property Relationships. Wiley, Chichester (Mar (2016)
4. Wells, H.L.: über die cäsium- und kalium-bleihalogenide. Z. Anorg. Allg. Chem. **3**(1), 195–210 (1893)
5. Goldschmidt, V.M.: Die gesetze der krystallochemie. Die Naturwissenschaften **14**(21), 477–485 (1926) (introduction of tolerance factor concept)
6. Wul, B., Goldman, I.M.: Dielectric constants of titanates of metals of the second group. Comptes Rendus de l'Académie des Sciences de l'URSS **46**, 139–142 (1945)
7. von Hippel, A., Breckenridge, R.G., Chesley, F.G., Tisza, L.: High dielectric constant ceramics. Ind. Eng. Chem. **38**(11), 1097–1109 (1946)
8. Megaw, H.D.: Crystal structure of barium titanate. Nature **155**(3938), 484–485 (1946)
9. Devonshire, A.F.: Theory of barium titanate: Part I. Phil. Mag. **40**(309), 1040–1063 (1949)
10. Devonshire, A.F.: Theory of barium titanate: Part II. Lond. Edinburgh Dublin Philos. Mag. J. Sci. **42**(333), 1065–1079 (1951)
11. Johnsson, M., Lemmens, P.: Crystallography and chemistry of perovskites. In: Handbook of Magnetism and Advanced Magnetic Materials, pp. 1–11 (2007)
12. Jaffe, B., Cook, W.R., Jaffe, H.: Piezoelectric Ceramics. Academic, London (1971)
13. Cross, L.E.: Relaxor ferroelectrics. Ferroelectrics **76**(1), 241–267 (1987)
14. Lines, M.E., Glass, A.M.: Principles and Applications of Ferroelectrics and Related Materials. Oxford University Press (2001). https://doi.org/10.1093/acprof:oso/9780198507789.001.0001
15. Møller, C.K.: Crystal structure and photoconductivity of cæsium plumbohalides. Nature **182**(4645), 1436 (1958)
16. Fruchart, D., Bertaut, E.F.: Magnetic studies of the metallic perovskite-type compounds of manganese. J. Phys. Soc. Jpn. **44**(3), 781–791 (1978)
17. Takenaka, K., Ichigo, M., Hamada, T., Ozawa, A., Shibayama, T., Inagaki, T., Asano, K.: Magnetovolume effects in manganese nitrides with antiperovskite structure. Sci. Technol. Adv. Mater. **15**(1), 015009 (2014)
18. Kamishima, K., Goto, T., Nakagawa, H., Miura, N., Ohashi, M., Mori, N., Sasaki, T., Kanomata, T.: Giant magnetoresistance in the intermetallic compound Mn_3GaC. Physical Review B 63(2), 024426 (2001)
19. Kojima, A., Teshima, K., Shirai, Y., Miyasaka, T.: Organometal halide perovskites as visible-light sensitizers for photovoltaic cells. J. Am. Chem. Soc. **131**(17), 6050–6051 (2009)
20. Green, M.A., Ho-Baillie, A., Snaith, H.J.: The emergence of perovskite solar cells. Nat. Photonics **8**(7), 506–514 (2014)

21. Saparov, B., Mitzi, D.B.: Organic-inorganic perovskites: structural versatility for functional materials design. Chem. Rev. **116**(7), 4558–4596 (2016)
22. Stranks, S.D., Eperon, G.E., Grancini, G., Menelaou, C., Alcocer, M.J.P., Leijtens, T., Herz, L.M., Petrozza, A., Snaith, H.J.: Electron-hole diffusion lengths exceeding 1 micrometer in an organometal trihalide perovskite absorber. Science **342**(6156), 341–344 (2013)
23. Borriello, I., Cantele, G., Ninno, D.: Ab initio investigation of hybrid organic-inorganic perovskites based on tin halides. Phys. Rev. B **77**(23), 235214 (2008)
24. Zhang, X., Zhao, L.D.: Thermoelectric materials: energy conversion between heat and electricity. Journal of Materiomics **1**(2), 92–105 (2015)
25. Hu, S., Ren, Z., Djurišić, A., Rogach, A.: Metal halide perovskites as emerging thermoelectric materials. ACS Energy Lett. **6**(11), 3882–3905 (2021)
26. Nyayban, A., Panda, S., Chowdhury, A.: Structural, electronic and optical properties of lead free Rb based triiodide for photovoltaic application: an ab initio study. J. Phys.: Condens. Matter **33**(37), 375702 (2021)
27. Khechoubi, M., Bendani, A., Chanh, N., Courseille, C., Duplessix, R., Couzi, M.: Thermal conformational changes in a bidimensional molecular composite material: A thermodynamic and crystallographic study of $NH_3–(CH_2)_4–NH_3$ $CdCl_4$. J. Phys. Chem. Solids **55**(11), 1277–1288 (1994). https://www.sciencedirect.com/science/article/pii/0022369794902100
28. Reiss, G.J.: Bis[(dimethylphosphoryl)methanaminium] tetrachloridopalladate(ii). Acta Crystallographica Section E **69**(11), m614–m615 (2013). https://doi.org/10.1107/S1600536813028067
29. Ziouani, H., Mazouar, S., Tchapet Njafa, J.P., Abdelilah, T., Ettakni, M., Khechoubi, E.M.: Structural and electronic properties of hybrid organic inorganic perovskite [nh3-(ch2)4-nh3]mcl4 compound with (m = cu, mn, cd): Ab initio calculations. In: Elkhattabi, E.M., Boutahir, M., Termentzidis, K., Nakamura, K., Rahmani, A. (eds.) Advanced Materials for Sustainable Energy and Engineering, pp. 265–273. Springer Nature, Cham (2024)
30. Gonze, X., Beuken, J.M., Caracas, R., Detraux, F., Fuchs, M., Rignanese, G.M., Sindic, L., Verstraete, M., Zerah, G., Jollet, F., Torrent, M., Roy, A., Mikami, M., Ghosez, P., Raty, J.Y., Allan, D.: First-principles computation of material properties: The ABINIT software project. Comput. Mater. Sci. **25**(3), 478–492 (2002). https://doi.org/10.1016/s0927-0256(02)00325-7
31. Gonze, X., Rignanese, G.M., Verstraete, M., Beuken, J.M., Pouillon, Y., Caracas, R., Jollet, F., Torrent, M., Zerah, G., Mikami, M., Ghosez, P., Veithen, M., Raty, J.Y., Olevano, V., Bruneval, F., Reining, L., Godby, R., Onida, G., D.C. Allan, D.H.: A brief introduction to the ABINIT software package. Zeitschrift für Kristallographie - Crystalline Mater. 220(5/6), 558–562 (2005)
32. Anisimov, V.I., Zaanen, J., Andersen, O.K.: Band theory and mott insulators: Hubbarduinstead of Stoneri. Phys. Rev. B **44**(3), 943–954 (1991)
33. Anisimov, V.I., Aryasetiawan, F., Lichtenstein, A.I.: First-principles calculations of the electronic structure and spectra of strongly correlated systems: thelda+umethod. J. Phys.: Condens. Matter **9**(4), 767–808 (1997)
34. Liechtenstein, A.I., Anisimov, V.I., Zaanen, J.: Density-functional theory and strong interactions: orbital ordering in Mott-Hubbard insulators. Phys. Rev. B **52**(8), R5467–R5470 (1995). https://doi.org/10.1103/physrevb.52.r5467
35. Perdew, J.P., Burke, K., Ernzerhof, M.: Generalized gradient approximation made simple [Phys. Rev. Lett. 77, 3865 (1996)]. Phys. Rev. Lett. **78**(7), 1396–1396 (1997). https://doi.org/10.1103/physrevlett.78.1396
36. Blöchl, P.E.: Projector augmented-wave method. Phys. Rev. B **50**(24), 17953–17979 (1994)
37. Holzwarth, N., Tackett, A., Matthews, G.: A projector augmented wave (PAW) code for electronic structure calculations, Part I: Atompaw for generating atom-centered functions. Comput. Phys. Commun. **135**(3), 329–347 (2001). https://doi.org/10.1016/s0010-4655(00)00244-7
38. Lejaeghere, K., Bihlmayer, G., Björkman, T., Blaha, P., Blügel, S., Blum, V., Caliste, D., Castelli, I.E., Clark, S.J., Dal Corso, A., de Gironcoli, S., Deutsch, T., Dewhurst, J.K., Di Marco, I., Draxl, C., Dułak, M., Eriksson, O., Flores-Livas, J.A., Garrity, K.F., Genovese, L., Giannozzi, P., Giantomassi, M., Goedecker, S., Gonze, X., Grånäs, O., Gross, E.K.U., Gulans, A., Gygi,

F., Hamann, D.R., Hasnip, P.J., Holzwarth, N.A.W., Iuşan, D., Jochym, D.B., Jollet, F., Jones, D., Kresse, G., Koepernik, K., Küçükbenli, E., Kvashnin, Y.O., Locht, I.L.M., Lubeck, S., Marsman, M., Marzari, N., Nitzsche, U., Nordström, L., Ozaki, T., Paulatto, L., Pickard, C.J., Poelmans, W., Probert, M.I.J., Refson, K., Richter, M., Rignanese, G.M., Saha, S., Scheffler, M., Schlipf, M., Schwarz, K., Sharma, S., Tavazza, F., Thunström, P., Tkatchenko, A., Torrent, M., Vanderbilt, D., van Setten, M.J., Van Speybroeck, V., Wills, J.M., Yates, J.R., Zhang, G.X., Cottenier, S.: Reproducibility in density functional theory calculations of solids. Science **351**(6280) (2016)
39. Monkhorst, H.J., Pack, J.D.: Special points for Brillouin-zone integrations. Phys. Rev. B **13**(12), 5188–5192 (1976)
40. Houmad, M., Zaari, H., Benyoussef, A., El Kenz, A., Ez-Zahraouy, H.: Optical conductivity enhancement and band gap opening with silicon doped graphene. Carbon **94**, 1021–1027 (2015)
41. Murnaghan, F.D.: The compressibility of media under extreme pressures. Proc. Natl. Acad. Sci. U.S.A. **30**, 244–247 (1944)
42. Perdew, J.P., Burke, K., Ernzerhof, M.: Generalized gradient approximation made simple. Phys. Rev. Lett. **77**(18), 3865–3868 (1996). https://doi.org/10.1103/physrevlett.77.3865
43. Payne, M.C., Teter, M.P., Allan, D.C., Arias, T.A., Joannopoulos, J.D.: Iterative minimization techniques for ab initio total-energy calculations: molecular dynamics and conjugate gradients. Rev. Mod. Phys. **64**(4), 1045–1097 (1992)
44. Kresse, G., Furthmüller, J.: Efficient iterative schemes for ab initio total-energy calculations using a plane-wave basis set. Phys. Rev. B **54**(16), 11169–11186 (1996)
45. Martin, R.M.: Electronic Structure: Basic Theory and Practical Methods. Cambridge University Press (2020)
46. Himmetoglu, B., Floris, A., de Gironcoli, S., Cococcioni, M.: Hubbard-corrected dft energy functionals: The lda+u description of correlated systems. Int. J. Quantum Chem. **114**(1), 14–49 (2013)
47. Cohen, A.J., Mori-Sańchez, P., Yang, W.: Insights into current limitations of density functional theory. Science **321**(5890), 792–794 (2008)
48. Ding, J., Zhao, P., Chen, H., Fu, H.: Ultraviolet photodetectors based on wide bandgap semiconductor: a review. Appl. Phys. A **130**(5) (2024)
49. Hafner, J.: Ab-initio simulations of materials using VASP: density-functional theory and beyond. J. Comput. Chem. **29**(13), 2044–2078 (2008)
50. Becke, A.D.: Density-functional thermochemistry. III. the role of exact exchange. J. Chem. Phys. **98**(7), 5648–5652 (1993)
51. Walsh, A.: Principles of chemical bonding and band gap engineering in hybrid organic–inorganic halide perovskites. J. Phys. Chem. C **119**(11), 5755–5760 (2015)
52. Khenata, R., Bouhemadou, A., Sahnoun, M., Reshak, A.H., Baltache, H., Rabah, M.: Elastic, electronic and optical properties of zns, znse and znte under pressure. Comput. Mater. Sci. **38**(1), 29–38 (2006). https://www.sciencedirect.com/science/article/pii/S0927025606000024
53. Houmad, M., Dakir, O., Abbassi, A., Benyoussef, A., El Kenz, A., Ez-Zahraouy, H.: Optical properties of sic nanosheet. Optik **127**(4), 1867–1870 (2016). https://www.sciencedirect.com/science/article/pii/S0030402615016253
54. Rizwan, M., Ayub, A., Shakil, M., Usman, Z., Gillani, S., Jin, H., Cao, C.: Putting DFT to trial: For the exploration to correlate structural, electronic and optical properties of m-doped (m = group i, ii, iii, xii, xvi) lead free high piezoelectric c-bialo3. Mater. Sci. Eng.: B **264**, 114959 (2021). https://www.sciencedirect.com/science/article/pii/S0921510720304669
55. Absike, H., Labrim, H., Hartiti, B., Douhou, K., Ez-Zahraouy, H.: Ab initio calculations on electronic, optical, and thermoelectric properties of (Si, Pb) (Co)-doped ZnS for solar cell device applications. J. Phys. Chem. Solids **132**, 10–17 (2019), https://www.sciencedirect.com/science/article/pii/S0022369718335108
56. Rizwan, M., Gul, S., Mahmood, T., Shakil, M., Majid, A., Rafique, M., Zafar, A., Jin, H., Cao, C.: Tailoring electronic and optical properties of laalo3 by cu inclusion: a DFT study. Can. J. Phys. **99**(1), 38–43 (2021). https://doi.org/10.1139/cjp-2019-0558

57. Madsen, G.K.H., Carrete, J., Verstraete, M.J.: BoltzTraP2, a program for interpolating band structures and calculating semi-classical transport coefficients. Comput. Phys. Commun. **231**, 140–145 (2018)
58. Alrahamneh, M.J., Mousa, A.A., Khalifeh, J.M.: First principles study of the structural, electronic, magnetic and thermoelectric properties of zr2rhal. Physica B: Condens. Matter **552**, 227–235 (2019). https://www.sciencedirect.com/science/article/pii/S0921452618306422
59. Snyder, G.J., Toberer, E.S.: Complex thermoelectric materials. Nat. Mater. **7**(2), 105–114 (2008)
60. Dresselhaus, M., Chen, G., Tang, M., Yang, R., Lee, H., Wang, D., Ren, Z., Fleurial, J., Gogna, P.: New directions for low-dimensional thermoelectric materials. Adv. Mater. **19**(8), 1043–1053 (2007)
61. CRC Handbook of Thermoelectrics. CRC Press (2018)

Thermal Performance of Non-Newtonian Water–Ethylene Glycol Nanofluids in Enclosed Natural Convection

Bilal El hadoui, Youness Ighris, Mourad Kaddiri, and Jamal Baliti

Abstract This study investigates natural convection in shallow rectangular cavities filled with non-Newtonian Al_2O_3 nanofluids based on water–ethylene glycol (W/EG) mixtures. Analytical and numerical approaches are developed to study the impact of nanoparticle volume fraction ($0 \le \varphi \le 0.05$), power-law index ($0.6 \le n \le 1$), and base fluid mixture on heat transfer performance. The governing equations are solved numerically using the finite difference method, and analytically using a parallel flow approximation. Both approaches demonstrate excellent agreement of results, which validates the analytical model for high-aspect-ratio cavities. It is found that the increase in φ or ethylene glycol concentration results in significant viscosity improvement that deteriorates both the flow and heat transfer rate despite moderate improvements in thermal conductivity. The Nusselt number decreases drastically with φ, especially for EG-rich mixtures. Convection is also reduced by increasing n, but to a lesser extent at already high viscosity.

Keywords Heat transfer · Nanofluid · Non-Newtonian · Natural convection · Parallel flow approximation · Rectangular enclosure

1 Introduction

Natural convection in enclosures has been a subject of thermal sciences research due to its intrinsic importance in a large variety of engineering systems [1]. A few examples of applications that rely heavily on buoyancy-induced heat transfer in cavities of different geometries are the thermal regulation of building [2], passive solar collectors [3], cooling of electronic components [4], and heat exchangers [5]. Natural convection has been investigated thoroughly by researchers in the past under different

B. El hadoui (✉) · M. Kaddiri
Faculty of Sciences and Technologies, University of Sultan Moulay Slimane, Beni Mellal, Morocco
e-mail: bilal.elhadoui@usms.ac.ma

Y. Ighris · J. Baliti
Polydisciplinary Faculty, University of Sultan Moulay Slimane, Beni Mellal, Morocco

E. El Mehdi and M. Boutahir (eds.), *Advanced Materials for Sustainable Energy and Engineering*, Engineering Materials, https://doi.org/10.1007/978-3-032-10069-6_9

thermal boundary conditions and aspect ratios, particularly in square cavities, which have practical use for compact devices [6–10]. Despite significant progress, the optimization of heat transfer in such systems remains an open issue, particularly for constrained geometries or small driving forces.

With the intent to enhance thermal performance, nanofluids which are solid nanoparticle suspensions in traditional base fluids emerged as a novel class of heat transfer fluid. Nanofluids as investigated by many researchers [11, 12], provide higher effective thermal conductivity and heat transfer coefficient compared to traditional fluids. Many numerical studies confirmed that the introduction of nanoparticles such as Al_2O_3, CuO, or TiO_2 into water or ethylene glycol can improve convective performance for both natural and forced convection flows [13–17]. However, the influence of such nanoparticles on thermal behavior is not always favorable [18, 19]. At high volume fractions or in certain regimes of flow, the increased viscosity and potential for particle agglomeration can suppress convective motion, resulting in a net decrease in heat transfer [20–24]. These contradictory findings indicate the need for more targeted studies considering not only the particle loading but also the flow structure and fluid rheology.

Among the most critical but often overlooked factors in nanofluid modeling is the rheological nature. While the majority of works implicitly take nanofluids to be Newtonian by default for simplicity [22], experimental studies have shown that certain nanofluids, most prominently those made up of water-ethylene glycol solutions, exhibit non-Newtonian behavior [25–27], with the most common type being shear-thinning (pseudoplastic). This is attributed to complex interactions between the base fluid molecules and nanoparticles that affect apparent viscosity for any shear rate. As a result, the use of power-law model to describe the non-Newtonian nanofluid behavior is gaining attention in numerical simulations [28, 29]. However, combined impact of the shear-dependent viscosity and buoyancy forces for natural convection is still not fully explored, especially for shallow enclosures when viscous forces can dominate.

In this context, the use of ethylene glycol (EG)–water mixtures as base fluids is beneficial and model complexities. EG can readily be used in thermal fluid management due to its low volatilization, high boiling point, and solubility in metal oxides [30, 31]. When water is added to it, the mixture possesses adjustable thermophysical properties and proved to stabilize the nanoparticles' dispersion more than pure water [32]. However, EG-water mixtures are inherently Newtonian [33], and when nanoparticle-blended, they possess non-Newtonian properties [34]. Natural convective behavior of non-Newtonian EG-water nanofluids in shallow enclosures is not yet well researched despite their widespread occurrence in industrial applications.

The current work is designed to close this gap through an examination of natural convection in shallow rectangular enclosures filled with non-Newtonian Al_2O_3 nanofluids using EG-water mixtures. The enclosures are subjected to equal and uniform heat fluxes on the opposing sidewalls, a configuration common in most passive thermal systems. Both numerical and analytical methods are employed: the parallel flow approximation provides theoretical solutions, and extensive flow and temperature fields are given by a finite difference solution. Thermophysical properties like viscosity and thermal conductivity are handled with experimentally validated correlations. This work presents new knowledge regarding nanoparticle volume fraction, power-law index, and base fluid composition impacts on the structure of natural convective flow and the resulting heat transfer performance in shallow enclosures.

2 Geometry and Governing Equations

The physical model consists of a two-dimensional shallow rectangular enclosure of height *H* and width *W*, filled with a non-Newtonian nanofluid composed of Al_2O_3 nanoparticles dispersed in a water–ethylene glycol (W/EG) base fluid as shown in Fig. 1. The right and left vertical walls are subjected to equal and uniform heat fluxes *q*, while the horizontal bottom and top walls are insulated. The nanofluid is assumed to have incompressible properties, constant thermophysical properties except for variations in the density term used for buoyancy, and a laminar flow condition. Local thermal equilibrium is assumed between nanoparticles and base fluid, using the single-phase model [19, 22].

The governing equations for mass, momentum (in the vorticity-stream function form), and energy under those assumptions are as follows [35]:

Continuity equation:

$$\frac{\partial u'}{\partial x'} + \frac{\partial v'}{\partial y'} = 0 \tag{1}$$

Momentum equation:

The momentum equation is expressed in terms of vorticity Ω', where $\Omega' = \frac{\partial v'}{\partial x'} - \frac{\partial u'}{\partial y'}$. The transport equation for vorticity is:

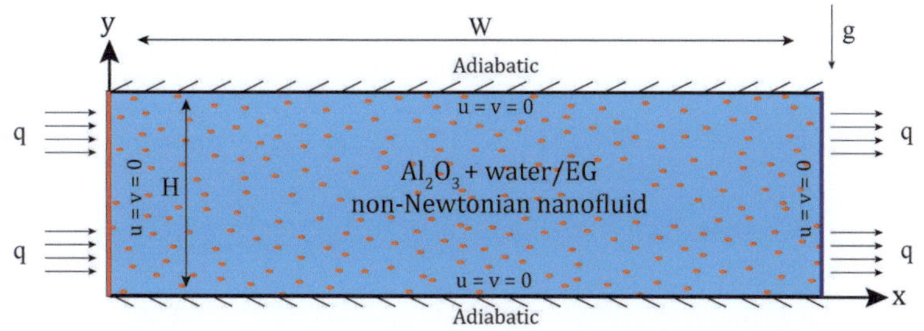

Fig. 1 Sketch of the physical domain

$$v'\frac{\partial \Omega'}{\partial y'} + u'\frac{\partial \Omega'}{\partial x'} = \vartheta_{eff}\left(D'_f\left(\frac{\partial^2 \Omega'}{\partial x'^2} + \frac{\partial^2 \Omega'}{\partial y'^2}\right) + S'_m\right) + \beta_{eff}\frac{\partial T'}{\partial x'} \tag{2}$$

Here ϑ_{eff} is the effective kinematic viscosity of the nanofluid, β_{eff} is the effective thermal expansion coefficient, and S'_m is the source term accounting for non-Newtonian viscosity gradients.

The stream function Ψ' is related to velocity components by:

$$u' = \frac{\partial \Psi'}{\partial y'},\ v' = -\frac{\partial \Psi'}{\partial x'} \tag{3}$$

The Poisson equation for the stream function is:

$$-\Omega' = \frac{\partial^2 \Psi'}{\partial x'^2} + \frac{\partial^2 \Psi'}{\partial y'^2} \tag{4}$$

Energy equation:

$$u'\frac{\partial T'}{\partial x'} + v'\frac{\partial T'}{\partial y'} = \alpha_{eff}\left(\frac{\partial^2 T'}{\partial x'^2} + \frac{\partial^2 T'}{\partial y'^2}\right) \tag{5}$$

where *T'* is the temperature field and α_{eff} is the effective thermal diffusivity.

Non-Newtonian viscosity (power-law fluid model):

The apparent viscosity D'_f of the power-law is defined as [36]:

$$D'_f = \dot{\gamma}^{n-1} \tag{6}$$

where n is the power-law index ($n < 1$: shear-thinning, $n = 1$: Newtonian, $n > 1$: shear-thickening) and $\dot{\gamma}$ is the shear rate, approximated as:

$$\dot{\gamma} = \left(\left(\frac{\partial v'}{\partial x'} + \frac{\partial u'}{\partial y'}\right)^2 + 2\left(\left(\frac{\partial v'}{\partial y'}\right)^2 + \left(\frac{\partial u'}{\partial x'}\right)^2\right)\right)^{\frac{1}{2}} \tag{7}$$

Adopting the following scaling analysis [37]:

$$(x, y) = \frac{(x', y')}{H},\ (u, v) = \frac{H}{\alpha_f}(u', v'),\ T = \frac{(T' - T'_0)}{\Delta T} \tag{8}$$

The non-dimensional governing equations are:

$$v\frac{\partial \Omega}{\partial y} + u\frac{\partial \Omega}{\partial x} = Pr\vartheta_r\left(D_f\nabla^2\Omega + S_m\right) + RaPr\beta_r\frac{\partial T}{\partial x} \tag{9}$$

$$u\frac{\partial T}{\partial x} + v\frac{\partial T}{\partial y} = \alpha_r\left(\frac{\partial^2 T}{\partial x^2} + \frac{\partial^2 T}{\partial y^2}\right) \tag{10}$$

$$\begin{aligned} S_m = {} & 2\left(\frac{\partial D_f}{\partial x}\frac{\partial \Omega}{\partial x} + \frac{\partial D_f}{\partial y}\frac{\partial \Omega}{\partial y}\right) \\ & - 2\frac{\partial^2 D_f}{\partial x \partial y}\left(\frac{\partial u}{\partial x} - \frac{\partial v}{\partial y}\right) \\ & + \left(\frac{\partial^2 D_f}{\partial x^2} - \frac{\partial^2 D_f}{\partial y^2}\right)\left(\frac{\partial u}{\partial y} + \frac{\partial v}{\partial x}\right) \end{aligned} \tag{11}$$

The boundary conditions in non-dimensional form are as follows: the horizontal boundaries are heated uniformly, which implies $\partial T/\partial x = -1/k_r$, while the vertical sidewalls are adiabatic $\partial T/\partial y = 0$, and all the boundaries are immobile $v = u = 0$.

The model introduces several key dimensionless parameters [38], namely: the Prandtl number, $Pr = \frac{\vartheta_f H^{2-2n}}{\alpha_f^{2-n}}$, the Rayleigh number, $Ra = \frac{qH^{2+2n} g \beta_f}{k_f \vartheta_f \alpha_f^n}$, the aspect ratio, $A = W/H$, and the properties ratio, $f_r = f_{eff}/f_f$.

The thermophysical properties used for base fluids and Al_2O_3 nanoparticles are presented in Table 1. To incorporate realistic physical behavior, thermophysical properties of the nanofluid are modeled using experimentally validated correlations. The effective density, specific heat, and thermal expansion coefficient are computed following standard mixture rules [19, 39].

$$\rho_{eff} = \rho_{np}\varphi + \rho_f(1-\varphi) \tag{12}$$

$$\left(\rho c_p\right)_{eff} = \left(\rho c_p\right)_{np}\varphi + \left(\rho c_p\right)_f(1-\varphi) \tag{13}$$

$$(\rho\beta)_{eff} = (\rho\beta)_{np}\varphi + (\rho\beta)_f(1-\varphi) \tag{14}$$

To model the viscosity and thermal conductivity ratios, different empirical correlations are applied depending on the base fluid composition and nanoparticle volume fraction. These correlations, extracted from experimental data [32, 41], are:

$$100:0\%\ \mathrm{W/EG}\ k_r = 4.97\varphi^2 + 2.72\varphi + 1;\ \mu_r = 123\varphi^2 + 7.3\varphi + 1 \tag{15}$$

Table 1 Thermo-physical properties of W/EG mixture and Al_2O_3 [32, 40]

	100:0%	80:20%	60:40%	40:60%	Al_2O_3
c_p	4179	3815	3468	3084	765
β	21×10^{-5}	27.6×10^{-5}	34.2×10^{-5}	40.5×10^{-5}	0.85×10^{-5}
ρ	997.1	1029.72	1059.68	1086.27	3970
k	0.613	0.492	0.404	0.334	40

$$0:20\%\ \mathrm{W/EG}\ k_r = 8.0175\varphi + 1.1236;\ \mu_r = 0.9396e^{24.16\varphi} \quad (16)$$

$$60:40\%\ \mathrm{W/EG}\ k_r = 10.164\varphi + 1.0806;\ \mu_r = 0.9299e^{67.43\varphi} \quad (17)$$

$$40:60\%\ \mathrm{W/EG}\ k_r = 10.448\varphi + 1.0618;\ \mu_r = 1.1216e^{77.56\varphi} \quad (18)$$

These expressions enable precise modeling of nanofluid behavior across a range of operating conditions, accounting for the strong nonlinear dependence of viscosity and conductivity on nanoparticle concentration and base fluid composition.

Figure 2 illustrates the impact of nanoparticle concentration φ on the viscosity and thermal conductivity ratios, for different water/ethylene glycol (W/EG) mixtures. These trends are crucial for understanding the competing effects of nanoparticle addition on heat transfer performance. The viscosity increases nonlinearly with the nanoparticle volume fraction, and this effect is amplified with higher ethylene glycol content showing an enhancement highly sensitive to base fluid composition. While, the thermal conductivity increases almost linearly with φ across all base fluids, but the rate of increase is less sensitive to the EG content.

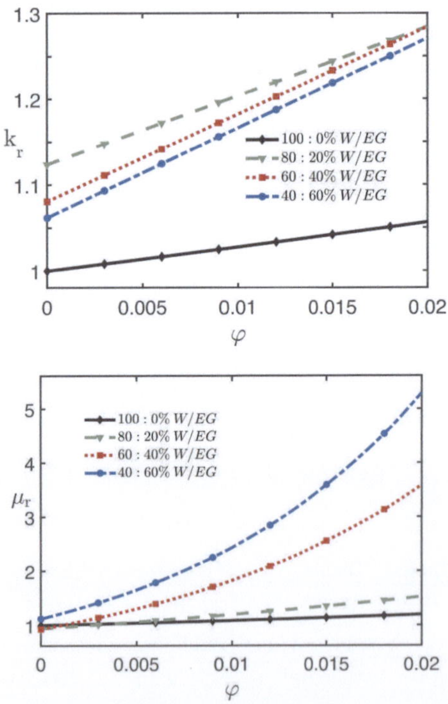

Fig. 2 Variations of viscosity and thermal conductivity ratios for different compositions

3 Numerical and Analytical Approaches

In order to solve the dimensionless governing equations of the flow of non-Newtonian nanofluids inside the cavity, a finite difference technique is employed. Spatial derivatives are approximated using a second-order central difference scheme with the aim of preserving accuracy in momentum and energy equations. The iterative solution is achieved using the Alternating Direction Implicit (ADI) method that offers improved stability and computational efficiency. A full description of this numerical method, including convergence behavior and implementation structure, is given in [35, 37].

The primary output of thermal performance assessment in this study is the average Nusselt number, calculated along the vertical centerline of the cavity (at $x = A/2$) to quantify the heat exchange in the enclosure. It is defined as:

$$\overline{\text{Nu}} = -\int_0^1 1/\left(\frac{\delta T}{\delta x}\right)_{x=\frac{A}{2}} \text{dy} \tag{19}$$

To ensure the reliability of the numerical code, a validation process is conducted by comparing the results of the current solver with benchmark studies for both Newtonian and non-Newtonian fluids. As summarized in Tables 2 and 3, comparisons with Davis and Jones [42] in the case of natural convection for different *Ra* and Ghasemi et al. [43] in the case of nanofluid natural convection for different φ. Further validation with the work of Turan et al. [44], presented in Table 4, confirms the accuracy of the solution across a range of power-law indices.

In order to balance computational efficiency and accuracy, a preliminary grid sensitivity analysis was performed. A uniform grid size of 301 × 101 nodes was adopted based on convergence and solution independence tests, providing a reliable resolution of both thermal and hydrodynamic fields throughout the domain as

Table 2 Validation of the numerical code in the case of natural convection for different *Ra*

	10^3		10^4		10^5	
	$\lvert\Psi_{max}\rvert$	$\overline{\text{Nu}}$	$\lvert\Psi_{max}\rvert$	$\overline{\text{Nu}}$	$\lvert\Psi_{max}\rvert$	$\overline{\text{Nu}}$
Davis [42]	1.174	1.118	5.071	2.243	9.111	4.519
Our code	1.173	1.111	5.070	2.242	9.156	4.603

Table 3 Validation of the numerical code in the case of nanofluid natural convection for different φ

	0.0		0.02		0.04		0.06	
	$\lvert\Psi_{max}\rvert$	$\overline{\text{Nu}}$	$\lvert\Psi_{max}\rvert$	$\overline{\text{Nu}}$	$\lvert\Psi_{max}\rvert$	$\overline{\text{Nu}}$	$\lvert\Psi_{max}\rvert$	$\overline{\text{Nu}}$
Ghasemi [43]	11.053	4.738	11.313	4.820	11.561	4.896	11.801	4.968
Our code	11.025	4.722	11.287	4.804	11.536	4.880	11.776	4.951

Table 4 Validation of the numerical code in the case of non-Newtonian fluids for different n

	n		
	0.6	1	1.8
Our code	34.52	9.37	2.56
Turan et al. [44]	33.68	9.17	2.54

tabulated in Table 5. Furthermore, it was found that an aspect ratio $A \geq 24$ satisfies the infinitely long cavity approximation, which allows for flow development in the central zone without end effects.

Qualitative examination of the streamline and isotherm patterns, Fig. 3, for representative conditions $Ra = 10^4$, $n = 0.6$, $\varphi = 0.05$ reveals key flow features that support simplifying assumptions. In the central region of the cavity, away from the boundary effects, the flow becomes predominantly unidirectional, parallel to the horizontal walls, and the temperature field is linearly stratified in the horizontal direction. These observations align with previous works [9, 29, 35, 38] and justify the application of an analytical approximation, parallel flow approximation, in the core of the domain.

Based on these findings, a simplified analytical model is formulated by applying the following assumptions in the cavity core:

$$u = u(y), \psi = \psi(y), v = 0, T = \theta_T(y) + C_T\left(x - \frac{A}{2}\right) \tag{20}$$

where $C_T = \partial T/\partial x$ is an unknown constant gradient in the x direction.

Table 5 Grid sensitivity analysis for $n = 0.6$

	161 × 61	241 × 61	241 × 81	281 × 81	301 × 101	341 × 121
$\lvert\Psi_{max}\rvert$	6.458	6.463	6.463	6.463	6.463	6.463
$\overline{Nu}$	15.758	15.751	15.713	15.713	15.693	15.680

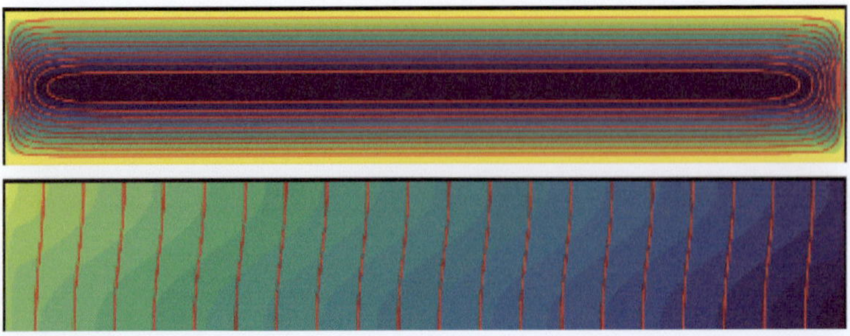

Fig. 3 Streamlines and isotherms $Ra = 10^4$, $\varphi = 0.05$, $n = 0.6$ and different base fluids (colorful pattern: 100:0% W/EG, red line: 40:60% W/EG)

The nonlinearity of the nanofluid's power-law behavior and the presence of a return flow region, where du/dy = 0, necessitate the division of the vertical domn into three intervals: $[0, y_1]$, $[y_1, y_2]$, and $[y_2, 1]$ where $y_1 + y_2 = 1$. These intervals are separated by points at which the shear rate vanishes. To simplify the integration across these regions, an auxiliary function is introduced:

$$b(y) = \frac{y^2}{2} - \frac{y}{2} + \frac{y_2 y_1}{2} \tag{21}$$

Substituting these simplifications into the dimensionless governing Eqs. (9)–(11), and integrating under the relevant boundary conditions, leads to a reduced set of differential equations describing the velocity and temperature profiles.

$$u(y) = D_n \Psi_c F(y) \tag{22}$$

$$\theta_T(y) = \frac{C_T}{\alpha_r} D_n \Psi_c H(y) + cst \tag{23}$$

$$\Psi(y) = D_n \Psi_c G(y) \tag{24}$$

Ψ_c is the central stream function and has the following expression:

$$(D_n \Psi_c)^n = \frac{\beta_r}{\vartheta_r} C_T Ra \tag{25}$$

$$D_n = \left[\left(\frac{1}{2} - y_0 \right) \int_0^{y_1} b(y)^{\frac{1}{n}} dy + \int_{y_1}^{\frac{1}{2}} \left[\int_y^{y_1} [-b(y)]^{\frac{1}{n}} \right] dy \right. \\ \left. + \int_0^{y_1} \left[\int_0^y b(y)^{\frac{1}{n}} dy \right] dy \right]^{-1} \tag{26}$$

The functions *F(y)*, *G(y)*, and *H(y)* represent cumulative integrals of $b(y)^{\frac{1}{n}}$, each tailored to the corresponding interval. To determine the constant C_T a procedure, consists of integrating the energy equation, is used and it leads to:

$$C_T = \frac{1}{\overline{Nu}} = -\frac{\alpha_r}{\mathrm{k}_r} \left(\frac{1}{\alpha_r} (D_n \Psi_c)^2 A_n - \alpha_r \right)^{-1} \tag{27}$$

After some mathematical operations, we obtain the following transcendent equation:

$$\left((D_n\Psi_c)^4\left(\frac{A_n}{\alpha_r}\right)^2 - \left[\frac{1}{\alpha_r^2}+1\right](D_n\Psi_c)^2A_n + 1\right)$$

$$(D_n\Psi_c)^n = \frac{\beta_r Ra}{\vartheta_r \mathrm{k}_r}\left[1-(D_n\Psi_c)^2A_n\right] \tag{28}$$

The coefficients A_n and D_n depend only on n. Equation 28 is solved by the Newton–Raphson method. It must be mentioned that this analytic method is valid for the case of very elongated enclosures ($A >> 1$) as discussed by [9, 35, 38].

4 Results and Discussion

This work presents and analyzes both numerical and analytical results for natural convection within the shallow rectangular enclosure, focusing on the average Nusselt number and the stream function magnitude at the center. The simulations are performed for a fixed Rayleigh number $Ra = 10^4$, aspect ratio $A = 22$, and Prandtl number $Pr = 6.5$, while systematically varying the nanoparticle volume fraction $0 \leq \varphi \leq 0.05$, power-law index $0.6 \leq n \leq 1$ and different base fluid mixtures (from pure water to ethylene glycol–rich mixtures). The comparison between numerical results (symbols) and analytical predictions (lines) demonstrates an excellent match in both $|\Psi_c|$ and $\overline{Nu}$, which confirms the validity of the parallel flow approximation used in the analytical formulation.

Figure 4 demonstrates the effect of increasing nanoparticle concentration φ on $|\Psi_c|$ and $\overline{Nu}$, particularly for a shear-thinning fluid $n = 0.6$, with different base fluids. As φ increases, a consistent deterioration in heat transfer is observed. This deterioration in $|\Psi_c|$ and $\overline{Nu}$ results primarily from the sharp increase in viscosity ratio where it reaches 5 (Fig. 2), which affects the convective motion and diminishes buoyancy-driven flow in addition to the decrease in the thermal expansion coefficient. Although the addition of nanoparticles increases thermal conductivity, its positive effect is dominated by the dominant viscous damping. The degree of this deterioration greatly differs depending on the base fluid: in water, $\overline{Nu}$ decreases linearly to 15.69 from 24.82; but for a 40:60 W/EG solution, it drops drastically to 1.73. This sudden decline is due to the synergistic effect of ethylene glycol's inherently high viscosity and its interaction with nanoparticle-induced thickening. Notably, this trend holds true for both Newtonian and non-Newtonian conditions.

The effect of base fluid composition is also witnessed in qualitative flow and temperature field observations, as shown in Fig. 3. When the high-viscosity mixtures like 40:60 W/EG are utilized, the streamlines near the heated sidewalls become thin, and the isotherms become nearly vertical. This behavior indicates transition from convection-dominated to conduction-dominated heat transport, where stagnation of the flow at boundaries and loss of thermal mixing leads to a dramatic reduction of $\overline{Nu} \approx 1$ and $|\Psi_c| \approx 0$. Low-viscosity liquids (pure water), by contrast, permit the

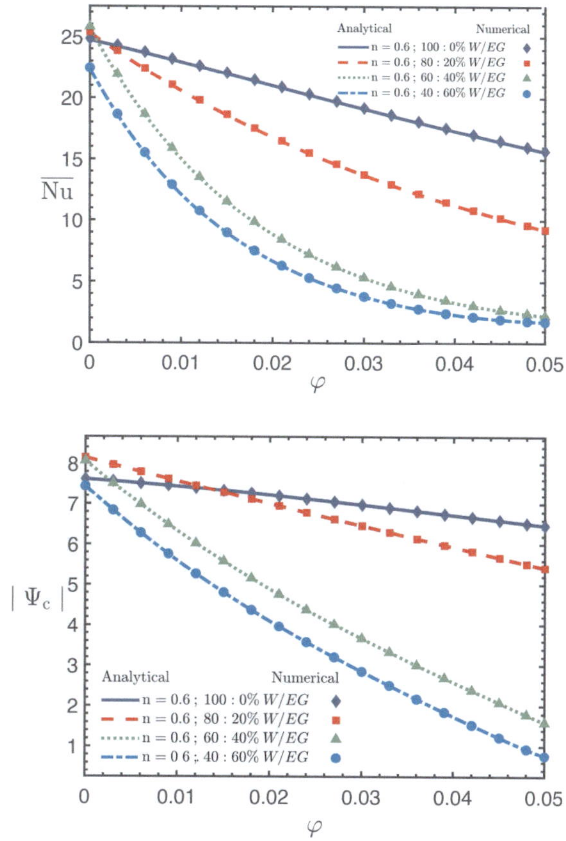

Fig. 4 Variation of $\overline{Nu}$ and $|\Psi_c|$ with φ

presence of strong convective cells and steeper isotherms, which indicate efficient thermal transport.

Figure 5 analyzes the effect of the power-law index n, highlighting the contribution of fluid rheology to convective performance for various fluid mixtures and concentrations of nanoparticles. As n increases, transitioning fluid behavior away from shear-thinning ($n < 1$) towards Newtonian ($n = 1$), both $\overline{Nu}$ and $|\Psi_c|$ decrease as the resultant rise in apparent viscosity. However, its value depends on the base fluid and nanoparticle concentration. For instance, at $\varphi = 0$ and pure water, $\overline{Nu}$ decreases from 24.82 (at $n = 0.6$) to 6.88 (at $n = 1$). Similarly, at $\varphi = 0.03$ in 60:40 W/EG solution, $\overline{Nu}$ decreases from 5.32 to 2.18. Interestingly, with an increase of either φ or the ethylene glycol concentration, the influence of n vanishes. This is likely a reflection of the fact that under very viscous conditions, the rheological sensitivity of the flow is dampened by an already-low intensity of convection. In general, the three parameters, volume fraction of the nanoparticle, base fluid composition, and power-law index, all affect flow behavior in a similar manner: they increase viscosity, which in turn degrades convective efficiency.

Fig. 5 Variation of $\overline{Nu}$ and $|\Psi_c|$ with n

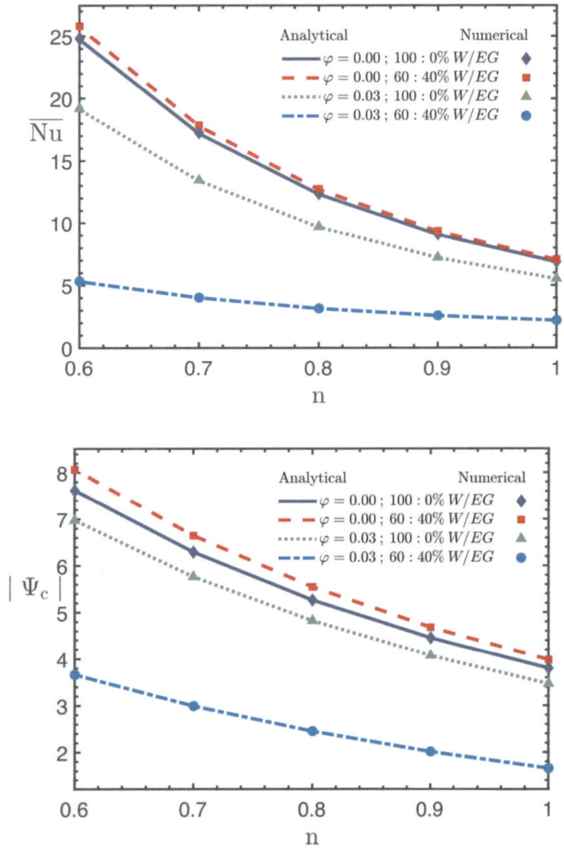

5 Conclusion

This study analyzed natural convection in shallow rectangular cavities filled with non-Newtonian Al_2O_3–W/EG nanofluids using both analytical and numerical approaches. The parallel flow assumption showed excellent agreement with the numerical results, validating its use for high-aspect-ratio enclosures. Key findings include:

- The developed analytical solution exhibits a good agreement with the numerical simulation for a wide range of the studied parameters.
- The use of nanofluids of different base fluids in long enclosures affects negatively the convective thermal exchange.
- Increasing nanoparticle volume fraction and ethylene glycol content significantly increased viscosity, reducing flow strength and Nusselt number despite modest gains in thermal conductivity.
- The flow intensity and Nusselt number decrease with the augmentation of the power-law index and its effect diminished in already viscous fluids.

Nomenclature

A	Aspect ratio;
C_T	Thermal gradients in the x direction;
cp	Specific heat capacity ($J.kg^{-}1.K^{-1}$);
H	Height of the cavity (m);
D_f	Dimensionless apparent viscosity;
g	Gravitational acceleration ($m.s^{-2}$);
n	Power-law index;
k	Thermal conductivity ($W.m^{-1}.K^{-1}$);
Pr	Prandtl number;
Nu	Nusselt number;
q	Constant heat flux ($W.m^{-2}$);
Ra	Rayleigh number;
T	Dimensionless temperature;
(u, v)	Dimensionless components of velocity;
W	Width of the cavity (m);
α	Thermal diffusivity ($m^2.s$);
μ	Dynamic viscosity ($kg.m^{-1}.s^{-1}$);
(x, y)	Dimensionless coordinates;
β	Thermal expansion coefficient (K^{-1});
ΔT	Characteristic temperature difference (K);
φ	Nanoparticle volume fraction;
Ω	Dimensionless vorticity;
ν	Kinematic viscosity ($m^2.s$);
ρ	Fluid density ($kg.m^{-3}$)

Subscripts

f	Fluid;
np	Nanoparticles;
nf	Nanofluid;
r	Ratio value

Superscripts

‘	Dimensional variable

References

1. Baïri, A., Zarco-Pernia, E., García de María, J.-M.: A review on natural convection in enclosures for engineering applications. The particular case of the parallelogrammic diode cavity. Appl. Therm. Eng. **63**, 304–322 (2014). https://doi.org/10.1016/j.applthermaleng.2013.10.065
2. SA Mikhailenko IV Miroshnichenko MA Sheremet 2021 Thermal radiation and natural convection in a large-scale enclosure heated from below: building application Build. Simul. 14 681 691 https://doi.org/10.1007/s12273-020-0668-4
3. AL Hernández JE Quiñonez 2018 Experimental validation of an analytical model for performance estimation of natural convection solar air heating collectors Renew. Energy 117 202 216 https://doi.org/10.1016/j.renene.2017.09.082
4. LA Florio A Harnoy 2007 Combination technique for improving natural convection cooling in electronics Int. J. Therm. Sci. 46 76 92 https://doi.org/10.1016/j.ijthermalsci.2006.03.007
5. B Zhao 2024 Fundamental heat and entropy transfer mechanisms of convection underlying the first and second laws for an unsteady multiport system and their applications in parallel flow heat exchangers Int. J. Heat Fluid Flow 107 109400https://doi.org/10.1016/j.ijheatfluidflow.2024.109400
6. Hamid, M., Usman, M., Khan, W.A., Haq, R.U., Tian, Z.: Natural convection and multidirectional magnetic field inside a square shaped cavity with sinusoidal temperature and heated/cold blocks. Int. Commun. Heat Mass Transf. **152**, 107291 (2024). https://doi.org/10.1016/j.icheatmasstransfer.2024.107291
7. Issakhov, A., Sabyrkulova, A., Abylkassymova, A.: Influence of tilt angles and different models of fluid viscosity on coupled natural convection in a differentially heated closed square cavity with a partition. Numer. Heat Transf. Part A: Appl. 1–19 (2024). https://doi.org/10.1080/10407782.2023.2299287
8. AR Al-Badri AAY Al-Waaly G Saha T Saha SC Saha 2025 Improving thermal performance in building heating, ventilation, and air conditioning systems: a study of natural convection and entropy in plus-shaped cavity Heat Transf. https://doi.org/10.1002/htj.23288
9. El Hadoui, B., Kaddiri, M., Lamsaadi, M., El Harfi, H.: Effect of the Thermal Rayleigh Number on Natural Double Diffusive Convection in Shallow Rectangular Cavities Filled With Nanofluids (2023). https://doi.org/10.1063/5.0172067
10. Ighris, Y., Qaffou, M., El hadoui, B., Baliti, J., Elguennouni, Y., Hssikou, M., Abbassi, Z.: Numerical simulation of fluid flows using the lattice boltzmann method. In: Boutahir, M., El Mehdi, E. (eds.) Advanced Materials for Sustainable Energy and Engineering: Volume 1: Novel Nanomaterials for Sustainable Energy, pp. 413–433. Cham, Springer Nature Switzerland (2025). https://doi.org/10.1007/978-3-031-75032-8_28
11. I Razzaq W Xinhua G Rasool T Sun AS Shflot MY Malik K Abbas S Ali A Ali 2025 Nanofluids for advanced applications: a comprehensive review on preparation methods, properties, and environmental impact ACS Omega 10 5251 5282 https://doi.org/10.1021/acsomega.4c10143
12. RB Ganvir PV Walke VM Kriplani 2017 Heat transfer characteristics in nanofluid—a review Renew. Sustain. Energy Rev. 75 451 460 https://doi.org/10.1016/j.rser.2016.11.010
13. Ighris, Y., Qaffou, M., Baliti, J., Elguennouni, Y., Hssikou, M.: Thermal management optimization of natural convection in a triangular chamber: role of heating positions and ternary hybrid nanofluid. Phys. Fluids **36** (2024). https://doi.org/10.1063/5.0226427
14. Ighris, Y., El hadoui, B., Baliti, J., Elguennouni, Y., Hssikou, M.: Optimizing thermal management of convective heat transfer in a complex nanofluid-filled cavity using the lattice Boltzmann method. Int. J. Numer. Methods Heat Fluid Flow (2025). https://doi.org/10.1108/HFF-01-2025-0044
15. MÖ Korukçu 2024 Heat transfer and entropy generation for mixed convection of Al_2O_3–water nanofluid in a lid-driven square cavity with a concentric square blockage Processes 12 1079 https://doi.org/10.3390/pr12061079
16. N Faraz M Shemyal Nisar Y Khan A Hussain K Iqbal 2023 Natural convection of Cu-H_2O nanofluid inside hexagonal enclosure fitted with a square cavity with a non-uniformly heated wall(s) Results Phys. 51 106648https://doi.org/10.1016/J.RINP.2023.106648

17. Çiçek, O., Sheremet, M.A., Baytaş, A.C.: Effect of natural convection hybrid nanofluid flow on the migration and deposition of MWCNT-Fe_3O_4 in a square enclosure. Int. J. Therm. Sci. **190**, 108318 (2023). https://doi.org/10.1016/j.ijthermalsci.2023.108318
18. SZ Heris MB Pour O Mahian S Wongwises 2014 A comparative experimental study on the natural convection heat transfer of different metal oxide nanopowders suspended in turbine oil inside an inclined cavity Int. J. Heat Mass Transf. 73 231 238 https://doi.org/10.1016/j.ijheatmasstransfer.2014.01.071
19. CJ Ho WK Liu YS Chang CC Lin 2010 Natural convection heat transfer of alumina-water nanofluid in vertical square enclosures: an experimental study Int. J. Therm. Sci. 49 1345 1353 https://doi.org/10.1016/j.ijthermalsci.2010.02.013
20. El hadoui, B., Kaddiri, M.: Enhancement of carbon nanotubes/kerosene nanofluids on mixed convective heat transfer in rectangular enclosures. Int. J. Thermofluids **24**, 100932 (2024). https://doi.org/10.1016/j.ijft.2024.100932
21. El hadoui, B., Kaddiri, M.: Double diffusive natural convection with variable properties of nanofluid using lattice boltzmann method. In: Ali-Toudert, F, Draoui, A., Halouani, K., Hasnaoui, M., Jemni, A., Tadrist, L. (eds.) Advances in Thermal Science and Energy, pp. 22–32. Cham, Springer Nature Switzerland (2024)
22. MZ Saghir A Ahadi A Mohamad S Srinivasan 2016 Water aluminum oxide nanofluid benchmark model Int. J. Therm. Sci. 109 148 158 https://doi.org/10.1016/j.ijthermalsci.2016.06.002
23. M Torki N Etesami 2020 Experimental investigation of natural convection heat transfer of SiO_2/water nanofluid inside inclined enclosure J. Therm. Anal. Calorim. 139 1565 1574 https://doi.org/10.1007/s10973-019-08445-9
24. P Mondal TR Mahapatra R Parveen 2021 Entropy generation in nanofluid flow due to double diffusive MHD mixed convection Heliyon. 7 e06143https://doi.org/10.1016/j.heliyon.2021.e06143
25. M Hojjat SG Etemad R Bagheri J Thibault 2011 Rheological characteristics of non-Newtonian nanofluids: experimental investigation Int. Commun. Heat Mass Transf. 38 144 148 https://doi.org/10.1016/j.icheatmasstransfer.2010.11.019
26. R Sivaraj S Banerjee 2021 Transport properties of non-Newtonian nanofluids and applications Eur. Phys. J. Spec. Top. 230 1167 1171 https://doi.org/10.1140/epjs/s11734-021-00031-1
27. Selimefendigil, F., Şenol, G., Öztop, H.F., Abu-Hamdeh, N.H.: A review on non-Newtonian nanofluid applications for convection in cavities under magnetic field. Symmetry (Basel) **15**, 41 (2022). https://doi.org/10.3390/sym15010041
28. Tizakast, Y., Kaddiri, M., Lamsaadi, M.: Double-diffusive mixed convection in rectangular cavities filled with non-Newtonian fluids. Int. J. Mech. Sci. **208**, 106667 (2021). https://doi.org/10.1016/j.ijmecsci.2021.106667
29. Bihiche, K., Lamsaadi, M.: Soret Driven Thermosolutal Convection of Power-Law Fluids in a Shallow Cavity Uniformly Heated From Below: Case of Cooperating Flows (2023). https://doi.org/10.1063/5.0171814
30. MdM Molla MF Hasan MdM Islam 2025 Elucidating thermal phenomena of non-Newtonian experimental data based copper-alumina-ethylene glycol hybrid nanofluid in a cubic enclosure with central heated plate by machine learning validations of D3Q27 MRT-LBM Int. J. Thermofluids 26 101033https://doi.org/10.1016/j.ijft.2024.101033
31. H Chen Y Ding A Lapkin X Fan 2009 Rheological behaviour of ethylene glycol-titanate nanotube nanofluids J. Nanopart. Res. 11 1513 1520 https://doi.org/10.1007/s11051-009-9599-9
32. Syam Sundar, L., Venkata Ramana, E., Singh, M.K., Sousa, A.C.M.: Thermal conductivity and viscosity of stabilized ethylene glycol and water mixture Al_2O_3 nanofluids for heat transfer applications: an experimental study. Int. Commun. Heat Mass Transf. **56**, 86–95 (2014). https://doi.org/10.1016/J.ICHEATMASSTRANSFER.2014.06.009
33. AV Brikov AN Markin SV Sukhoverkhov VA Avramenko 2016 Rheological properties of poly(ethylene glycol) solutions and gels Dokl. Phys. Chem. 469 121 123 https://doi.org/10.1134/S0012501616080030

34. Minakov, A.V., Rudyak, V.Y., Pryazhnikov, M.I.: Rheological behavior of water and ethylene glycol based nanofluids containing oxide nanoparticles. Colloids Surf. A: Phys.Chemical Eng. Asp. **554**, 279–285 (2018). https://doi.org/10.1016/j.colsurfa.2018.06.051
35. M Lamsaadi M Naïmi M Hasnaoui 2006 Natural convection heat transfer in shallow horizontal rectangular enclosures uniformly heated from the side and filled with non-Newtonian power law fluids Energy Convers. Manage. 47 2535 2551 https://doi.org/10.1016/J.ENCONMAN.2005.10.028
36. A Hossain MdM Molla 2023 MHD mixed convection of non-Newtonian power-law ferrofluid in a wavy enclosure J. Therm. Anal. Calorim. 148 11871 11892 https://doi.org/10.1007/s10973-023-12485-7
37. El Hadoui, B., Kaddiri, M.: Comparing two numerical methods in the case of aiding and opposing natural double diffusion in a square enclosure. In: 2024 4th International Conference on Innovative Research in Applied Science, Engineering and Technology (IRASET), pp. 1–5. IEEE (2024). https://doi.org/10.1109/IRASET60544.2024.10548769
38. Tizakast, Y., Kaddiri, M., Lamsaadi, M.: Rayleigh-Bénard double-diffusive mixed convection in two-dimensional rectangular cavities filled with non-Newtonian fluids. Int. J. Mech. Sci. **227**, 107448 (2022). https://doi.org/10.1016/j.ijmecsci.2022.107448
39. Zhou, S.Q., Ni, R.: Measurement of the specific heat capacity of water-based Al_2O_3 nanofluid. Appl. Phys. Lett. **92**(9) (2008). https://doi.org/10.1063/1.2890431
40. BK Saha JI Jihan G Barai NJ Moon G Saha SC Saha 2025 Exploring natural convection and heat transfer dynamics of Al_2O_3-H_2O nanofluid in a modified tooth-shaped cavity configuration Int. J. Thermofluids 25 101005https://doi.org/10.1016/j.ijft.2024.101005
41. Maïga, S.E.B., Nguyen, C.T., Galanis, N., Roy, G.: Heat transfer behaviours of nanofluids in a uniformly heated tube. In: Superlattices and Microstructures, pp. 543–557 (2004). https://doi.org/10.1016/j.spmi.2003.09.012
42. G Vahl Davis De IP Jones 1983 Natural convection in a square cavity: a comparison exercise Int. J. Numer. Methods Fluids 3 227 248 https://doi.org/10.1002/fld.1650030304
43. B Ghasemi SM Aminossadati A Raisi 2011 Magnetic field effect on natural convection in a nanofluid-filled square enclosure Int. J. Therm. Sci. 50 1748 1756 https://doi.org/10.1016/j.ijthermalsci.2011.04.010
44. O Turan A Sachdeva N Chakraborty RJ Poole 2011 Laminar natural convection of power-law fluids in a square enclosure with differentially heated side walls subjected to constant temperatures J. Nonnewton Fluid Mech. 166 1049 1063 https://doi.org/10.1016/j.jnnfm.2011.06.003

Microwave Absorption Properties of TiO_2/Al_2O_3 Ceramic Composite Material at X-band Frequencies

Youssef Ouhassan, Ali Ait Ali, Seddik Bri, Fouad Boughanzai, Mbarek Chahboun, and Mohamed Habibi

Abstract In this paper, we investigated the microwave absorption properties of TiO_2/Al_2O_3 ceramic composite as a function of three different concentrations of titanium dioxide TiO_2 in the alumina matrix Al_2O_3. The aim of this study is to analyze the impact of TiO_2 concentration on the complex dielectric properties as well as the microwave absorption capabilities of the composite in the X-band, using electromagnetic simulations. The results obtained show that the complex permittivity of the composite increases with increasing TiO_2 concentration and decreases with increasing frequency, in the frequency range from 8.2 to 12.4 GHz. This variation indicates a significant dependence for the complex permittivity of the composite on TiO_2 content and frequency. When the TiO_2 concentration reaches 15% by volume in the composite, microwave absorption performance is optimized. More specifically, the composite exhibits a reflection loss bandwidth of less than − 10 dB (absorption greater than 90%) of 1.45 GHz, with a minimum reflection loss of -37.20 dB observed at the frequency of 10.70 GHz. The frequency band over which this optimum absorption occurs extends from 10.15 to 11.60 GHz. These results indicate that the TiO_2/Al_2O_3 composite material with 15% TiO_2 by volume possesses excellent performance characteristics for microwave absorption, making this material an excellent candidate for various technological applications. The composites studied may be particularly useful in the fields of electromagnetic shielding, electromagnetic pollution reduction and other applications requiring effective microwave control. Therefore, this study confirms that the addition of TiO_2 in the Al_2O_3 matrix significantly improves microwave absorption properties, paving the way for new applications in electromagnetics.

Y. Ouhassan (✉) · F. Boughanzai · M. Habibi
Laboratory of Electronic Systems, Information Processing, Mechanics and Energy, Department of Physics, Faculty of Sciences, Ibn Tofail University, Kenetra, Morocco
e-mail: ouhassan.youssef@gmail.com

A. A. Ali · M. Chahboun
Laboratory of Engineering, Systems and Applications, The National School of Applied Sciences, Sidi Mohamed Ben Abdellah University, Fez, Morocco

S. Bri
Department of Electrical Engineering, Materials and Instrumentations Group, Superior School of Technology, Moulay Ismail University, Meknes, Morocco

E. El Mehdi and M. Boutahir (eds.), *Advanced Materials for Sustainable Energy and Engineering*, Engineering Materials, https://doi.org/10.1007/978-3-032-10069-6_10

Keywords Composite materials · Complex permittivity · Microwave absorption properties · Alumina · TiO_2 · Transmission/Reflection (T/R) · X-band

1 Introduction

Electromagnetic pollution is a major problem, affecting both the environment and human health [1, 2]. This phenomenon is notably increased by the increasing use of wireless technologies, in particular cell phones, Wi-Fi networks and other communication equipment operating on microwave frequencies [3]. At the same time, the environmental impact of this pollution is also increasingly recognized. Indeed, many animal species, especially those that navigate using natural electromagnetic fields, may find their habitats disturbed by artificial waves. Faced with these concerns, research into materials capable of absorbing these microwaves has developed [4]. By reducing their propagation in the environment, the aim is to develop effective materials to limit the harmful effects of these waves on both human health and life on earth. In this context, alumina (Al_2O_3) has become one of the most widely used ceramic materials in microwave absorbing composites [5]. Due to its particular dielectric properties [6], alumina has a relatively high permittivity compared with other ceramics, enabling it to interact effectively with electromagnetic waves and dissipate them in the form of heat. However, although alumina has good dielectric properties, its permittivity may sometimes be insufficient for certain applications, particularly in the field of electronic communications, where higher performance is necessary. To help overcome this limitation, a great deal of research has focused on improving the dielectric properties of alumina. One solution is to reinforced the alumina matrix with semiconducting and ceramic particles, such as titanium carbide (TiC) [5, 7], silicon carbide (SiC) [8], zirconium (ZrB_2) [9, 10], as well as metal alloys such as NiCrAlY [11], chromium (Cr) [6] or FeSiAl [3]. Integrating these materials into the alumina matrix not only improves its dielectric properties, but also enhances its ability to absorb electromagnetic waves, creating composites that are more efficient. Titanium dioxide (TiO_2) is one of the best performing materials for the design of these composites. This material is characterized by highly specific properties such as low density, resistance to high-temperature oxidation and good electrical properties, making it an excellent candidate for improving the conductivity of composite materials [12]. When combined with alumina, TiO_2 forms a hybrid TiO_2/Al_2O_3 composite, which benefits from the advantages of both materials. This combination offers not only improved thermal stability, but also enhanced microwave-absorbing capacity, while remaining lightweight and durable. This means that the TiO_2/Al_2O_3 composite is the material of choice for applications requiring both optimized dielectric properties and good resistance to environmental conditions.

The main objective of this study is to analyze the influence of TiO_2 concentration on the dielectric properties of TiO_2/Al_2O_3 composite. This analysis is crucial because the concentration of TiO_2 can significantly influence the ability of the composite to absorb microwaves. In addition, the study also looks at the effect of microwave frequency, particularly in the 8.2–12.4 GHz frequency range, which corresponds to the bands used in many wireless communication technologies. By studying these frequency ranges, it is possible to better understand how materials react to different microwave exposure conditions and thus optimize their performance. To characterize the dielectric properties of the TiO_2/Al_2O_3 composite, the transmission and reflection (T/R) method was used. This method enables precise measurement of the transmission and reflection coefficients of electromagnetic waves, offering a detailed assessment of the materials performance. A numerical electromagnetic simulation was used to extract transmission and reflection coefficients for different configurations of the composite matrix. The results were then processed using a MATLAB program to calculate the real and imaginary parts of the composites dielectric properties. These values are essential for understanding the materials behavior when exposed to microwaves, and for determining its efficiency in absorbing the waves. Using this data, the reflection loss coefficient (RL) was calculated, which allows to determine the performance of the TiO_2/Al_2O_3 composite to reduce electromagnetic pollution. This detailed approach opens up interesting prospects for the design of new microwave absorbing materials that are more efficient and better adapted to today's environmental and health challenges.

2 Theory and Method

In this study, we chose to adopt the Transmission/Reflection method, which was successfully used in a previous work we published in [13]. This method is widely acknowledged for its ability to provide accurate measurements of the electromagnetic properties of dielectric materials. The fundamental principle of this method is based on the measurement of the interaction of electromagnetic waves with the sample under test, enabling its complex dielectric characteristics, such as complex relative permittivity, to be calculated. We used a WR90 rectangular waveguide, with transverse dimensions of $(22.86 \times 10.16)\ mm^2$ (Fig. 1).

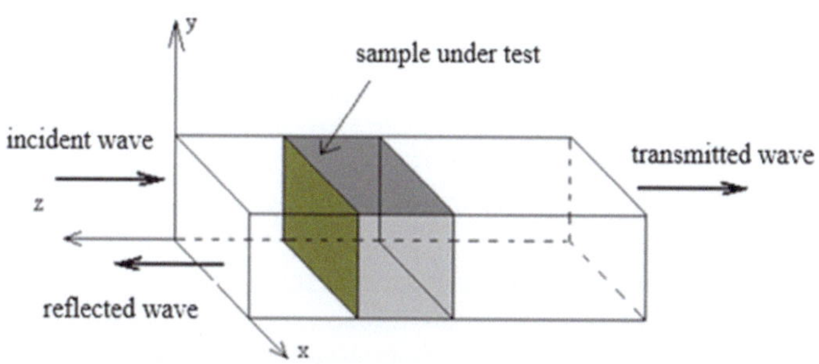

Fig. 1 WR90 rectangular waveguide, with transverse dimensions of $(22.86 \times 10.16)\ mm^2$

This waveguide is used to propagate electromagnetic waves in a specific direction, while retaining control over the frequency and properties of the waves passing through it. The WR90 waveguide is particularly well suited to the microwave frequency range, which is essential for this study, as it enables materials to be tested under conditions close to those encountered in real wireless communication or detection applications. In our experimental configuration, the waveguide is loaded at its center with the dielectric sample to be tested. This sample has a diameter of 10 mm, which is a standard dimension that allows proper insertion into the waveguide, while ensuring correct wave propagation through the material. The sample placed in the waveguide influences the way electromagnetic waves are transmitted and reflected, enabling us to measure transmission and reflection coefficients directly. The Transmission/Reflection method involves measuring these coefficients to assess how the sample affects wave propagation. The transmission coefficient indicates the proportion of energy that passes through the sample, while the reflection coefficient indicates the amount of energy that is reflected back to the source. From these measurements, it is possible to calculate the complex permittivity of the sample, which is made up of two parts: the real permittivity (which describes the material's ability to store electromagnetic energy) and the imaginary permittivity (which represents the loss of energy due to absorption and dissipation within the material). This method is particularly effective for characterizing dielectric materials in the microwave frequency range, such as those used in wireless communication technologies, radar, or electromagnetic protection devices. It offers the advantage of being relatively simple to implement, while providing highly reliable and accurate results. Furthermore, the use of electromagnetic simulation software and data processing tools such as MATLAB makes it possible to analyse in detail the impact of different frequencies and material compositions on their electromagnetic properties. This method provides crucial information about a materials performance under real-world conditions, which is essential for developing more efficient microwave-absorbing materials tailored to specific applications. In the X-band frequency range, which generally extends from 8.2 to 12.4 GHz, one of the most commonly used methods for determining the dielectric properties of materials is the Nicholson-Ross-Weir technique. This approach is widely recognized in the field of materials characterization for its accuracy, reliability, and ability to provide detailed results with a minimum of approximation. It is particularly effective for the study of materials in microwave applications, and is widely used in communication systems, radar, detection and surveillance technologies [14, 15]. When applying this method, the dielectric properties of materials are extracted by analyzing the values of parameters S_{11} and S_{21}, which are lattice parameters describing respectively the reflection and transmission of electromagnetic waves as they pass through the material under test. The parameter S_{11} represents the reflection coefficient, i.e. the amount of energy reflected by the sample at the frequency under study, while S21 corresponds to the trans-mission coefficient, indicating the energy passing through the sample. These two parameters are fundamental for characterizing the electromagnetic properties of a material, as they provide information on the attenuation, propagation and absorption of electromagnetic waves by the material. To calculate reflection coefficients (Γ) and transmission coefficients (T) from these parameters S11 and S21, specific mathematical relationships exist. These relationships transform the measured values of S11 and S21 into physical quantities directly related to the dielectric properties of the material, such as the complex relative permittivity. The Γ and T coefficients thus enable precise quantification of how the material reacts to electromagnetic waves in the X frequency range. The exact mathematical expressions of these coefficients are given by the following equations [16], which are based on the principles of electromagnetism and the incident plane wave:

$$S_{21} = \frac{\mathrm{T}(1-\Gamma^2)}{1-\Gamma^2\mathrm{T}^2} \tag{1}$$

$$S_{21} = \frac{\mathrm{T}(1-\Gamma^2)}{1-\Gamma^2\mathrm{T}^2} \tag{2}$$

The following relations give the reflection Γ and transmission T coefficients:

$$\Gamma = K \pm \sqrt{K^2-1} \tag{3}$$

$$\mathrm{T} = \frac{S_{11}+S_{21}-\Gamma}{1-(S_{11}+S_{21})\Gamma} \tag{4}$$

with the condition that $|\Gamma| < 1$ is verified and K given by:

$$K = \frac{S_{11}^2 - S_{21}^2 + 1}{2S_{11}} \tag{5}$$

Equation (6) gives the relationship for calculating μ_r:

$$\mu_r = \frac{\lambda_{0g}(1+\Gamma)}{\Lambda(1-\Lambda)} \tag{6}$$

with μ_r: relative permeability of composite and λ_{0g}: the guided wavelength in the vacuum cell is given below:

$$\lambda_{0g} = \frac{1}{\sqrt{\left(\frac{1}{\lambda_0}\right)^2 - \left(\frac{1}{\lambda_c}\right)^2}} \tag{7}$$

where the cutoff and free-space wavelengths are denoted by λ_0 and λ_C respectively.

According to the following relation, Λ is the normalized wavelength associated with the transmission coefficient:

$$\frac{1}{\Lambda^2} = \left[\frac{j}{2\pi d}\ln(\mathrm{T})\right]^2 \tag{8}$$

where d represents the material's length to be characterized.

The relative permittivity is calculated by the following relationship:

$$\varepsilon_r = \frac{\lambda_0^2}{\mu_r}\left(\frac{1}{\Lambda^2} + \frac{1}{\lambda_c^2}\right) \tag{9}$$

The RL values were calculated from the complex permittivity values of the composite using transmission line theoretical methods. The formulae below express relationship between reflection loss RL, relative material permittivity, normalized material input impedance Z_{in} and free space characteristic impedance Z_0 [6]:

$$RL(dB) = 20\log\left|{(Z_{in} - Z_0)}/{(Z_{in} + Z_0)}\right| \tag{10}$$

with $Z_0 = \sqrt{{\mu_0}/{\varepsilon_0}}$

Z_{in} can be given by the equation below:

$$Z_{in} = Z_0\sqrt{\frac{\mu_r}{\varepsilon_r}}\tanh\left[j\frac{2\pi fd}{c}\sqrt{\mu_r\varepsilon_r}\right] \tag{11}$$

In Eq. (11):

- ε_r and μ_r are the relative permittivity of the composite and its relative permeability respectively. Because TiO_2 has poor magnetism, in our case, the composite's permeability is assumed to be 1.

3 Results and Discussion

Using the characterization method detailed above, we were able to determine the dielectric properties of the TiO_2/Al_2O_3 composite at different frequencies. The results obtained show a significant variation in the real and imaginary parts of the complex permittivity of the material, as demonstrated in Fig. 2. These variations were examined as a function of frequency and the three different concentrations of TiO_2 in the Al_2O_3 matrix, enabling us to understand the influence of the composite's composition on its electromagnetic properties.

The values measured for real and imaginary permittivity show distinct behaviors depending on frequency, a phenomenon frequently observed for composite materials in the microwave range. In general, real permittivity, which determines the material's ability to store electromagnetic energy, varies according to a curve specific to each TiO_2 concentration. Similarly, imaginary permittivity, related to energy losses by dissipation in the material, also shows frequency-dependent fluctuations, which is particularly relevant in the context of electromagnetic shielding and microwave absorption applications. The values measured for real and imaginary permittivity show distinct behaviors depending on frequency, a phenomenon frequently observed

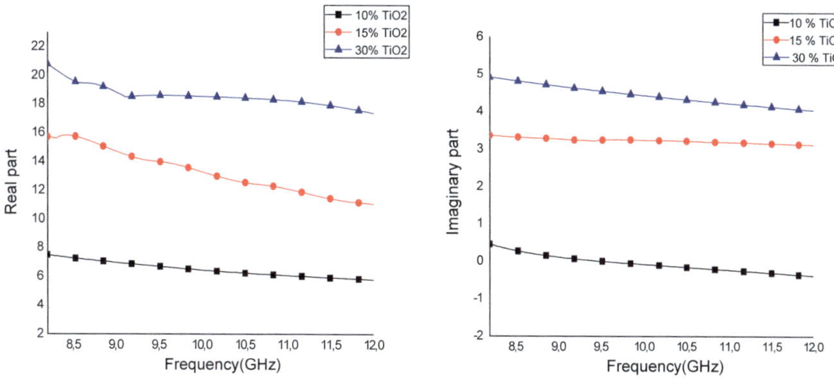

Fig. 2 The real parts and imaginary parts of de dielectric property in the X band for the TiO_2/Al_2O_3 composite with three TiO_2 inclusion contents

for composite materials in the microwave range. In general, real permittivity, which determines the material's ability to store electromagnetic energy, varies according to a curve specific to each TiO_2 concentration. Similarly, imaginary permittivity, related to energy losses by dissipation in the material, also shows frequency-dependent fluctuations, which is particularly relevant in the context of electromagnetic shielding and microwave absorption applications.

Particular attention was given to the effect of TiO_2 concentration on these properties. Indeed, the three TiO_2 incorporation rates showed marked differences, with improved permittivity and losses for higher TiO_2 concentrations. This phenomenon can be explained by the increased density of electronic states in the material, promoted by the presence of TiO_2 particles, which alters the composite's response to electromagnetic fields. It is therefore possible that the addition of TiO_2 optimizes microwave absorption properties, particularly in the frequency range from 8.2 to 12.4 GHz. Furthermore, the frequency-dependent variation in dielectric properties could be attributed to interactions between TiO_2 particles and the Al_2O_3 matrix, which influence the scattering and absorption of electromagnetic waves. These complex interactions play a key role in the overall behavior of the material and enable the permittivity of the composite to be modulated as a function of its constituents and experimental conditions. For example, the effects of electron polarization and dielectric loss can be amplified or reduced depending on TiO_2 concentrations and incident wave frequencies.

The experimental results obtained open up new prospects for optimizing the properties of TiO_2/Al_2O_3 composite materials in a variety of applications, including the design of microwave-absorbing materials, electromagnetic shielding and protection against electromagnetic pollution. The results also show that precise adjustments to TiO_2 concentration can offer finer control over the electromagnetic properties of the composite, enabling tailor-made materials to be designed for specific applications.

In conclusion, this study highlights the importance of TiO_2/Al_2O_3 composite composition and excitation frequency in determining dielectric properties. A thorough understanding of these effects will enable these materials to be better exploited in increasingly common applications in wireless communication and sensing technologies, as well as in electromagnetic protection devices.

Figure 2 details the experimental results, demonstrating the significant influence of the TiO_2 inclusion rate on the dielectric properties of the composites studied. As can be observed, as the TiO_2 concentration increases in the Al_2O_3 matrix, the complex permittivity values of the composite also increase. This phenomenon indicates that the addition of TiO_2, a material with good conductivity, significantly improves the dielectric characteristics of the composite. In particular, the composite containing 30% by volume of TiO_2 displays the best performance, with a high complex permittivity. This suggests that the inclusion of TiO_2, a semiconducting material, creates a better interaction with electromagnetic waves in the frequency range studied, making the material more suitable for applications requiring good microwave absorption, such as electromagnetic shielding and detection systems.

The properties of the imaginary part of the complex permittivity, measured at microwave frequencies, are mainly influenced by dielectric losses, which can be divided into conduction losses and losses due to polarization relaxation. As mentioned in reference [5], conduction losses are directly related to the presence of free carriers in the material, while polarization relaxation is linked to the material's ability to respond to variations in the electromagnetic field. When increasing the concentration of TiO_2 in the composite, the density of free carriers in the material increases, leading to an increase in conduction losses. At the same time, the addition of TiO_2 promotes better dipole response in the Al_2O_3 matrix, improving losses due to polarization relaxation. Both phenomena result in an increase in the imaginary part of permittivity, which is beneficial for applications where microwave absorption and electromagnetic energy dissipation are crucial. Moreover, experimental curves clearly show that frequency has a direct impact on the complex permittivity of composites. Both the real and imaginary parts of the complex permittivity decrease as frequency increases in the X-band. This observation is typical of dielectric materials and indicates that composite properties are highly dependent on irradiation frequency. At higher frequencies, the polarization mechanisms in the material become less effective, as the dipoles cannot follow changes in the electromagnetic field as quickly. As a result, at higher frequencies, the response of the material becomes weaker, resulting in lower permittivity values. This phenomenon is important in microwave applications, as it provides a better understanding of how the material reacts under operating conditions at different frequencies.

These results also show the possibility of optimizing the properties of TiO_2/Al_2O_3 composites by adjusting the TiO_2 concentration. Such optimization enables the design of composite materials whose dielectric properties and microwave absorption capabilities can be finely controlled according to specific application requirements. For example, composites with a high concentration of TiO_2 could be used in devices requiring high electromagnetic energy dissipation, while composites with a lower concentration could be chosen for applications where less absorption is required.

In conclusion, the experimental results obtained show that TiO_2 concentration and irradiation frequency play a crucial role in the electromagnetic properties of TiO_2/Al_2O_3 composites. A thorough understanding of these effects makes it possible to design tailor-made materials for specific applications in microwave technology, electromagnetic shielding, and other fields where electromagnetic energy management is essential.

To investigate the influence of TiO_2 concentration on the microwave absorption characteristics of the composite under study, we calculated the reflection loss coefficient (RL) of the materials as a function of different TiO_2 concentrations. These RL values were determined from the complex permittivity values measured for each sample. The reflection loss coefficient (RL) is a crucial parameter for assessing the microwave absorption efficiency of a material, as it indicates the amount of electromagnetic waves reflected by the material surface. The higher the RL value, the greater the material's absorption capacity, making it a good candidate for applications such as electromagnetic shielding.

To calculate RL, we used the complex values of permittivity, which include both the real and imaginary parts. The real part of permittivity is related to the material's ability to store energy, while the imaginary part is associated with energy dissipation losses in the material, which directly influences microwave absorption. By studying these parameters for various concentrations of TiO_2, we were able to identify the relationship between the proportion of TiO_2 in the composite and its efficiency in absorbing electromagnetic waves.

The results show that, as the TiO_2 concentration increases, the composite's ability to absorb microwaves improves, which is directly related to the increase in conduction losses and losses due to polarization relaxation, as mentioned above. The addition of TiO_2, with its semiconducting properties, improves microwave attenuation, making the composite more effective for applications where microwave absorption is essential. In this way, this analysis sheds light on the impact of TiO_2 concentration on microwave absorption properties, providing valuable information for the design of optimized composite materials in fields such as electromagnetic shielding and protection against electromagnetic pollution.

The reflection loss (RL) of the material studied, measured over the frequency range from 8.2 to 12.4 GHz, is shown in Fig. 3. This frequency range corresponds to the microwave X-band, which is commonly used in various communication and radar applications.

The reflection loss coefficient is a key indicator of a material's effectiveness in absorbing microwaves, as it measures the amount of electromagnetic waves reflected by the material's surface. The more negative the RL coefficient, the greater the material's ability to absorb electromagnetic energy, which is crucial for applications such as electromagnetic shielding and the reduction of electromagnetic pollution.

The results presented in Fig. 3 show how the RL value varies with frequency for the different composites studied. By analyzing these curves, we can observe variations in absorption behavior as a function of excitation frequency. In general, for many dielectric materials, the ability to absorb microwaves tends to decrease with increasing frequency, due to the weaker response of the material at higher frequencies.

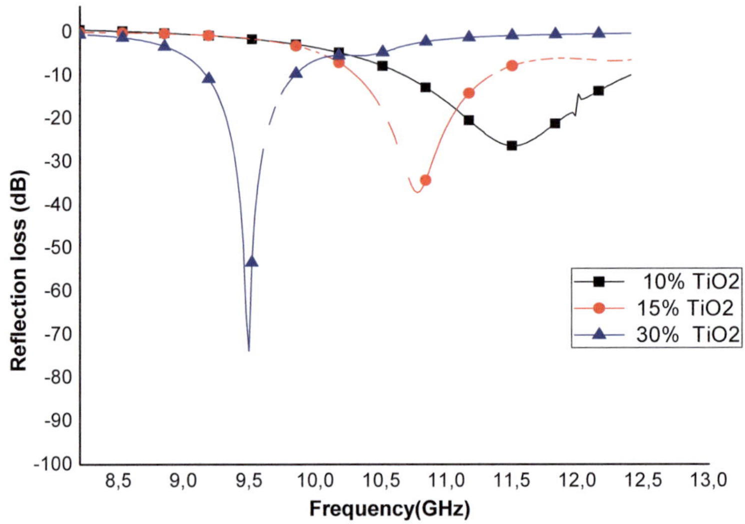

Fig. 3 Frequency evolution of the reflection loss (RL) of TiO_2/Al_2O_3 ceramic composite for three TiO_2 concentrations

However, specific variations can be observed as a function of TiO_2 concentration, allowing microwave absorption to be optimized at given frequencies. This is directly related to changes in complex permittivity, as indicated by previous calculations. In addition, the RL curves also allow us to visualize the impact of composite structure and material properties on microwave absorption efficiency. Composites with a higher concentration of TiO_2 tend to display more negative RL values, indicating a better ability to dissipate electromagnetic waves. These observations underline the importance of a judicious choice of materials and their concentration to design effective composites for specific microwave applications.

Figure 3 shows the evolution of the reflection loss (RL) of the TiO_2/Al_2O_3 composite as a function of frequency, for three different TiO_2 concentrations. The results show that for composites containing less than 15% TiO_2, the reflection loss remains above − 10 dB. This indicates that these composites are not efficient for microwave absorption, as more than 90% of the microwaves are reflected. On the other hand, when the TiO_2 concentration exceeds 15% in the composite, the reflection loss decreases and reaches values below − 10 dB. This means that microwave absorption improves considerably as the amount of TiO_2 in the Al_2O_3 matrix increases.

The results also show that composites containing 10, 15 and 30% TiO_2 exhibit a reflection loss of less than − 20 dB, meaning that these materials absorb more than 90% of incident microwaves. At a specific frequency of 10.70 GHz, the composite with 15% TiO_2 by volume displays the best microwave absorption performance, with a particularly low reflection loss of − 37.20 dB. At this frequency, the material is

Table 1 The values of the minimum reflection loss coefficient RL_{min} and their corresponding frequencies for three composites studied

	RL_{min} (dB)	$f_{m(GHz)}$ of RL_{min}	Bandwidth for RL < − 10 dB
Al_2O_3-10% TiO_2	− 26.30	11.50	10.90 GHz at 12.02 GHz (1.12 GHz)
Al_2O_3-15% TiO_2	− 37.20	10.70	10.15 GHz at 11.60 GHz (1.45 GHz)
Al_2O_3-30% TiO_2	− 73.74	9.49	9.18 GHz at 9.82 GHz (0.64 GHz)

able to absorb almost all microwaves, making it an excellent candidate for applications in electromagnetic protection systems and the reduction of electromagnetic interference.

Additionally, for this 15% TiO_2 composite, the bandwidth for which reflection loss remains below − 10 dB (indicating over 90% microwave absorption) covers a frequency range of 1.45 GHz, from 10.15 to 11.60 GHz. This wide bandwidth is an important asset, as it enables the material to absorb microwaves efficiently over a wide frequency range.

Table 1 shows the values of the minimum reflection loss coefficient RLmin and the corresponding frequencies (fm), as well as the bandwidth RL < 10 dB for the three compounds studied.

This table shows that the TiO_2-reinforced composites have significant bandwidths, especially the composite with 10% inclusion, reaching up to 1.6 GHz, confirming that the composites studied may be potential candidates for microwave absorption applications.

In conclusion, the results of this study demonstrate that TiO_2/Al_2O_3 composites, particularly those with TiO_2 concentrations above 15%, have strong potential for microwave absorption applications. These materials could be used in electromagnetic shielding devices, electromagnetic pollution reduction, as well as in other applications requiring effective microwave control.

4 Conclusion

This study examined the effect of titanium dioxide (TiO_2) inclusion concentration on the complex dielectric properties and microwave absorption characteristics of the TiO_2/Al_2O_3 composite. We first determined the complex dielectric properties of this material in the X-band using the transmission/reflection method. Using the obtained complex permittivity values, return loss (RL) values were then calculated.

The results clearly show that the dielectric properties and microwave absorption characteristics of the composite material are significantly improved by the addition of TiO_2 inclusions. In particular, the composite containing 15% by volume of TiO_2 inclusions demonstrated the best microwave absorption performance. With this composite, the absorption bandwidth (RL < − 10 dB) is 1.45 GHz, covering the frequency range from 10.15 to 11.60 GHz. This improvement is significant because it demonstrates the effectiveness of TiO_2 inclusion in increasing microwave absorption capacity across a wide frequency range.

The results obtained confirm that the TiO_2/Al_2O_3 composites studied have strong potential for applications in the field of microwave absorption. Due to their excellent absorption properties, these materials are promising for applications such as electromagnetic shielding, electromagnetic pollution reduction, and other areas requiring efficient microwave management. This study paves the way for further research on optimizing inclusion concentrations and exploring new applications for these composites.

References

1. Yao, H., Qing, Y., Li, Y., Wen, J., Li, W., Cao, Y., Luo, F.: Effect of B_4C content and annealing on complex permittivity and microwave-absorption properties of B_4C/Al_2O_3 coatings. J. Eur. Ceram. Soc. **43**(4), 1450–1458 (2023)
2. Ouhassan, Y., Bri, S., El Boubakraoui, M.C., Habibi, M.: Dielectric and micro-wave absorption properties of ZrB_2/Al_2O_3 ceramic composite. Adv. Mater. Process. Technol. **9**(3), 1091–1101 (2022)
3. Zhou, L., Huang, J., Wang, X., Su, G., Qiu, J., Dong, Y.: Mechanical, dielectric and microwave absorption properties of $FeSiAl/Al_2O_3$ composites fabricated by hot-pressed sintering. J. Alloys Compd. **774**, 813–819 (2019)
4. Jafarian, M., Kashani-Bozorg, S.F., Amadeh, A.A., Atassi, Y.: Hierarchical brain-coral-like structure (3D) vs rod-like structure (1D): Effect on electromagnetic wave loss features of $SrFe_{12}O_{19}$ and $CoFe_2O_4$. Ceram. Int. **47**(21) (2021)
5. Wang, Y., Luo, F., Zhou, W., Zhu, D.: Dielectric and microwave absorption properties of $TiC\text{-}Al_2O_3$/Silica coatings at high temperature. J. Electron. Mater. **46**, 5225–5231 (2017)
6. Zhou, L., Cui, S., Zhai, Y., Luo, F., Dong, Y.: Dielectric and microwave absorption properties of plasma sprayed Cr/Al_2O_3 composite coatings. Ceram. Int. **41**(10), 14908–14914 (2015)
7. Ouhassan, Y., Bri, S., boubakraoui, M.C.E., Habibi, M.: Dielectric properties of ceramic composite materials based on alumina reinforced by titanium carbide. FME Trans. **48**(4), 908–913 (2020)
8. Nakahira, A., Niihara, K.: Sintering behaviors and consolidation process for Al_2O_3/SiC Nanocomposites. J. Ceram. Soc. Japan **100**(1160), 448–453 (1992)
9. Liu, Y., Su, X., He, X., Xu, J., Wang, J., Qu, Y., Fu, C., Wang, Y.: Dielectric and microwave absorption properties of ZrB_2/Al_2O_3 composite ceramics. J. Mater. Sci. Mater. Electron. **30**, 2630–2637 (2019)
10. Qing, Y., Yao, H., Li, Y., Luo, F.: Plasma-sprayed ZrB_2/Al_2O_3 ceramics with excel-lent high temperature electromagnetic interference shielding properties. J. Eur. Ceram. Soc. **41**(2), 1071–1075 (2021)
11. Zhou, L., Luo, F., Cui, S., Zhang, Y.: Dielectric and microwave absorption proper-ties of low power plasma sprayed $NiCrAlY/Al_2O_3$ coatings, J. Mater. Sci.: Mater. Electron **26**(6), 3853–3860 (2015)
12. Zhang, H., Banfield, J.F.: Structural characteristics and mechanical and thermodynamic properties of nanocrystalline TiO_2. Chem. Rev. **114**, 9613–9644 (2014)
13. Ouhassan, Y., Bri, S., Boubakraoui, M.C.E., Habibi, M.: Permittivity of SiC ceramic matrix composites in Ku band. In: E3S Web Conferences, vol. 229, p. 01012. EDP Sciences (2021)
14. Nicolson, A.M., Ross, G.F.: Measurement of the intrinsic properties of materials by time-domain techniques. IEEE Trans. Instrum. Meas. **19**(4), 377–382 (1970)

15. Challa, R.K., Kajfez, D., Gladden, J.R., Elsherbeni, A.Z.: Permittivity measurement with a non-standard waveguide by using TRL calibration and fractional linear data fitting. Prog. Electromagn. Res. B **2**, 1–13 (2008)
16. Mabhouti, K., Karamirad, M., Norouzzadeh, P., Golzan, M., Naderali, R.: Measurement of electric and magnetic properties of ZnO nanoparticles in the X-band using Nicolson–Ross–Weir analysis. J. Electron. Mater. **49**, 3668–3676 (2020)

First-Principles Study of the Electronic Structure and Optical Properties of $CH_3NH_3PbX_3$ (X = I, Br, Cl) for Perovskite Silicon Tandem Solar Cell Applications

Fatima Oubihi, El Hassan El Harouny, Jamal El Khamkhami, and El Mahdi Assaid

Abstract Enhancing the power conversion efficiency is crucial for achieving affordable photovoltaic systems. Tandem solar cells, which fuse wide-bandgap perovskite top cells with silicon bottom cells, are one of the most promising methods to improve the efficiency and utilization of the solar spectrum. Besides, a cheap and effective method of increasing the performance of small band gap solar technologies is to put wide band gap polycrystalline semiconductor on the top to form a tandem structure. Methylammonium lead halides (MALH) whose chemical formula is $CH_3NH_3PbX_3$ (where X = Br, I, or Cl) have been extensively discussed as solar energy conversion materials due to their superior absorption of light. This paper explores the electronic structure and optical properties of MALH perovskite films tandem on silicon substrates through first-principles calculations that are based on Density Functional Theory (DFT). The calculated energy band structures and density of states (DOS) indicate that these tandem con-figurations exhibit a direct bandgap, which is tunable-approximately 1.64 eV for X = I, 2.43 eV for X = Br, 1.50 eV for X = Cl, and 1.12 eV

F. Oubihi (✉) · E. H. El Harouny · J. El Khamkhami
Laboratoire de Physique de La Matière Condensée, Département de Physique, Faculté Des Sciences, Université Abdelmalek Essaadi, Tétouan, Morocco
e-mail: fatima.oubihi@etu.uae.ac.ma

E. H. El Harouny
e-mail: elharouny.elhassan-etu@uae.ac.ma

J. El Khamkhami
e-mail: jelkhamkhami@uae.ac.ma

E. M. Assaid
Equipe d'Optique Et Electronique du Solide, Département de Physique, Faculté Des Sciences, Université Chouaïb Doukkali, El Jadida, Morocco

Laboratoire d'Instrumentation, Mesure Et Contrôle, Département de Physique, Faculté Des Sciences, Université Chouaïb Doukkali, El Jadida, Morocco

E. M. Assaid
e-mail: assaid.e@ucd.ac.ma

E. El Mehdi and M. Boutahir (eds.), *Advanced Materials for Sustainable Energy and Engineering*, Engineering Materials, https://doi.org/10.1007/978-3-032-10069-6_11

for silicon. The results suggest that the tandem design strategy can effectively narrow the band gap, and thus improve the solar energy conversion efficiency. The authors' statements are in agreement with older new experimental and theoretical literature, reiterating confidence in the tandem approach for the future market of photovoltaic technologies.

Keywords Tandem · Methylammonium lead halide perovskite · Density functional theory · Electronic structure and · Optical properties

1 Introduction

Without any doubt, the most pressing problem of modern times is the search for fitting materials to produce and store energy. The renewable energy sources are predominantly solar-based that constitute a key resource because of its global availability and low environmental impact. For decades, silicon-based photovoltaic technologies have led the solar energy market, which is currently dominating about 95% of the market. They do this on the basis of their high module efficiency, plentiful supply of non-toxic ingredients, and long-term reliability. However, the maximum theoretical limit of the efficiency of a single silicon photovoltaic cell is 33%. In practice this value is lower, because the construction and operation of cells inevitably lead to small energy losses. The limitations and deficiencies arise from the material properties of silicon. Its band gap means that only photons with a specific energy can be converted into electricity. The photon with higher energy cannot be fully used by the solar cell. To achieve higher performances, many efforts in the latest years have been devoted by recent research has focused on developing advanced prototypes, and promising results have been reached in this context. So the most powerful solution for increasing the efficiency of modules PV is to make tandem solar cells [1].

Metal halide perovskites are a semiconductor material category that has unique properties, such as the presence of tunable, high efficiency absorption characteristics, broad absorption spectrums, band gaps, extended charge diffusion lengths, elevated charge carrier mobility, these qualities make them suitable for a wide range of photovoltaic and optoelectronic applications. Inorganic–organic perovskite solar cells have ABX3 as their basic structure [2]. Normally position A typically contains organic cations organic [MA, FA, …], while position B contains bivalent metal cations [Pb, Sn, Ge,…], and halogen anions [I, Br, Cl,…] are occupy the X site. B and X ions are forming a regular octahedron structure in this unit, B ion is at the center of the octahedron, X ion is at the vertex of the octahedron, and A ion fills in the void formed by the three-dimensional octahedron network [3]. Perovskite solar cells have shown a power conversion efficiency of 3.8% since their first demonstration, performance has rapidly improved, reaching a certified efficiency of 22.1% within just a few years [4]. However, what gives a distanced $CH_3NH_3PbX_3$ over other materials and makes it an attractive alternative to silicon is its low-cost, solution-based synthesis and easy processing. Since its introduction as a solar absorber, extensive research

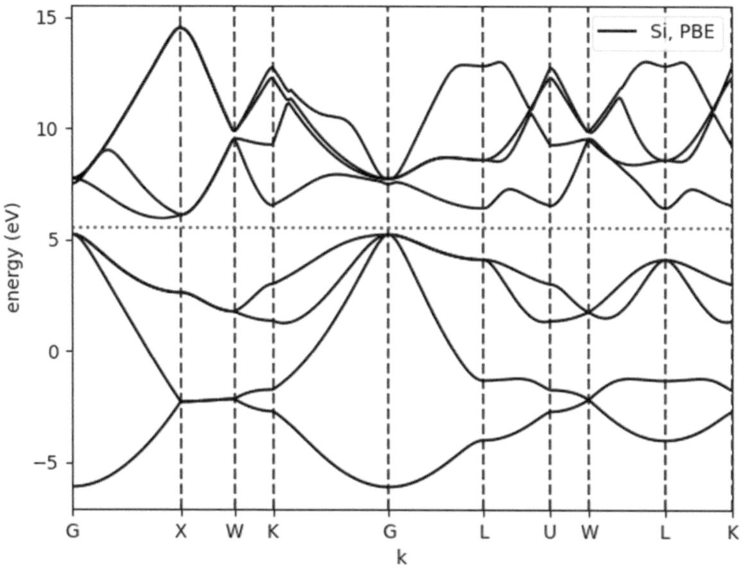

Fig. 1 Electronic band structure of silicon in the diamond crystal structure, calculated using density functional theory with the PBE exchange–correlation functional

has been devoted to device fabrication, efficiency optimization, and the development of scalable architectures for industrial application. This classification makes metal halide perovskite ideal for double junction tandem cells. Because of its outstanding attributes, Including a high absorption coefficient, extended carrier lifetime, direct band gap, long, and high balanced mobilities for both hole and electron [4]. So to improve the PCE, carefully designing the integration of high-efficiency wide-bandgap top solar cells (Perovskite) with low-bandgap (Silicon, see Fig. 1) bottom solar cell will form tandem solar cells.

2 Computational Methods

In all ab initio methods used to describe an N-electron system, the goal is to approximate the exact wave function, which contains all the accessible information about the system. However, such calculations can be extremely demanding in terms of computational time and memory, particularly for larger systems. It is reasonable to consider coming up with new ways to approach this problem, even if this means some detail is sacrificed. The authors studied the structural and electronic characteristics of $CH_3NH_3PbX_3$ compounds (where X = Br, I, and Cl) by applying the DFT method. To characterize and predict the properties of crystalline materials, Density Functional Theory (DFT) stands as the benchmark computational approach. The

Generalized Gradient Approximation (GGA) as well as Local Density Approximation (LDA) were the basis for our calculations. To treat the exchange–correlation interactions we used the Perdew-Burke-Ernzerhof (PBE) functional in the Projector Augmented Wave (PAW) framework. In order to get trustworthy results, the HSE hybrid functional was combined with precise electronic band structure calculations.

3 Various Properties of $CH_3NH_3PbX_3$ (X = I, Br, Cl)

3.1 Structural Properties

The crystallographic structures of different hybrid perovskite are among the most studied fundamental aspects so far, especially since they are subject to radical changes as a function of temperature. Studies exploiting the temperature dependent variations in the behavior of $CH_3NH_3PbX_3$ have led to a classification of the crystal structures of this perovskite according to the temperature region. At high temperature, this perovskite adopts a cubic structure, with symmetry group (P_{m3m}) [5]. The $CH_3NH_3PbX_3$ perovskites (X = Cl: T > 179 K, X = Br: T > 236 K, X = I: T > 330 K) have also been observed to exhibit a weak ferroelectric displacement of the Pb^{2+} cation, which induces the formation of a tetragonal unit cell with space group ($P|4mm$) [4, 6]. When the temperature is lowered, a first transition from the cubic phase to the tetragonal phase, with space group ($I|4/mcm$),, occurs at room temperature and results in a tetragonal unit cell with a volume double that of the cubic phase. By further decreasing the temperature, a second transition from the tetragonal symmetry to an orthorhombic symmetry (X = Cl: T < 173 K, X = Br: T < 149 K, X = I: T < 161 K) reduces the disordering of (CH_3NH_3) occupation [6]. In the orthorhombic phase of $CH_3NH_3PbBr_3$, the cation acquires a fixed position and ($P|na2$) whereas $CH_3NH_3PbI_3$ the orthorhombic phase of $CH_3NH_3PbCl_3$ ($P|222$) now offers a twofold disordered position for the (CH_3NH_3) cation [4]. In this investigation, $CH_3NH_3PbX_3$ compounds display perovskite structures with an orthorhombic crystal phase. At room temperature, they belong to the (P_{nma}) space group [7]. In order to remain stable. An orthorhombic phase with a 48 atom unit cell exists at low temperatures (Fig. 2).

The compounds of $CH_3NH_3PbX_3$ crystallize in an orthorhombic structure in perovskite. they adopt the (P_{nma}) space group to maintain structural stability [8]. A 48-atom Fig. 2 illustrates the atomic unit cell of the orthorhombic phase, which remains stable under low temperature conditions. Each structure comprises four CH_3NH_3 molecules incorporated into the $PbBr_3$, PbI_3 and $PbCl_3$ frameworks, Each Pb^{2+} ion is coordinated by six Br^-, I^-, or Cl^- anions to form corner sharing $PbBr_6$ PbI_6 and $PbCl_6$ octahedra. The Pb–Br bond lengths are found to range between approximately 3.312 Å and 3.314 Å, While Pb–I bond lengths vary from 3.175 Å to 3.199 Å, And Pb–Cl bond lengths vary from 2.386 Å to 2.684 Å. furthermore, the

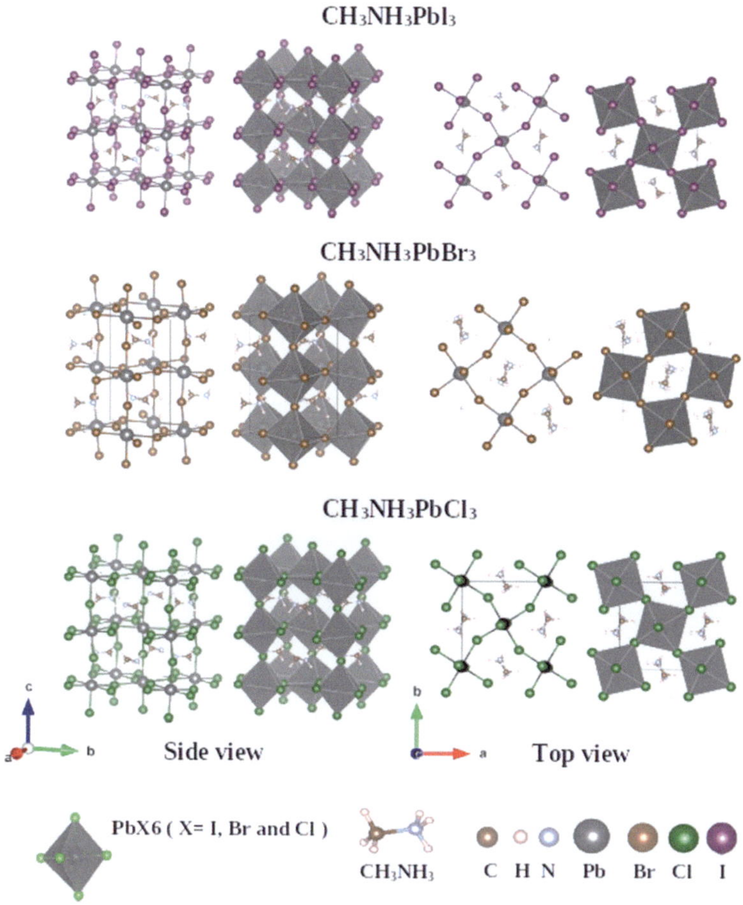

Fig. 2 Side and top views, along and chemical bonds in the CH_3NH_3 molecules along with PbI_6, $PbBr_6$, and $PbCl_6$ groups, associating to $CH_3NH_3PbI_3$, $CH_3NH_3PbBr_3$ and $CH_3NH_3PbCl_3$

bond lengths for C–N, C–H, and N–H within the organic component are measured to be 1.501 Å–1.617 Å, 1.017 Å–1.12 Å, and 1.70–1.180 Å respectively.

3.2 *Electronic Properties*

3.2.1 The Band Structure

Using "first-principles pseudo potential calculations" method, in 2004, Chang and Park [9] studied the electronic properties of inorganic–organic lead halide perovskites $CH_3NH_3PbX_3$ where X is (Br, I and Cl), According to this study, it turns out that the electronic levels of the organic molecule are located deep in the conduction and valence bands. While the edge states of the bands mainly come from the Pb–X bonds. The authors conclude that the impact of the organic cation on the opto-electronic properties of the material is negligible. On the other hand, I. Borriello, G. Cantele, and D. Ninno show in 2008 [9] that the electronic properties, specifically band gap, can be adjusted by an appropriate choice of organic cation, and more precisely its size. Furthermore, this study, based on ab initio calculations, by analyzing the energy band diagrams and density of states, the orthorhombic phase electronic structures of $CH_3NH_3PbX_3$ were calculated using PBE. The band structures obtained for $CH_3NH_3PbX_3$, X (I, Br and Cl) are shown in Fig. 3. The highest occupied bands and lowest unoccupied bands are highlighted in red and blue on the graph for each structure. Each of the calculated structures display a direct band gap located at the Γ-point in the Brillouin zone. These findings suggest that $CH_3NH_3PbI_3$, $CH_3NH_3PbBr_3$ and $CH_3NH_3PbCl_3$ possess a direct band gap, with values of energy 1.64 eV, 2.43 eV and 1.50 eV respectively (Fig. 3).

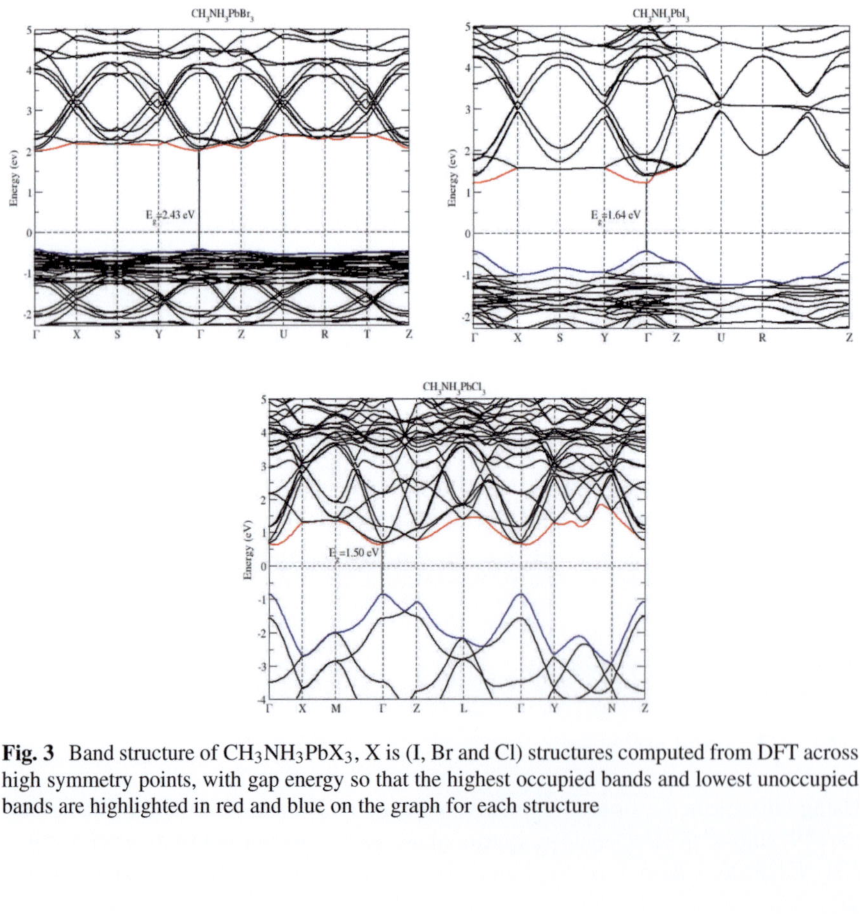

Fig. 3 Band structure of $CH_3NH_3PbX_3$, X is (I, Br and Cl) structures computed from DFT across high symmetry points, with gap energy so that the highest occupied bands and lowest unoccupied bands are highlighted in red and blue on the graph for each structure

Table 1 Lattice constant (Å) of $CH_3NH_3PbX_3$, along with the corresponding experimental data, are also provided for comparison, along with band gap values Eg (eV) calculated using (PBE and HSE)

Compounds	a	b	c	DFT (eV)	HSE (eV)
$CH_3NH_3PbBr_3$	8.06[a] 7.95 [11] 8.07 [12] 7.97 [13] 8.05 [14]	8.77[a] 8.15 [11] 8.78 [12] 8.58 [13] 8.52 [14]	12.08[a] 11.55 [11] 12.09 [12] 11.84 [13] 12.34 [14]	2.43[a] 2.14 [12]	2.75[a] 2.81 [12] 2.30 [14]
$CH_3NH_3PbI_3$	8.55[a] 8.86 [15] 5.67 [13] 8.87 [14]	9.25[a] 8.58 [15] 5.62 [13] 8.62 [14]	12.96[a] 12.62 [15] 8.63 [14] 13.12 [14]	1.64[a] 1.77 [12] 1.68 [15]	2.13[a] 2.34 [12] 1.63 [14]
$CH_3NH_3PbCl_3$	8.21[a] 8.86 [13]	8.21[a] 8.58 [13]	12.34[a] 12.62 [13]	1.50[a]	

x^a: result from this work

3.2.2 The Density of State

The density of states plays essential role for determining the hybridization between the various states of a material's components as well as for understanding the electronic transitions between the conduction band and the valance band. Moreover, their analysis offers valuable insight into determining the energy band gap [10]. The density of states (DOS) diagram for this case is presented in Fig. 4 which is correlated with the band structure depicted in the same Fig. 3, and its analysis clearly that the p states of Br and I and Cl hybridize with the p states of pb, while the s states of CH_3NH_3 contribute minimally, This hybridization plays a important role in the energy interval under investigation. The van Hove singularities in the density of states around Fermi energy E_f, disappear within energy ranges of 2.43 eV, 1.64 eV and 1.50 eV, respectively, for $CH_3NH_3PbBr_3$, $CH_3NH_3PbI_3$, and $CH_3NH_3PbCl_3$,Suggesting that these materials display the proprieties typical of tight gap semiconducting properties (Table 1; Fig. 4).

3.2.3 The Effective Masse

The effective mass is a fundamental parameter for describing the electronic properties of photovoltaic materials as it direct impact charge carrier mobility and electronic transport. A thorough understanding of these transport mechanisms properties of all heterostructures studies and their influence on photovoltaic efficiency, in this context, analyzing the average effective masses of holes and electrons becomes particularly crucial. It's well known that a lower effective masse generally leads to higher carrier mobility and conversely, a higher effective mass results in reduced mobility, which in turn can increase the solar cells efficiency by enhancing charge transport and minimizing recombination losses. The average effective mass was determinate by

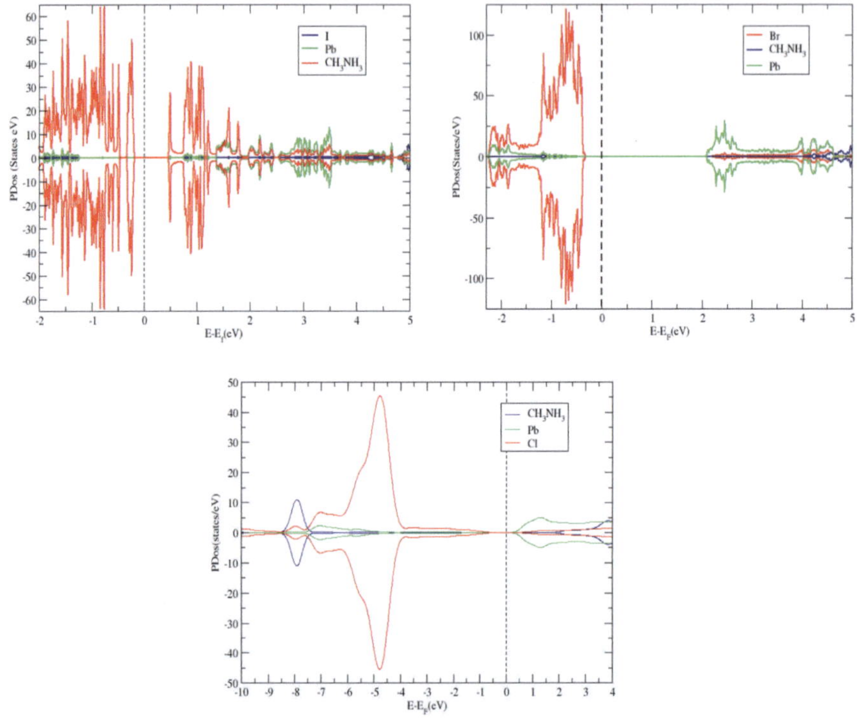

Fig. 4 Projected density of states of $CH_3NH_3PbX_3$ (X = I, Br and Cl) structures calculated with PBE functional

fitting the parabolic functions to the conduction band minimum and valence band maximum along the x and y directions respectively, using the following approach:

$$\frac{1}{m} = \frac{1}{\hbar^2}\frac{\partial^2 E_n(k)}{\partial k_i \partial k_j} \tag{1}$$

Table 2 is summarizes the effective masses of holes (h) and electrons (e) along the x and y axes, as well as the average effective mass m_d for $CH_3NH_3PbX_3$, X(I, Br and Cl), It's observed that along both the x and y directions, the effective mass of conduction electrons is relatively small, whereas the holes exhibit a significantly larger effective mass. This suggests that the mobility of electrons is more easily and higher than holes in the material. This disparity influences the carrier transport properties, affecting how efficiently charges are extracted and generate an electric current when an electric field is applied. The Variations in the effective masses affect recombination rates. When electron effective mass is lower, electrons are less likely to recombine with holes, which enhances charge separation and increases the likelihood of charge collection at the contacts, thereby improving the production of electrical energy.

Table 2 Calculated effective mass of electron and holes for $CH_3NH_3PbX_3$ (X = I, Br and Cl)

		$CH_3NH_3PbBr_3$	$CH_3NH_3PbI_3$	$CH_3NH_3PbCl_3$
m_e*	$m^*_{e\_x}$	0.18857	0.14023	0.10735
	$m^*_{e\_y}$	0.16736	0.13565	0.10088
	m_{e_d}	0.17765	0.13792	0.10406
	m_{e_d} [11]	0.16910	0.11876	0.21 [8]
m_h*	$m^*_{h\_x}$	– 1.08557	– 0.29269	– 0.12876
	$m^*_{h\_y}$	– 0.84527	– 0.23209	– 0.13215
	m_{h_d}	0.95791	0.26063	0.13044
	m_{h_d} [11]	0.12943	0.14716	0.19 [8]

3.3 Optical Properties

3.3.1 Dielectric Function

The investigation of the optical properties of perovskite materials $CH_3NH_3PbX_3$ (X = I, Br and Cl) is important to comprehending the characteristics of these materials and assessing their suitability for diverse utilizations [16], Therefore, a thorough analysis of its optical characteristics has been performed. The complex dielectric constant $\varepsilon(\omega)$ is defined by the following Eq. (2):

$$\varepsilon(\omega) = \varepsilon_1(\omega) + i\varepsilon_2(\omega) \tag{2}$$

This dielectric function is consists of two components: The imaginary parts can be theoretically derived from the calculation of momentum matrix elements between the occupied and unoccupied states as shown below [14]:

$$\varepsilon_2 = \frac{e^2\hbar}{\pi m^2\omega^2}\sum_{\upsilon,c}\int_{BZ}\lceil\langle ck|e\nabla|\upsilon k\rangle\rceil^2\delta\{\omega_{ck}(k)-\omega\}d^3k \tag{3}$$

And the real part $\varepsilon 1$ (ω), which is responsible for the polarization and dispersion of the incident light is determined from the imaginary part $\varepsilon_2(\omega)$ via the Kramers–Kronig relation:

$$\varepsilon_1 = 1 + \frac{2}{\pi}\int_0^\infty \frac{\varepsilon_2(\omega')}{\omega'^2-\omega^2}\omega' d\omega' \tag{4}$$

The both parts imaginary and real of dielectric functions with various directions, respectively, $CH_3NH_3PbI_3$, $CH_3NH_3PbBr_3$ and $CH_3NH_3PbCl_3$ are shown in Fig. 5a–f. $\varepsilon_2(\omega)$ is the imaginary part of the dielectric function which corresponds to the optical absorption at a given frequency ω.

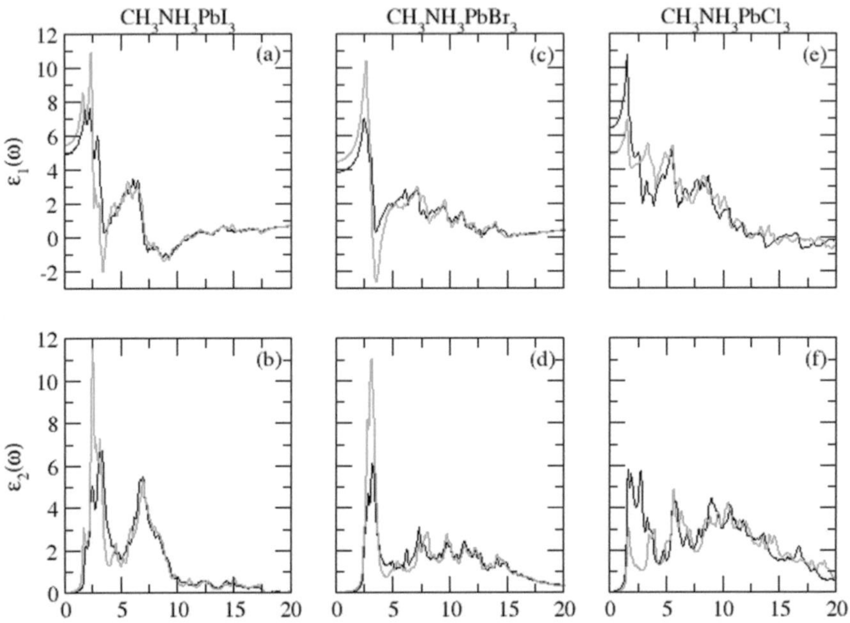

Fig. 5 Both the imaginary (b), (d) and (f) and real parts (a), (c) and (e) of dielectric functions following xx direction (black curves), zz direction (green curves) respectively, **a–b** for $CH_3NH_3PbI_3$, **c–d** for $CH_3NH_3PbBr_3$ and **e–f** for $CH_3NH_3PbCl_3$

3.3.2 Absorption Coefficient

The absorption coefficient was determined using the equation provided below:

$$\alpha(\omega) = \varepsilon_2 \frac{\omega}{nc} \tag{5}$$

n is the refractive index, c the speed of light in vacuum and ω is in energy unit and. In addition, there are others optical parameters, such as optical conductivity σ(ω) Both the energy loss function L(ω) and reflectivity R(ω) can be obtained based on the refractive index n(ω) and extinction coefficient K(ω). These, in turn, are derived from the real and imaginary parts of the dielectric function.

$$n(\omega) = \left(\frac{\sqrt{2\varepsilon_1^2 + 2\varepsilon_2^2} + 2\varepsilon_1}{\varepsilon_i^2} \right)^{\frac{-1}{2}} \tag{6}$$

$$K(\omega) = \left(\frac{\sqrt{\varepsilon_1^2 + \varepsilon_2^2} - \varepsilon_1}{2} \right)^{\frac{1}{2}} \tag{7}$$

$$\sigma(\omega) = \frac{\alpha(\omega)n(\omega)c}{4\pi} \tag{8}$$

$$R(\omega) = \frac{(n-1)^2 + K^2}{(n+1)^2 + K^2} \tag{9}$$

$$L(\omega) = \frac{\varepsilon_i}{\varepsilon_1^2 + \varepsilon_2^2} \tag{10}$$

We also considered the both parts of the dielectric functions real and imaginary to compare the presence/absence of isotropy effects under various directions, as illustrated in Fig. 5. Our analysis indicates that the $CH_3NH_3PbI_3$, $CH_3NH_3PbBr_3$ and $CH_3NH_3PbCl_3$ perovskite structures exhibit anisotropy along both the xx/zz and yy/zz axes. However, the anisotropy in the xx/yy directions is negligible (on the order of 10^{-5}), prompting us to concentrate our discussion on the xx/zz directions. Figure 5(a) reveals a peak at about 2 eV for $CH_3NH_3PbI_3$, in Fig. 5(c) at 2.5 eV for $CH_3NH_3PbBr_3$ and in Fig. 5(e) at 1.8 eV, suggesting notable electronic transitions occurring at the corresponding energies. These peaks, corresponding to interband transitions, demonstrate that electrons are excited from the valence bands to the conduction bands, in accordance with theoretical calculations. Regarding Fig. 5(b–f), which indicates the imaginary part ε(ω) which indicates the imaginary part ε(ω), exhibits a prominent peak in the 2–3 eV region for both compound. But, $CH_3NH_3PbI_3$ shows a slightly higher absorption in the energy region between 5 and 10 eV than $CH_3NH_3PbBr_3$ and $CH_3NH_3PbCl_3$. This suggests that $CH_3NH_3PbI_3$ may absorb more light in this region enhancing its potential for photovoltaic applications. The observed peaks are likely the result of electronic transitions between the valence and the conduction bands.

The absorption coefficient α(ω), which characterizes the degree to which light is absorbed by these semiconductors, is presented in Fig. 6. This parameter is critical for solar cell applications, as high absorption in the visible range allows these chlorine, bromine containing or iodine perovskite halides to efficiently capture sunlight for energy conversion. As evident in Fig. 6, when the photon energy is falls below than the onset frequency, the absorption coefficient α(ω) diminishes to negligible levels, as electronic excitation does not occur. However, beyond the threshold frequency, α(ω) exhibits significant variations, reflecting an enhanced sensitivity to electronic excitation processes. Concerning α(ω), Specifically, $CH_3NH_3PbI_3$ demonstrates light absorption within the visible spectrum, initiating at approximately (~ 1.7 eV). The absorption spectrum within this visible range plays a significant role in contributing to the outstanding performance of $CH_3NH_3PbI_3$ in photovoltaic applications. The capability to absorbing light in the visible spectrum is particularly ideal for improving solar cells functionality, as it enables efficient conversion of sunlight into electrical energy, making it highly suitable for photovoltaic applications. In comparison, $CH_3NH_3PbBr_3$, exhibits a blue-shifted absorption onset, associated with a wide band gap. A distinct absorption peak is appears at approximately (~ 2.5 eV). Indicating a blue-shifted absorption spectrum. This shift suggests that $CH_3NH_3PbBr_3$ processes

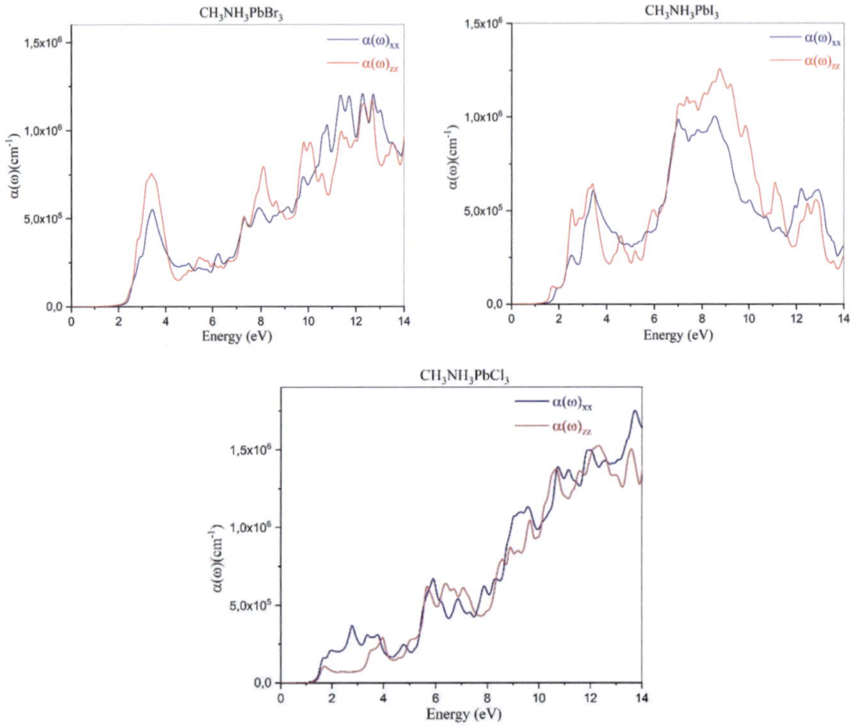

Fig. 6 Absorption coefficient of $CH_3NH_3PbBr_3$, $CH_3NH_3PbI_3$ and $CH_3NH_3PbCl_3$as a function of energy, with xx-(blue curves) and zz-(red curves) directions

a higher energy, and for the $CH_3NH_3PbCl_3$ the peak is appears at approximately (~ 1.5 eV).

3.3.3 Loss Function

To gain a deeper insight into the optical characteristics of $CH_3NH_3PbX_3$ compounds we also examine additional optical properties, such as the energy loss function (ELF). Each peak in the energy loss function (ELF) confirms the presence of the ω_p plasmon mode, which represents the quantization of charge oscillations. These values closely align with the zero points of the both parts of the dielectric functions the imaginary and real. As illustrated in Fig. 7, the ELFs exhibit pronounced maxima in the energy range from 4.50 eV to 4.90 eV for both materials, respectively. We noticed also that our structures exhibit anisotropy along both the xx and yy axes.

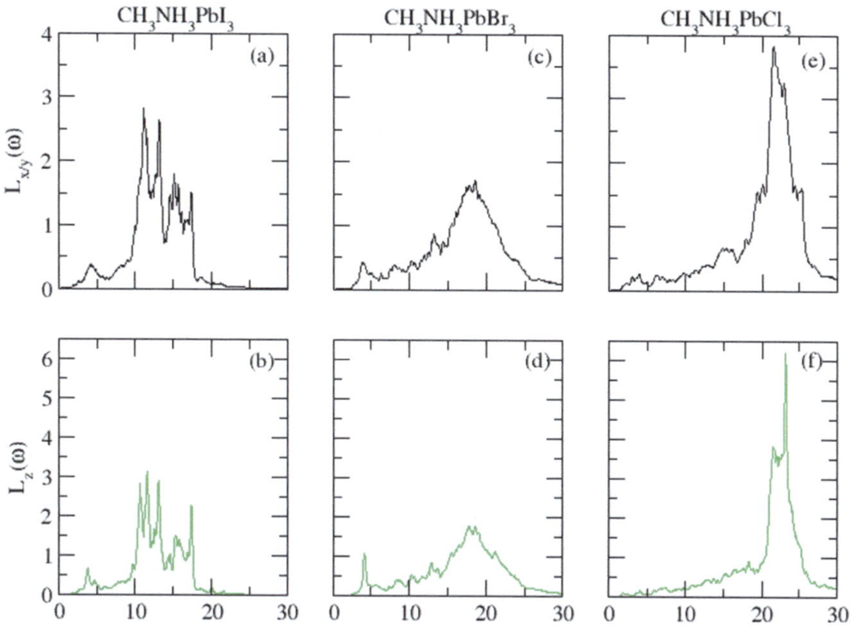

Fig. 7 Energy loss functions of $CH_3NH_3PbBr_3$ and $CH_3NH_3PbI_3$ as a function of energy, with xx and zz directions

4 Conclusion

In summary, this study investigates the electronic structure and optical properties of Methylammonium lead halides perovskite (MALH) stacking arrangements on silicon substrates using Density Functional Theory. The results of our calculations indicate that all the structural systems analyzed display a direct band gap for MALH, with both the valence band maximum and conduction band minimum positioned at the Γ-point. Furthermore, the optical characteristics of these systems have been evaluated over the energy range [0–20 eV]. Their absorption coefficients in the visible and ultraviolet regions are particularly notable, making them promising candidates for photovoltaic applications. Our findings suggest that these structures could be leveraged in the development of optoelectronic and photovoltaic devices. As a result, these results pave the way for further research into the creation of more efficient solar cells using these materials. Our upcoming work, we aim to integrate these materials into a tandem solar cell to assess its performance and determine a configuration that allows us to attain a higher energy conversion efficiency and improved form factor (FF), so we can prove with the help of the results that we will obtain the high potential of this type of materials for advanced photovoltaic applications.

Acknowledgements The authors would like to express their gratitude to the National Center for Scientific and Technical Research (CNRST), Rabat, High Performance Computing (HPC) Marwan (www.marwan.ma/hpc) for granting permission and providing support for our computational research.

References

1. Fu, F., Li, J., Yang, T.C., Liang, H., Faes, A., Jeangros, Q., Ballif, C., Hou, Y.: Monolithic perovskite-silicon tandem solar cells: from the lab to fab? Adv. Mater. **34**, 2106540 (2022). https://doi.org/10.1002/adma.202106540
2. El-Demsisy, H., Selmy, W.: Perovskite-based tandem/multi-junction solar cells. Int. J. Mater. Technol. Innov. **1**(2), 67–80 (2021). https://doi.org/10.21608/ijmti.2021.204571
3. Yin, T., Fang, Y., Fan, X., Zhang, B., Kuo, J.-L., White, T.J., Chow, G.M., Yan, J., Shen, Z.X.: Hydrogen-bonding evolution during the polymorphic transformations in $CH_3NH_3PbBr_3$: experiment and theory. Chem. Mater. **29**, 5974–5981 (2017). https://doi.org/10.1021/acs.chemmater.7b01630
4. Zhang, W., Eperon, G.E., Snaith, H.J.: Metal halide perovskites for energy applications. Nat. Energy **1**, 16048 (2016). https://doi.org/10.1038/nenergy.2016.48
5. Liao, X., Lu, R., Xia, L., Liu, Q., Wang, H., Zhao, K., Wang, Z., Zhao, Y.: Density functional theory for electrocatalysis. Energy Environ. Mater. **5**, 157–185 (2022). https://doi.org/10.1002/eem2.12204
6. Stoumpos, C.C., Malliakas, C.D., Kanatzidis, M.G.: Semiconducting tin and lead iodide perovskites with organic cations: phase transitions, high mobilities, and near-infrared photoluminescent properties. Inorg. Chem. **52**, 9019–9038 (2013). https://doi.org/10.1021/ic401215x
7. Bikerouin, M., Balli, M., Farkous, M., El-Yadri, M., Dujardin, F., Abdellah, A.B., Feddi, E., Correa, J.D., Mora-Ramos, M.E.: Effect of lattice deformation on electronic and optical properties of $CuGaSe_2$: ab-initio calculations. Thin Solid Films **696**, 137783 (2020). https://doi.org/10.1016/j.tsf.2019.137783
8. Chang, Y., Park, C.H., Matsuishi, K.: First-principles study of the structural and the electronic properties of the lead-halide-based inorganic-organic perovskites (CH~ 3NH~ 3) PbX~ 3 and CsPbX~ 3 (X= Cl, Br, I). J.-Korean Phys. Soc. **44**, 889–893 (2004)
9. Belardinelli, R.E., Pereyra, V.D.: Fast algorithm to calculate density of states. Phys. Rev. E **75**, 046701 (2007). https://doi.org/10.1103/PhysRevE.75.046701
10. Jayakrishnan, R., Raj, A., Varma, S.J.: Self-assembled methyl-ammonium lead bromide thin films with blue photoluminescence. Appl. Nanosci. **11**, 2095–2102 (2021). https://doi.org/10.1007/s13204-021-01933-1
11. Thi Han, N., Le Manh, T., Khuong Dien, V.: First-principles calculations into electronic, and excitonic effects of CH3NH3PbX3 (X = Br, I) perovskite solar cells. Phys. B Condens. Matter **682**, 415915 (2024). https://doi.org/10.1016/j.physb.2024.415915
12. Umari, P., Mosconi, E., De Angelis, F.: Relativistic GW calculations on $CH_3NH_3PbI_3$ and $CH_3NH_3SnI_3$ perovskites for solar cell applications. Sci. Rep. **4**, 4467 (2014). https://doi.org/10.1038/srep04467
13. Borriello, I., Cantele, G., Ninno, D.: *Ab initio* investigation of hybrid organic-inorganic perovskites based on tin halides. Phys. Rev. B **77**, 235214 (2008). https://doi.org/10.1103/PhysRevB.77.235214
14. Liu, Y., Yang, Z., Cui, D., Ren, X., Sun, J., Liu, X., Zhang, J., Wei, Q., Fan, H., Yu, F., Zhang, X., Zhao, C., Liu, S.F.: Two-inch-sized perovskite $CH_3NH_3PbX_3$ (X = Cl, Br, I) crystals: growth and characterization. Adv. Mater. **27**, 5176–5183 (2015). https://doi.org/10.1002/adma.201502597
15. Feng, J., Xiao, B.: Crystal structures, optical properties, and effective mass tensors of $CH_3NH_3PbX_3$ (X = I and Br) phases predicted from HSE06. J. Phys. Chem. Lett. **5**, 1278–1282 (2014). https://doi.org/10.1021/jz500480m

16. Chargui, T., Lmai, F., Rahmani, K.: Comprehensive analysis of optoelectronic and photovoltaic properties of Cs_2ScAgX_6 (X = Br, I) double perovskites for tandem cells using DFT and SCAPS-1D methods. Sol. Energy **283**, 113016 (2024). https://doi.org/10.1016/j.solener.2024.113016

Theoretical Study of the Physical Properties of XH_2 Hydrides (X = Mg and Sr) for Hydrogen Storage

Adil Bardane, El Mahjoub Chakir, Al Ibrahmi El Mehdi, Hassane El Bekkouri, and Mohamed Sayouti

Abstract The search for environmentally friendly energy sources amid the challenges posed by petroleum usage and diminishing fossil fuel reserves has propelled hydrogen to the forefront. Hydrogen, with its high calorific value and versatile applications, is a promising solution. Metal hydrides, particularly magnesium hydride (MgH_2) and strontium hydride (SrH_2), offer efficient hydrogen storage options. Density Functional Theory (DFT) simulations, provide insights into the structural and electronic properties of these metal hydrides. Despite challenges such as energy-intensive production and slow-release kinetics, research continues to enhance metal hydride storage, with alkaline earth metals like magnesium holding potential for future hydrogen storage solutions. DFT serves as a valuable computational tool in materials science, facilitating the exploration of electronic properties and contributing to advancements in hydrogen storage technologies.

Keywords Density functional theory (DFT) · Hydrogen · Energy storage

1 Introduction

The environmental and economic problems caused using petroleum products and the shortage of these fossil fuels have led to a search for other abundant and environmentally friendly energy sources. Many of these sources are intermittent and require storage solutions. Hydrogen appears to be a good candidate to fulfil this function. Dihydrogen, the molecular form of this element, which is gaseous under normal temperature and pressure conditions, has a higher calorific value of 120 MJ/kg, i.e. 2.5 times that of petrol. The hydrogen industry is based on 3 pillars: production, storage/distribution and use.

A. Bardane (✉) · E. M. Chakir · A. I. El Mehdi · H. El Bekkouri
Laboratory of Materials and Subatomic Physics, Physics Department, Faculty of Sciences, Ibn Tofail University, Kenitra, Morocco
e-mail: adil.bardane@uit.ac.ma

M. Sayouti
Innovative Technologies Laboratory, Fez School of Technology, Fez, Morocco

E. El Mehdi and M. Boutahir (eds.), *Advanced Materials for Sustainable Energy and Engineering*, Engineering Materials, https://doi.org/10.1007/978-3-032-10069-6_12

Hydrogen is the lightest atom under normal conditions, with a density of 0.09 kg/m^3 [1]. It can be liquefied (70 g/L) at − 253.8 °C, a temperature that is difficult to achieve, or compressed using modern techniques to 700 bars, the pressure at which the density becomes 42 kg/m^3. This method presents significant safety concerns when it comes to designing tanks to precise standards. Solid hydrogen storage by chemical means in metal alloys is therefore particularly appropriate for certain applications.

Hydrogen storage is therefore likely to play a fundamental role in the future energy production and transport chain. There are three methods of storing hydrogen:

- Gaseous hydrogen storage.
- Storage in liquid form.
- Storage in solid form (hydrides).

Metal hydrides make it possible to release hydrogen in a very pure state by retaining the impurities present in the hydrogen, which are liable to damage the fuel cell membrane. What's more, they offer a safe storage solution because, on the one hand, they absorb and release hydrogen at moderate pressures of around 1 MPa and, on the other, the endothermic hydrogen release reaction is self-limiting: if hydrogen leaks into a metal hydride tank, the temperature of the desorbing material will drop rapidly until it reaches reaction equilibrium conditions. At equilibrium, the reaction stops, and no more hydrogen is released from the material [2].

Numerical simulation has now become a strategic issue in terms of scientific research, with no prior experimental need for the systems under study. It is now possible to characterize the electronic and structural properties of the most complex systems with a high degree of accuracy by applying methods based on the fundamental laws of quantum mechanics, in particular ab initio methods using the Density Functional Theory (DFT) formalism developed by Hohenberg-Kohn [3] and Kohn-Sham [4]. These methods are adapted to the different types of problems encountered and enable several physico-chemical properties of matter to be explored and modelled with precision [5].

The present work is devoted to the study of the structural and electronic properties of the metal hydrides MgH_2 and SrH_2 using the first-principles method implemented in the Quantum Espresso code [6], which is based on the plane wave (PW) method within the DFT framework.

Alkaline earth metals, such as magnesium, calcium and barium, are promising candidates for hydrogen storage because of their ability to form stable metal hydrides that can store large quantities of hydrogen. These metal hydrides can be produced from the reaction of alkaline earth metals with hydrogen under controlled temperature and pressure.

Magnesium is particularly attractive for hydrogen storage because it is abundant, inexpensive and has a high energy density. This means it can store a large quantity of hydrogen in a small unit volume. What's more, magnesium can be recycled and reused many times, making it a sustainable option for hydrogen storage.

However, storing hydrogen in the form of metal hydrides also presents challenges. The reaction of magnesium with hydrogen requires high temperatures and high pressure to produce metal hydrides, which can make the process energy intensive. In addition, the production of metal hydrides can be slow and requires heat treatment to release the stored hydrogen, which can lead to energy losses.

Despite these challenges, metal hydrides remain a promising technology for hydrogen storage and continue to be investigated and developed. With continued improvements in the production and use of metal hydrides, it is possible that magnesium and other alkaline earth metals could play an important role in the future storage and use of hydrogen as a sustainable energy source.

Density functional theory (DFT) is a computational method in quantum physics and chemistry that allows the electronic properties of molecular systems and solid materials to be described on the basis of electron density. Unlike traditional calculation methods, which solve Schrödinger's equations for each electron individually, DFT uses the electron density as the central variable.

This method has many applications in research in chemistry, materials physics and materials science. It is widely used to predict the properties of new materials or compounds, such as their binding energy and electronic structure.

2 Hydrides

2.1 Types of Hydrides

Hydrides are chemical compounds formed by the combination of hydrogen with another element. There are several types of hydrides, depending on the nature of the element with which the hydrogen combines and the nature of the bond between the atoms.

- **Ionic hydrides**: Ionic hydrides are compounds formed by combining hydrogen with highly electropositive metals such as sodium, potassium or calcium. In these hydrides, the hydrogen is in the form of a hydride ion (H-), which binds to the metal ion to form an ionic salt. Ionic hydrides are generally solid at room temperature and are often used as chemical reagents.
- **Covalent hydrides**: Covalent hydrides are compounds formed by combining hydrogen with non-metallic elements such as carbon, nitrogen, oxygen or sulphur. In these hydrides, the hydrogen bonds to the non-metallic element in a covalent bond, sharing electrons to form a molecule. Covalent hydrides can be gases, liquids or solids, depending on their molecular structure.
- **Metal hydrides**: Metal hydrides are compounds formed by combining hydrogen with transition metals such as titanium, zirconium or vanadium. In these hydrides, the hydrogen bonds to the metal via a metallic bond, which involves the sharing of several electrons between the atoms. Metal hydrides are often solids at room temperature and can have interesting properties such as high electrical conductivity or hydrogen adsorption.

- **Molecular hydrides**: these are compounds formed by the bonding of several hydrogen molecules together, such as H_2S or NH_3. They are often used in chemical reactions and energy production processes.

Alkaline earth metals

The alkaline earth metals are the six chemical elements in the second group of the periodic table: beryllium 4Be, magnesium 12Mg calcium 20Ca, strontium 38Sr, barium 56Ba and radium 88Ra. Their properties are very similar: they are silvery white, shiny, and chemically quite reactive at ambient temperature and pressure. Their electronic configuration contains a saturated s sublayer with two electrons, which they easily lose to form a divalent cation (+2 oxidation state).

Their name comes from the term "earth metals" used in alchemy to describe metals that are fire-resistant, as alkaline earth metal oxides remain solid at high temperatures.

Alkaline-earth metals are characterized by a silvery luster, low density, a melting temperature only slightly higher than those of poor metals (and a boiling temperature lower than some of them), great malleability, as well as a certain reactivity with halogens, leading to ionic salts (with the exception of beryllium chloride $BeCl_2$, which is covalent), and with water (with the exception of beryllium), although less readily than with alkali metals, to form strongly basic hydroxides. The reactivity of these elements increases with their atomic number.

Beryllium and magnesium are rather grey because they are covered by a protective, passivating film of BeO and MgO oxide, while calcium, strontium, barium and radium are brighter and softer. The surface of these metals' tarnishes rapidly in the open air.

For example, while sodium and potassium react with water at room temperature, calcium only reacts with hot water, and magnesium only with steam:

$$\mathbf{Mg + 2H_2O \rightarrow Mg(OH)_2 + H_2}$$

Beryllium is an exception to these behaviors: it does not react with either liquid water or water vapor, and its halides are covalent. For example, beryllium fluoride BeF_2, a priori the most ionic of the beryllium halides, is essentially covalent, with a melting temperature of just 553.85 °C and low electrical conductivity in the liquid state.

M^{2+} ions from the alkaline earths Ca, Sr and Ba can be characterized qualitatively by a flame test: When a salt of an alkaline earth is treated with concentrated hydrochloric acid (giving a volatile metal chloride) and heated strongly in the non-illuminating flame of a Bunsen burner, a characteristic flame color is observed. This flame is orange red for Ca (but pale green through blue glass), purple for Sr (but violet through blue glass), and apple-green for Ba Crystal structure of MgH_2:

The rutile structure of α-MgH_2 is present in its room temperature state [7] (Fig. 1). At least four forms exist at high pressure: γ-MgH_2 with α-PbO_2 structure, for example [8]. Additionally, whereas bulk MgH_2 is essentially stoichiometric in that it only permits a modest concentration of hydrogen vacancies [9], a non-stoichiometric

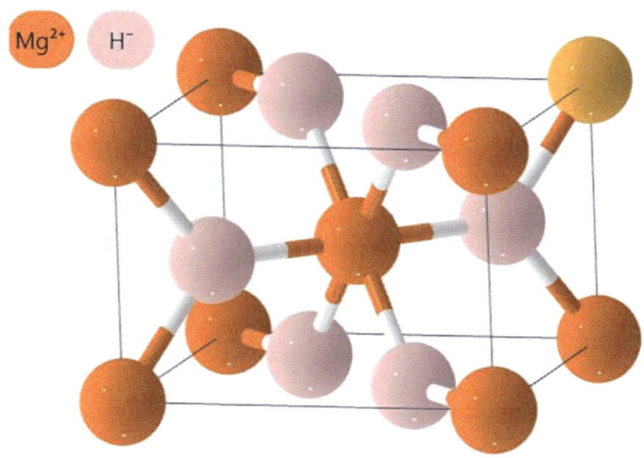

Fig. 1 Rutile structure of MgH_2

Table 1 Mesh parameters for MgH_2 positions of atoms

Hydrides	Architecture	Space group	Mesh parameter (a)		Atoms positions	
			Theoretical study	Experimental study	Theoretical study	Experimental study
MgH_2	Tetragonal	P42/ mnm N° 136	a = 4.498[c] c = 3.004[c]	a = 4.501[d] c = 3.01[d]	Mg(2a): 0.0, 0.0, 0.0[c] H(4f):0.3044, 0.3044,0.0[c]	M_g(2a): 0.0, 0.0, 0.0[b] H(4f): 0.304, 0.304, 0.0[b]

[d]Ref. [13], [b]Ref. [14], [c]Ref. [15]

MgH (2-δ) has been characterised and seems to exist only for very small particles [10].

A study of the charge density by synchrotron X-ray diffraction shows that the magnesium atom is fully ionised and spherical in shape, and the hydride ion is elongated. The link in rutile form is sometimes regarded as partially covalent rather than purely ionic in nature [11]. Magnesium hydroxide (MgH, MgH_2, Mg_2H, Mg_2H_2, Mg_2H_3, and Mg_2H_4) molecules are among the potential molecular forms. The bridging structure of the Mg_2H_4 molecule is comparable to that of the dimeric aluminium hydride dimer Al_2H_6 [12] (Table 1).

2.2 *Crystalline Structure of SrH_2*

SrH_2, or strontium hydride, is a crystalline chemical compound that is a metal hydride. The possible crystal structure of SrH_2 is calcium fluoride (CaF_2) and belongs to the face-centered cubic (FCC) family of structures.

In the SrH_2 crystal structure, strontium atoms (Sr) occupy the sites of the face-centered cubic lattice, while hydrogen atoms (H) occupy the tetrahedral sites between the strontium atoms. Each hydrogen atom is surrounded by four strontium atoms, while each strontium atom is surrounded by eight hydrogen atoms. The crystalline structure of SrH_2 is relatively stable and can store large quantities of hydrogen. Because of its hydrogen storage properties, SrH_2 is a material of interest for hydrogen storage applications, such as fuel cells and hydrogen vehicles. Further research is needed to improve the hydrogen storage properties of SrH_2 to make them more efficient and sustainable on a large scale.

2.3 Importance of Hydrides

Hydrides play an important role in hydrogen storage because they can store hydrogen efficiently and safely in solid form at ambient pressures and temperatures. Metal hydrides are particularly useful for hydrogen storage because they can adsorb large quantities of hydrogen to form stable compounds. These metal hydrides can be used in fuel cells, hydrogen-powered vehicles, renewable energy storage systems and other hydrogen-related applications.

Covalent hydrides are also being investigated for their potential as hydrogen storage materials. Although they generally store less hydrogen than metal hydrides, they can be more stable and easier to produce. In general, hydrides are attractive for hydrogen storage because they allow hydrogen to be stored in solid form, making it easier to transport and store hydrogen. Hydrides are also relatively safe, as they are not flammable or explosive like hydrogen gas.

However, further research is needed to improve the performance and durability of hydrides for large-scale hydrogen storage.

3 Hydrogen Storage

Hydrogen storage is a promising technology for energy storage because hydrogen is a versatile, clean and renewable energy carrier. Hydrogen can be produced from a variety of renewable energy sources, such as solar, wind or hydro power, and stored for later use.

Hydrogen can be stored in compressed or liquefied gaseous form, but these methods require special storage conditions, such as high-pressure or cryogenic tanks. Hydrogen storage in the form of metal hydrides is a safer and more efficient alternative for long-term hydrogen storage. Metal hydrides can store large quantities of hydrogen at ambient pressures and temperatures and can be used to power fuel cells or internal combustion engines.

Hydrogen storage can be used for a wide range of applications, including transport, power generation, heating and cooling of buildings, and other industrial uses. In the field of renewable energy, hydrogen storage can help solve the intermittency problems of renewable energy sources such as solar and wind power. Stored hydrogen can be used to generate electricity during periods of low renewable energy production.

In summary, hydrogen storage is a promising technology for storing renewable energy. Metal hydrides are an efficient and safe method of storing hydrogen over the long term and can be used to power fuel cells and internal combustion engines for a variety of applications. Further research is needed to improve the performance and durability of metal hydrides for large-scale hydrogen storage.

4 Schrödinger Equation for a Crystalline Solid

A crystalline solid is composed of atomic nuclei (of mass *mn* and atomic number Z) interacting with each other, and electrons (of spin σ_j, mass m_e and charge electrical e) positioned at $R = \{R_i;\ i = 1,2,\ldots, N\}$ and $r = \{(r_j, \sigma_j);\ j = 1,2,\ldots, N\}$, respectively. The stationary state of the particles is described by the following Schrödinger equation:

$$H\psi = E\psi$$

With:

H the Hamiltonian operator,
ψ The wave function of the system.
E Its own energy.

The Hamiltonian is made up of the kinetic energy of all the particles, the potential energy of interaction between them, and if necessary, the potential energy of interaction with the external medium, to write this Hamiltonian in the absence of the external field in the following form:

$$H_{\text{total}} = T_e + T_N + V_{e-e} + V_{e-N} + V_{N-N}$$

With:
Electron kinetic energy: $T_e = \frac{-h^2}{2m_e}\sum_{i=1}^{N_e}\nabla_i^2$
Kinetic energy of nuclei: $T_N = \frac{-h^2}{2m_n}\sum_{i=1}^{N_n}\nabla_i^2$
Potential energy of electron–electron interaction: $V_{e\text{-}e}$.
Potential energy of nucleus-nucleus interaction:

$$V_{N-N} = \sum_{\alpha\neq\beta}\frac{1}{8\pi\varepsilon_0}\left(\frac{e^2Z_\alpha Z_\beta}{|R_\alpha - R_\beta|}\right)$$

Potential energy of electron-nucleus interaction:

$$V_{N-e} = \sum_{\alpha\neq\beta}\frac{-1}{4\pi\varepsilon_0}\left(\frac{e^2Z_\alpha}{|r_i - R_\beta|}\right)$$

Schrödinger's equation for the crystal structure contains a large number of unknowns (an infinite number of atoms and electrons), which makes it impossible to solve because it contains $\mathbf{3(Z+1)}\,N$ variables, where N represents the number of atoms. It is therefore impossible to find a general analytical or numerical solution to this equation, so several approximations have been made to simplify it.

4.1 Born–Oppenheimer Approximation

According to Born–Oppenheimer, treating electrons and nuclei separately is the only way to simplify the problem and solve Schrödinger's equation, i.e. a nuclear part and an electronic part. This approximation is based on the adiabatic approximation [16], which is based on the large difference in mass between electrons and nuclei [17].

Nuclei are very heavy compared to electrons (about 2000 times) so electrons can move in the solid much faster than nuclei. The motion of the nuclei is therefore negligible, so the potential energy of the interaction between the nuclei becomes constant and their kinetic energy is zero [18]. This approach leads to a Hamiltonian for which the electrons move in a field created by a static configuration of the nuclei [19]. In this case, Schrödinger's equation is decomposed into an electron Hamiltonian and a nuclear Hamiltonian in the form:

$$\mathrm{H} = \mathrm{H_e} + \mathrm{V_N}$$

With:

$$\mathrm{H_e} = \mathrm{T_e} + \mathrm{V_{e-e}} + \mathrm{V_{e-N}}$$

whereas the Schrödinger equation for electrons is:

$$H_e \psi_e(r, R) = (T_e + V_{e-e} + V_{e-N})\psi_e(r, R)$$

Further approximations are needed in order to solve this equation properly, even though this approximation greatly reduces the degree of complexity. The novel wave function of the system depends on N bodies.

4.2 Hartee Approximation

A method for determining the wave functions and approximate energy of atoms and ions was proposed by Douglas Hartree in 1927 [20]. This approximation's fundamental tenet is that electron motion is decorrelated, meaning it is independent of one another. Such a system's Hamiltonian can be expressed as follows:

with:

$$H = \sum_{i=1}^{N} H(i)$$

where H(i) is the single-electron Hamiltonian.

The electronic wave function used to solve this Hamiltonian consists of a single-electron product. It is called the Hartree product [21].

$$\psi^{HP}(x_1, x_2, \ldots, x_N) = \left|\varphi_i(x_1)\varphi_j(x_2)\ldots\varphi_k(x_N)\right|$$

Because it is based on the concept of free electrons, this approximation excludes interactions between electrons and spin states.

This approach's greatest asset is that it offers an internally consistent solution to the electronic system issue [22]. It has the following four significant effects:

There is an overestimation of the electron system's total coulombic repulsion, $V_{e\text{-}e}$. Although easy to solve, the outcome is subpar.

Every electron has a charge of its own.

Pauli's exclusion principle is not considered.

A more reasonable wave function must be antisymmetric when two electrons are exchanged [23]. This last consequence being more serious, the "Hartree–Fock" approximation [24] takes spin into account when solving the Schrödinger equation.

4.3 Hartee-Fock Approximation

To describe a system of electrons physically, the electrons must obey Pauli's exclusion principle (antisymmetry), i.e. the exchange of two electrons in the wave function must result in the appearance of a negative sign:

$$\psi(x_1; x_2) = -\psi(x_1; x_2)$$

First, we look for an approximate solution to the electronic Schrödinger equation. It is possible, in this case, to evaluate the best wave function for such a system by minimising the energy using the variational principle.

$$\psi_i(r_i) = \frac{1}{\sqrt{n_e!}} \begin{bmatrix} \psi_1(1) & \cdots & \psi_{n_e}(1) \\ \vdots & \ddots & \vdots \\ \psi_1(n_e) & \cdots & \psi_{n_e}(n_e) \end{bmatrix}$$

where $\frac{1}{\sqrt{n_e!}}$ *is the normalisation factor.*

The generalization of Hartree's method which takes these latter criteria into account was proposed in 1930 by Fock [25] and Slater is known as the Hartree–Fock method [26]. The Slater determinant, which is defined as a determinant of order N produced on N separate spin-orbitals, which are single-electron functions of the space and spin variables, is used to replace the Hartree wave functions in order to achieve this generalisation. The determinant comprises single-electronic wave functions as linear combinations of all Hartree functions. This theorem allows poly-electronic wave functions to be expressed in terms of linear combinations of Slater determinants.

These approximations are often compensated for by using more sophisticated methods such as DFT to solve the Schrödinger equation.

5 Density Functional Theory (DFT)

Density Functional Theory (DFT) forms the basis of a set of ab initio approaches currently used to explain and predict the electronic structures of materials. DFT originated in work by Hohenberg and Kohn [3], followed by Kohn and Sham [4]. The fundamental idea behind DFT is that electron density contains all the information needed to describe the electronic properties of a system. In other words, knowing the electron density of a system is equivalent to knowing all the electronic properties of that system. This considerably reduces the complexity of the calculations needed to describe the electronic properties of molecular systems and solid materials. See Fig. 2.

The DFT needs to be established once the electron density has been determined. These foundations, which form the basis of the entire DFT and may be summed up in two theorems, were first articulated by Hohenberg and Kohn [3] in 1964.

- **1st Theorem**

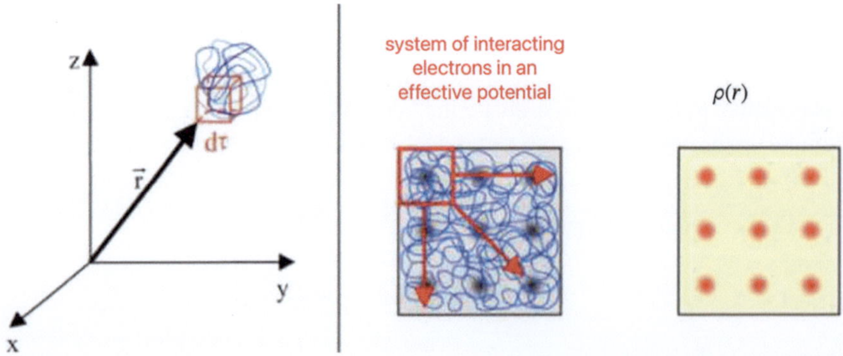

Fig. 2 Schematic representation of electron density

According to Hohenberg and Kohn, the basic principles can be expressed by considering a system of Ne interacting electrons subjected to an external potential V_{ext}. The Hamiltonian is then:

$$H_{el} = T + V_{ee} + \sum_{i \neq 1}^{N_e} V_{ext}$$

where T and V_{ee} are the kinetic energy and electron–electron interaction terms electron–electron interaction respectively. Recall that for an electronic system described by the Hel Hamiltonian, the energy and wave function of the ground state are determined by minimizing the E functional $[\psi]$. If we know the external potential and the number of electrons N in the system, we can uniquely determine the Hamiltonian and access the energy and wave function of the ground state.

An atomic system can be viewed either from the electron density through its electron cloud or through the nuclei via the external potential. These two quantities have an extremely obvious link. The outcome discovered by Hohenberg and Kohn indicates that the external potential and the electron density have a one-to-one correspondence, meaning that one appears to be the mirror image of the other.

$$E = E[\rho(r)]$$

This theorem immediately implies that the Hamiltonian operator is uniquely determined by the electron density. Therefore, the various properties of the molecule or material may be estimated by using the Ham-iltonian operator, which can be found by knowing the electron density.

- ***Second Theorem***

The second theorem of Hohenberg and Kohn [3] is a variational principle similar to that initially proposed in the Hartee-Fock approach for a wavefunction functional, but applied this time to an electron density functional:

$$\frac{\partial E[\psi]}{\partial \psi} = 0$$

There is a universal functional E[n] expressing the energy according to the electron density n(r), valid for any external potential $V_{ext}(r)$. Foreach specific $V_{ext}(r)$, the energy of the essential state of the device is the value that *ext* minimizes this functional, the density n(r) associated with it corresponds to the exact density $n_0(r)$ of the essential state.

5.1 The Kohn-Sham Equations

The equations of Kohn and Sham [27], published in 1965, make DFT a practical tool, and provide a practical method for obtaining the electron density of the ground state. They introduced the notion of a fictitious system of N non-interacting electrons with the same electron density in its ground state $\rho_0(\mathrm{r})$ as an interacting electron system. The formulations of the kinetic and potential energies for this hypothetical system are known, which is what makes it interesting. This choice of a system of N non-interacting electrons involves solving N single-electron Schrödinger equations. This leads us to rewrite the problem in the form of three interdependent equations.

These are the equations given by Kohn and Sham, which are formally equivalent to the variational principle of Hohenberg and Kohn:

$$\mathrm{V_{KS}}(\vec{\mathrm{r}}) = \mathrm{V_{ions}}(\vec{\mathrm{r}}) + \mathrm{V_H}(\vec{\mathrm{r}}) + \mathrm{V_{XC}}(\vec{\mathrm{r}})$$

$$\left[\frac{-\hbar^2}{2\mathrm{m}}\vec{\nabla}^2 + \mathrm{V_{RS}}(\vec{\mathrm{r}})\right]\phi_\mathrm{i}(\vec{\mathrm{r}}) = \varepsilon_\mathrm{i}\phi_\mathrm{i}(\vec{\mathrm{r}})$$

$$\rho(\vec{r}) = \sum_{a}^{0cc} |\phi_\mathrm{i}(\vec{\mathrm{r}})|^2$$

where: $\mathrm{V_{KS}}(\vec{\mathrm{r}}) = \frac{\delta V_{\mathrm{eff}}}{\delta\rho(\vec{r})}$

To solve such an equation, we introduce the notion of effective potential, such that:

$$V_{eff}\left[\rho(\vec{r})\right] = V_{\mathrm{ext}} + \int \frac{\rho(\vec{r}')}{|\mathrm{r}-\mathrm{r}'|}\mathrm{dr}' + V_{\mathrm{XC}}\left[\rho(\vec{r})\right]$$

With: $\int \frac{\rho(\vec{r}')}{|r-r'|}dr' = \mathrm{V_H}$ *is the Hartree potential.*

$\mathrm{V_{XC}}\left[\rho(\vec{r})\right] = \frac{\partial \mathrm{E_{XC}}\left[\rho(\vec{r})\right]}{\partial\rho(\vec{r})}$ *is the exchange and correlation potential.*

With (E_{XC}) is the exchange–correlation energy, which combines the effects of correlations caused by the quantum nature of electrons, and all other unknowns in the system.

The second uses the effective potential in the N single-electron Schrödinger equations with the aim of obtaining φ_i.

$$\left[-\frac{1}{2}\nabla^2 + V_{\mathrm{eff}}(\mathrm{r})\right]\varphi_i(r) = \varepsilon_i\varphi_i(r)$$

With: $\boldsymbol{V_{eff}(r) = V_{ext} + V_H + V_{xc}}$.

The third shows how to access the density from the N mono-functions φ_i.

$$\rho(r) = \sum_{i=1}^{N} |\varphi_i(r)|^2$$

There are many approximations to the exchange–correlation functional; they are designated in Gaussian by a group of letters. In general, the exchange–correlation functional is written as the sum of an exchange functional and a correlation functional, and is given by the following relation:

$$E_{XC}[\rho(r)] = E_X[\rho(r)] + E_C[\rho(r)]$$

With the exchange functional:

$$E_X[\rho(r)] = \frac{1}{2}\int\int \frac{\rho_1(r_1)\rho_X\rho_1(r_1, r_2)}{r_{12}} \mathrm{d}r_1 r_2$$

And the correlation function:

$$E_C[\rho(r)] = \frac{1}{2}\int\int \frac{\rho_1(r_1)\rho_C\rho_1(r_1, r_2)}{r_{12}} \mathrm{d}r_1 r_2$$

Its study and the search for the best possible approximations constitute an active branch of DFT.

5.2 *Solving the Kohn-Sham Equations*

To Kohn-Sham orbitals, which are expressed as linear combinations of orbitals and can be chosen as a basis for the wave functions to solve the Kohn-Sham equations, are expressed as follows:

$$\psi_i(r) = \sum C_{ij}\phi_j(r)$$

where:

$\Phi_i(k, r)$ are the base functions.
C_{ji} are the development coefficients.

Finding the coefficients C_{ji} for the occupied orbitals that minimise the total energy and make the calculations for the symmetry points in the first Brillouin Zone (BZ) simpler is the result of solving the Kohn and Sham equations.

Using a self-consistent iteration cycle. This is achieved by injecting the initial charge density $\boldsymbol{\rho_{in}}$ to diagonalise the secular equation:

$$H - \varepsilon_j S = 0$$

where:

H The Hamiltonian.
S represents the covering matrix.

The overall charge density, which can be derived by summing over all occupied orbitals, is used to generate the new charge density, $\boldsymbol{\rho_{out}}$, using the eigenvectors of this secular equation.

If convergence of the calculations is not obtained, the charge densities $\boldsymbol{\rho_{in}}$ and $\boldsymbol{\rho_{out}}$ are mixed as follows:

$$\rho_{out}^{i+1} = 1 - \alpha\rho_{in}^{i} + \alpha\rho_{out}^{i}$$

where:

i represents the $\boldsymbol{i}$th iteration.
$\boldsymbol{\alpha}$ a mixing parameter.

This process is iteratively repeated until convergence is achieved.

Figure 3 illustrates the flowchart for Solving the Kohn and Sham equations: self-consistent cycle.

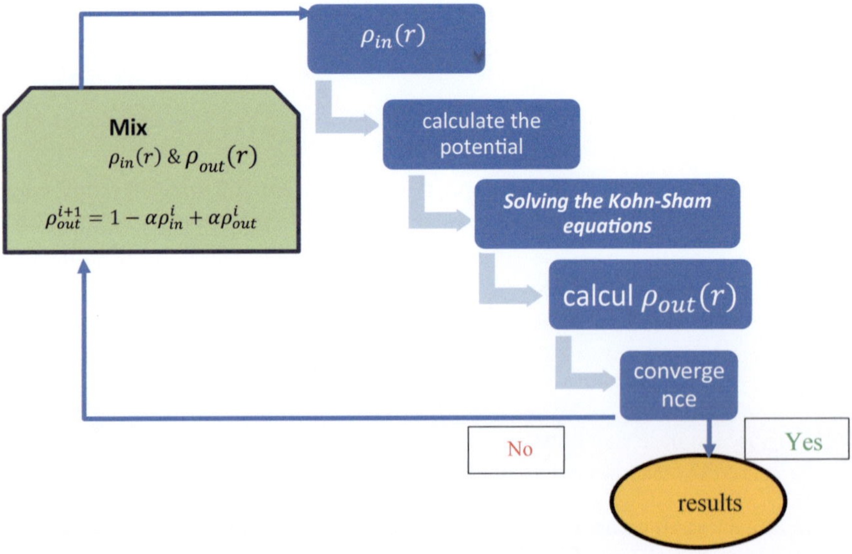

Fig. 3 Solving the Kohn and Sham equations: self-consistent cycle [27]

5.3 Exchange and Correlation Potential

Several approximations have been proposed to calculate the exchange and correlation potential, such as:

(a) **Local Density Approximation LDA**:

This approach considering the uniform electron gas model and forms the simplest approach to express the exchange–correlation energy. It is described as:

$$E_{XC}[n] = \int n(\vec{r})\varepsilon_{XC}[n]dr$$

where $\varepsilon_{xc}[n]$ is used to denote the exchange–correlation energy for a particle of a homogeneous electron gas. The function $\varepsilon_{xc}[n]$ can be decomposed into an exchange contribution $\varepsilon_x[n]$ and a correlation contribution $\varepsilon_c[n]$:

$$\varepsilon_{XC} = \varepsilon_X[n] + \varepsilon_C[n]$$

The contribution from electronic exchange in the local density approximation is known and comes from the exchange energy functional formulated by Dirac [28].

$$-\frac{3}{4}\left(\frac{3}{\pi}\right)^{\frac{1}{3}} \int\int\int n(\vec{r})^{\frac{4}{3}} d^3r$$

The LDA approximation can be formulated in a more general way by considering the electron spin in the expression of the functional, which is then referred to as the LSDA approximation (local spin density approximation). This approach was originally proposed by Slater (1900–1976) [29] and is used to solve certain problems associated with an LDA approach, especially the treatment of devices subjected to magnetic fields and devices where relativistic effects become important. Considering the LSDA approximation, the exchange functional is expressed as:

$$-2^{\frac{1}{3}}(-3/4)\left(\frac{3}{\pi}\right)^{\frac{1}{3}} \int\int\int n_\alpha(\vec{r})^{\frac{4}{3}} n_\beta(\vec{r})^{\frac{4}{3}} d^3r$$

(b) **Generalized gradient approximation (GGA)**

The generalized gradient approximation (GGA) is an improvement of the position density approximation in the treatment of exchange and correlation energy, which can be made dependent not only on the electron density, but also on its gradient $\nabla\rho(r)$, thus the function E_{xc} according to the non-uniform nature of the electron gas in this case [18], its contribution can be summed from the total energy.

For the system cumulatively from each part of an irregular gas as if it were local and is written:

$$E_{XC}^{GGA}[\rho \uparrow (r), \rho \downarrow (r)] = \int f(\rho \uparrow, \rho \downarrow, \nabla\rho \uparrow, \nabla\rho \downarrow)\rho(r)d^3r \ldots ()$$

- $\nabla\boldsymbol{\rho}(\boldsymbol{r})$: Represents the electron density gradient.
- $\boldsymbol{E_{xc}\rho}(\boldsymbol{r})$: It represents the energy of exchange and correlation in a system of non-uniform interaction system.
- f: Represents the local density function.

Improvements have been made with GGA-type functionals, and the use of this approximation increases the accuracy of calculations compared with LDA, especially for crystal lattice parameters and cohesive energies. However, due to its local nature, this improvement is not always systematic. GGA fails to treat correctly systems characterized by Van der Walls interactions, linked to long-range correlations.

To conclude this section, we can say that density functional theory is a very effective tool for studying systems of interacting electrons. It reduces the problem of N interacting bodies to that of N independent bodies moving in an effective potential. The introduction of this system of independent particles means that most of the kinetic energy can be considered. The neglected part of this energy comes from the fact that the total wave function of the system is not equal to the Slater determinant (otherwise the Hartree–Fock theory would be correct). The effort that must be made to get the right description of the kinetic energy is that instead of solving a single density equation, we must solve N.

6 Pseudo Potential and the Plane Wave Method

The pseudo potential method [30] has been useful in the density functional formalism, [18, 27] and has proved its worth in recent years by giving effective results in calculations and predictions of the ground state properties of solids [31, 32].

The basic concept of the pseudopotential is the frozen-core approximation, which assumes that the electronic states of the core electrons are insensitive to the neighboring electronic configuration. This assumption remains correct in many cases in the physical state of the solid.

The pseudo potential is constructed from the atomic potential. The eigenstate of the pseudo potential must therefore correspond to the valence state of the atomic potential. It is possible to use plane wave development in electronic structure calculations because of the presence of a smooth potential in the core region.

Replacing the ionic potential with a low potential enables electronic wave functions to be developed using a certain number of plane waves.

Then we need to develop the wave functions in a full ionic potential. The rapid oscillations of the valence wave functions in the core of the atom have been reduced and the states of the electrons in the small core are absent (Fig. 4).

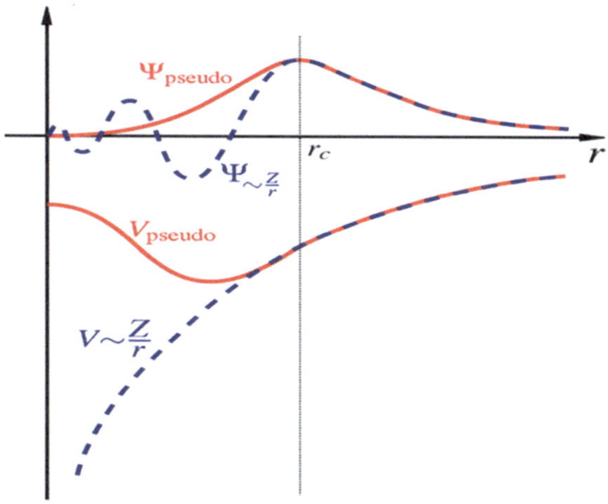

Fig. 4 Pseudo potential and pseudo wave (PP-PW) diagram

The pseudopotential approximation (Phillips 1958 [33], Heine and Cohen 1970 [32] and Yin and Cohen 1982 [28]) allows the development of electron wave functions using a reduced number of plane wave states.

Thus, the efficiency of the pseudopotential has evolved considerably since the Phillips-Kleinman construction. This has been made possible by the following conditions:

- The pseudopotential used must be as smooth as possible, as it must allow the expansion of valence pseudowave functions using the lowest possible number of plane waves.
- It must be as transferable as possible to ensure the reliability of the results in solid-state applications, where the crystalline potential is necessarily different from the atomic potential;
- The pseudo charge density constructed using the pseudo wave functions must reproduce the valence charge density as accurately as possible.

References

1. Wiberg, E., Wiberg, N., Holleman, A.F.: Inorganic Chemistry, 1st English edn. Academic Press, De Gruyter, San Diego. Berlin, New York (2001)
2. Klebanoff, L.E., Keller, J.O.: 5 Years of hydrogen storage research in the US DOE metal hydride center of excellence (MHCoE). Int. J. Hydrog. Energy **38**(11), 4533–4576 (2013)
3. Hohenberg, P., Kohn, W.: Phys. Rev. B **136**, 864 (1964)
4. Kan, M., Wang, B., Lee, Y. H., Sun, Q.: A density functional theory study of the tunable structure, magnetism and metal-insulator phase transition in VS2 monolayers induced by in-plane biaxial strain. Nano Res. **8**(4), 1348–1356 (2015)

5. Sundar, A., Huang, Y., Yu, J., Cinbiz, M.N.: The impacts of charge transfer, localization, and metallicity on hydrogen retention and transport capacity. Int. J. Hydrog. Energy **47**(46), 20194–20204 (2022)
6. Giannozzi, P., Andreussi, O., Brumme, T., Bunau, O., Buongiorno Nardelli, M., Calandra, M., Car, R., Cavazzoni, C., Ceresoli, D., Cococcioni, M., Colonna, N., Carnimeo, I., Dal Corso, A., de Gironcoli, S., Delugas, P., DiStasio, R.A., Ferretti, A., Floris, A., Fratesi, G., Fugallo, G., Gebauer, R., Gerstmann, U., Giustino, F., Gorni, T., Jia, J., Kawamura, M., Ko, H.-Y., Kokalj, A., Küçükbenli, E., Lazzeri, M., Marsili, M., Marzari, N., Mauri, F., Nguyen, N.L., Nguyen, H.-V., Otero-de-la-Roza, A., Paulatto, L., Poncé, S., Rocca, D., Sabatini, R., Santra, B., Schlipf, M., Seitsonen, A.P., Smogunov, A., Timrov, I., Thonhauser, T., Umari, P., Vast, N., Wu, X., Baroni, S.: Advanced capabilities for materials modelling with Quantum ESPRESSO. J. Phys.: Condens. Matter. **29**, 465901 (2017). https://doi.org/10.1088/1361-648X/aa8f79
7. Shinzato, Y., Komiya, K., Takahashi, Y., Yukawa, H., Morinaga, M., Orimo, S.: Characteristics of chemical bond in perovskite-type hydrides. Adv. Quantum Chem. **54**, 245–253 (2008)
8. El Kharbachi, A., Zavorotynska, O., Latroche, M., Cuevas, F., Yartys, V., Fichtner, M.: Exploits, advances and challenges benefiting beyond Li-ion battery technologies. J. Alloy. Compd. **817**, 153261 (2020)
9. Grau-Crespo, R., Smith, K.C., Fisher, T.S., de Leeuw, N.H., Waghmare, U.V.: Thermodynamics of hydrogen vacancies in MgH_2 from first-principles calculations and grand-canonical statistical mechanics. Phys. Rev. B. **80**, 174117 (2009). https://doi.org/10.1103/PhysRevB.80.174117
10. Schimmel, H.G., Huot, J., Chapon, L.C., Tichelaar, F.D., Mulder, F.M.: Hydrogen cycling of niobium and vanadium catalyzed nanostructured magnesium. J. Am. Chem. Soc. **127**, 14348–14354 (2005). https://doi.org/10.1021/ja051508a
11. Huang, Z., Cao, S., Feng, C., Li, Y., Liang, Y., Li, X., Mei, H., Fan, W., Xu, B., Yuan, S., Dai, F., Sun, D.: Polydopamine bridging strategy enables pseudomorphic transformation of multi-component MOFs into ultrasmall NiSe/WSe2@ NC heterojunctions for enhanced alkaline hydrogen evolution. Appl. Catal. B: Environ. **334**, 122769 (2023)
12. Bortz, M., Bertheville, B., Böttger, G., Yvon, K.: Structure of the high-pressure phase γ-MgH_2 by neutron powder diffraction. J. Alloy. Compd. **287**, L4–L6 (1999). https://doi.org/10.1016/S0925-8388(99)00028-6
13. San-Martin, A., Manchester, F.D.: The H−Mg (Hydrogen-Magnesium) system. JPE **8**, 431–437 (1987). https://doi.org/10.1007/BF02893152
14. Djellab, S., Bouhadda, Y., Bououdina, M., Fenineche, N., Boudouma, Y.: Structural, electronic and elastic properties of MgH_2, CaH_2 and $Ca_4Mg_3H_{14}$ for hydrogen storage materials. J. Electron. Mater. **45**, 3935–3942 (2016). https://doi.org/10.1007/s11664-016-4608-0
15. Calais, J.-L.: Density-Functional Theory of Atoms and Molecules. R.G. Parr and W. Yang, Oxford University Press, New York, Oxford, 1989. IX + 333 pp. https://doi.org/10.1002/qua.560470107
16. Singh, D.J., Schwarz, K., Blaha, P.: Electric-field gradients in $YBa_2Cu_3O_7$: Discrepancy between experimental and local-density-approximation charge distributions. Phys. Rev. B. **46**, 5849–5852 (1992). https://doi.org/10.1103/PhysRevB.46.5849
17. Jensen, F.: Introduction to computational chemistry, 2nd edn. John Wiley & Sons, Chichester, England (2007)
18. Thijssen, J.M.: Computational physics, 2nd edn. Cambridge University Press, Cambridge, UK, New York (2007)
19. Cramer, C.J.: Essentials of computational chemistry: theories and models, 2nd edn. Wiley, Chichester, West Sussex, England, Hoboken, NJ (2004)
20. Hartree, D.R.: The wave mechanics of an atom with a non-coulomb central field. Part I. theory and methods. Math. Proc. Camb. Phil. Soc. **24**, 89–110 (1928). https://doi.org/10.1017/S0305004100011919
21. Sahni, V., Qian, Z., Sen, K.D.: Atomic shell structure in Hartree theory. J. Chem. Phys. **114**, 8784–8788 (2001). https://doi.org/10.1063/1.1360202

22. Froese Fischer, C.: General Hartree-Fock program. Comput. Phys. Commun. **43**, 355–365 (1987). https://doi.org/10.1016/0010-4655(87)90053-1
23. Kohn, W., Sham, L.J.: Self-consistent equations including exchange and correlation effects. Phys. Rev. **140**, A1133–A1138 (1965). https://doi.org/10.1103/PhysRev.140.A1133
24. Martin, R.M.: Electronic Structure: Basic Theory and Practical Methods. Cambridge University Press, Cambridge, UK, New York (2004)
25. Slater, J.C.: A simplification of the Hartree-Fock method. Phys. Rev. **81**, 385–390 (1951). https://doi.org/10.1103/PhysRev.81.385
26. Hohenberg, P., Kohn, W.: Inhomogeneous electron gas. Phys. Rev. **136**, B864–B871 (1964). https://doi.org/10.1103/PhysRev.136.B864
27. Kohn, W.: Nobel lecture: electronic structure of matter—wave functions and density functionals. Rev. Mod. Phys. **71**, 1253–1266 (1999). https://doi.org/10.1103/RevModPhys.71.1253
28. Yin, M.T., Cohen, M.L.: Ground-state properties of diamond. Phys. Rev. B. **24**, 6121–6124 (1981). https://doi.org/10.1103/PhysRevB.24.6121
29. Nielsen, O.H., Martin, R.M.: First-principles calculation of stress. Phys. Rev. Lett. **50**, 697–700 (1983). https://doi.org/10.1103/PhysRevLett.50.697
30. Monkhorst, H.J., Pack, J.D.: Special points for Brillouin-zone integrations. Phys. Rev. B. **13**, 5188–5192 (1976). https://doi.org/10.1103/PhysRevB.13.5188
31. Perdew, J.P., Ruzsinszky, A., Csonka, G.I., Vydrov, O.A., Scuseria, G.E., Constantin, L.A., Zhou, X., Burke, K.: Restoring the density-gradient expansion for exchange in solids and surfaces. Phys. Rev. Lett. **100**, 136406 (2008). https://doi.org/10.1103/PhysRevLett.100.136406
32. Perdew, J.P., Ruzsinszky, A., Csonka, G.I., Vydrov, O.A., Scuseria, G.E., Constantin, L.A., Zhou, X., Burke, K.: Restoring the Density-Gradient Expansion for Exchange in Solids and Surfaces. Phys. Rev. Lett. **100**(13), 136406 (2008)
33. Phillips, J.C.: Energy-band interpolation scheme based on a pseudopotential. Phys. Rev. **112**, 685–695 (1958). https://doi.org/10.1103/PhysRev.112.685

Lattice Boltzmann Fluid Simulation with Neumann Boundary Conditions

Fouad Agoujil, Youssef Elguennouni, Youness Ighris, Jamal Baliti, and Mohamed Hssikou

Abstract This chapter offers an extended investigation of the lattice Boltzmann method (LBM) for simulating thermal flows in closed cavities, with particular focus on the implementation of Neumann-type boundary conditions. We present a theoretical overview of the method, highlighting its kinetic foundations and its ability to solve efficiently the Navier–Stokes and thermal transport equations. The treatment of hydrodynamic and thermal boundary conditions is analyzed with a focus on the classical approaches and recent methodology of Neumann conditions implementation. To validate the robustness and accuracy of this new methodology, a computational study is conducted on an air-filled square cavity, exposed to constant heat fluxes imposed on the vertical sidewalls according to the Neumann scheme. The results of this simulation are discussed in terms of temperature distributions, streamlines, as well as local and global Nusselt numbers, confirming the relevance of the model implemented. This work represents a significant contribution to advanced numerical modeling of heat transfer with more realistic boundary conditions.

Keywords Lattice Boltzmann method · Neumann conditions · Heat flux · Natural convection · Nusselt number · Grashof number

1 Introduction

Numerical simulation of thermofluidic flows is now a major area of research in many scientific and industrial fields [1, 2]. Whether in cooling systems for electronic equipment, microfluidic devices, air conditioning for smart buildings or energy conversion processes [3]. Therefore, an accurate understanding of heat transfer and fluid movement is a prerequisite for optimizing thermal and energy performance.

F. Agoujil (✉) · Y. Ighris · J. Baliti · M. Hssikou
Sultan Moulay Slimane University, Beni Mellal, Morocco
e-mail: fouad.agoujil@usms.ac.ma

Y. Elguennouni
Faculty of Sciences, Moulay Ismail University of Meknès, Meknes, Morocco

E. El Mehdi and M. Boutahir (eds.), *Advanced Materials for Sustainable Energy and Engineering*, Engineering Materials, https://doi.org/10.1007/978-3-032-10069-6_13

These phenomena, frequently interlinked, are governed by complex non-linear equations (Navier–Stokes, heat equation), which, in the majority of practical cases, are impossible to solve analytically. Hence the extensive use of numerical modelling, which enables thermal interactions to be predicted, analyzed and visualized in real geometries, under different limit conditions.

Among the classical numerical methods employed, finite volume (FVM) and finite element (FEM) methods have dominated the field of thermal flow simulation for several decades [4]. These methods are based on a macroscopic formulation of the conservation equations, discretized over often complex meshes [5]. Despite their considerable performance, they have certain limitations: they require heavy implicit resolution procedures, increased algorithmic complexity to deal with irregular geometries, and difficulty of implementation in multicomponent, multiphase or moving-interface cases. In addition, the consistent integration of boundary conditions (particularly of the imposed flux or Neumann type) often proves delicate and susceptible to discretization errors [6, 7]. In this context, the Lattice Boltzmann Method (LBM) has emerged as a promising new alternative. As opposed to conventional methods, the LBM adopts a mesoscopic standpoint: it does not directly solve the Navier–Stokes equations, but relies on the statistical evolution of particle distribution functions on a network of discrete velocities [8]. This method has many advantages as it is explicitly iterative, perfectly suitable for parallel computation, easy to implement even on complex geometries, and extensible to coupled phenomena (chemical reactions, multicomponent transfers, magnetic fields, etc.). In particular, for low-compressibility flows and laminar or transitional regimes, LBM offers much greater accuracy than traditional methods, while providing remarkable algorithmic flexibility [9–14].

However, despite its growing success, LBM also presents some challenges, particularly in the accurate consideration of boundary conditions. While the conditions of the Dirichlet type (imposed values) are relatively well controlled, the integration of Neumann conditions (imposed flux) remains a crucial challenge [15]. These conditions are common in real-life problems such as constant-flow heating, thermal isolation and imposed temperature gradients. Inadequate consideration of these elements can falsify simulation results, notably by affecting the calculation of heat flux (Nusselt) or temperature profiles. Several strategies have been proposed, including modified bounce-back schemes, boundary extrapolations or, more recently, the advanced scheme of Martins et al. [16], which introduces an explicit formulation of flow conditions based on cinetic theory.

In this context, the present chapter aims to analyze the implementation of Neumann conditions with respect to the Lattice Boltzmann method applied to heat transfer. More specifically, the purpose is to simulate, in a precise and stabilized way, a representative case of natural convection in a square cavity. The cavity is filled with air and exposed to differential lateral heating. This type of configuration allows the validation of the strength of the suggested design and the assessment of performance by comparing it with existing literature.

This section is structured as follows: first, we review the theoretical foundations of the LBM method, focusing on the modeling of heat transfer using a dual-distribution system (f_i, g_i). Then, particular attention is paid to the implementation of boundary conditions, first in general, then specifically for Neumann-type conditions, by comparing different approaches and detailing the chosen formulation. The third part is devoted to the computational study of a representative test case, including a description of the geometry, simulation conditions and physical parameters. In the fourth part, we report on the findings, examine the physical and numerical aspects, and discuss the limitations. Finally, we conclude by summarizing the contributions of this study and suggesting concrete future paths of research.

2 Theoretical Foundations of the LBM Method

The Lattice Boltzmann Method (LBM) is a numerical technique derived from the kinetic theory of gases, designed to simulate fluid flows more efficiently. It differs from classical approaches to fluid mechanics in that it does not directly solve the Navier–Stokes equations. Instead, it models the evolution of a discretized distribution function, describing the probability of particles being present at a given point in space, with a given velocity, at a given time. In this way, LBM operates on a mesoscopic scale, intermediate between molecular, microscopic description and continuous macroscopic modeling.

2.1 A Multi-scale Physics Approach

Fluid behavior can be viewed on three different scales:

- At the microscopic level, the fluids are considered as sets of individual molecules, with their motions governed by the Newton's laws. This is known as molecular dynamics (MD) simulations, which are accurate but need a long computation time for systems with large numbers of particles.
- At the macroscopic scale, the fluids are modeled as continuous systems, where their properties (density, velocity, temperature) are described by fields of variables governed by the Navier–Stokes equations. This approach is based on large-scale statistical models and is particularly suitable for high-dimensional flows.
- At the mesoscopic scale, which is the basis of the LBM method, the fluid is described by a distribution function $f(\vec{x}, \vec{v}, t)$, expressing the density of particles, where $\vec{v}$, $\vec{x}$, and t are velocity, position and time, respectively. This representation fully catches microscopic effects (collisions, diffusion) while remaining practical for complex macroscopic phenomena (Fig. 1).

The concept of using a statistical description of particles goes back to James Clerk Maxwell (1859), who introduced the notion of a velocity distribution function

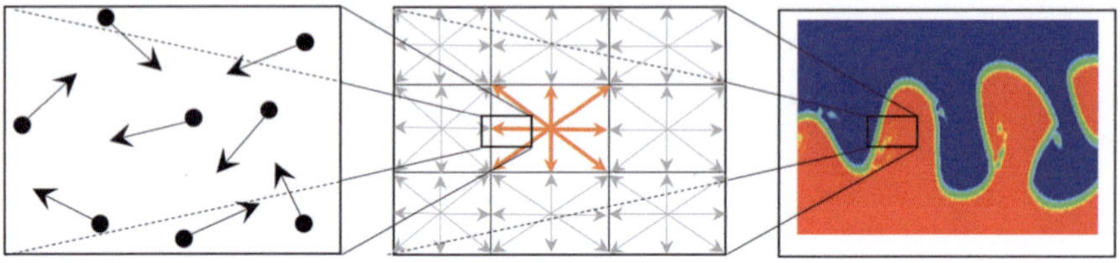

Fig. 1 Scales of fluid flows

in gases [8]. Rather than following every single particle, he proposed a function $f(\vec{x}, \vec{c}, t)$, where $\vec{x}$, $\vec{c}$, and t are position in real space, particle velocity, and time.

This function allows us to estimate the particle density in an infinitesimal volume $dx^3\ dc^3$ around $(\vec{x}, \vec{c})$,, at a given moment. The approach of Maxwell is behind the kinetic theory of gases, which establishes a bridge between molecular dynamics and the equations of fluid mechanics. The macroscopic properties of the fluid (such as density ρ and mean velocity $\vec{u}$) can be obtained by integrating this function over velocity space:

$$\rho(\vec{x}, t) = \int f(\vec{x}, \vec{c}, t)\, d\vec{c}, \quad \text{and} \quad \rho\vec{u} = \int \vec{c} f(\vec{x}, \vec{c}, t)\, d\vec{c}, \tag{1}$$

2.2 *Classical Boltzmann Equation*

The LBM method is based on a simplified form of the Boltzmann equation, which originates from the kinetic theory of gases. The general Boltzmann equation describes the evolution of the distribution function $f(\vec{x}, \vec{c}, t)$, which depends on the position $\vec{x}$, the particle velocity $\vec{c}$, and time t, within the phase space [17, 18].

$$\frac{\partial f}{\partial x} + \vec{c}.\vec{\nabla}_{\vec{x}} f + \vec{F}.\vec{\nabla}_{\vec{c}} f = B(f), \tag{2}$$

where $\vec{F}$ and $B(f)$ represent an external body force and the collision operator (which expresses the variations in f due to inter-particle interactions), respectively.

This equation is an integro-differential equation, because the collision operator depends on integrals over the velocity space; its analytical and even its numerical solution is complicated. For this reason, various approximations have been developed, including the Bhatnagar–Gross–Krook operator (BGK model), which will be introduced in the next section.

2.3 BGK Model

The collision operator $B(f)$ in the general Boltzmann equation defined from a complex integral taking into account all possible collisions between pairs of particles [19]. Theoretically, it defined as:

$$B = \frac{1}{m}\int \sigma\left|\vec{c_1} - \vec{c}\right| f^{(2)}\left(\vec{c}, \vec{c_1}\right) d\Omega\, d\vec{c_1}, \tag{3}$$

where $\left(\vec{c_1}, \vec{c}\right), f^{(2)}, \sigma, m,$ and $d\Omega$ are the velocities of the two particles, the distribution function of the two particles, the effective cross-section (probability of collision at a given angle), mass of a single particle and the solid angle element.

This formulation is highly accurate with physical reality, which includes major complication stemming from the fact that the operator is *integro-differential*. This makes its numerical treatment particularly difficult, if not impossible, in practical simulations. To overcome this issue, a simplified approach has proposed by Bhatnagar et al. [19], namely the BGK model. This model is based on the idea that the effects of collisions tend to bring the distribution f back to an equilibrium distribution $f^{(eq)}$, according to an exponential relaxation dynamic:

$$B_{BGK} = \vartheta\left(f - f^{(eq)}\right), \tag{4}$$

where ϑ is the relaxation time constant, which has the inverse of a time dimension. By convention, we note:

$$\tau = \frac{1}{|\vartheta|}, \tag{5}$$

For a stable fluid without any external force (zero gradient, homogeneous system), the equation becomes:

$$\frac{\partial f}{\partial t} = \vartheta\left(f - f^{(eq)}\right), \tag{6}$$

If $f > f^{(eq)}$, then relaxation to equilibrium requires a decrease in f over time, so:

$$\frac{\partial f}{\partial t} < 0 \Rightarrow \vartheta < 0, \tag{7}$$

We thus finally set:

$$B_{BGK}(f) = -\frac{1}{\tau}\left(f - f^{(eq)}\right), \tag{8}$$

The Boltzmann equation without the external force term then becomes:

$$\frac{\partial f}{\partial t} + \vec{c} \cdot \vec{\nabla} f = -\frac{1}{\tau}\left(f - f^{(eq)}\right), \tag{9}$$

In this respect, there are several notable advantages. First, it simplifies calculations considerably by replacing a complex integro-differential collision operator with an easily manipulated linear expression. It preserves the physical properties of the system (conservation of mass, quantity of motion, and energy). Moreover, it offers high efficiency in the modeling of local interactions, since the equilibrium distribution $f^{(eq)}$ depends only on local quantities such as density ρ, average velocity $\vec{u}$, and temperature T, which facilitates numerical calculations in a given spatial mesh.

2.4 Discretized Boltzmann Equation

To make the problem numerical, it is necessary to reformulate the Boltzmann equation in a discretized version [20–22]. Our starting point is to simplify the Boltzmann equation by using the BGK collision operator:

$$\frac{\partial f}{\partial t} + \vec{c} \cdot \vec{\nabla} f = -\frac{1}{\tau}\left(f - f^{(eq)}\right), \tag{10}$$

Such a continuous equation is based on an infinite number of possible molecular velocities $\vec{c}$, to make it numerically operational, we proceed to a discretization of the velocity: we replace the continuous set of velocities by a limited set of discrete directions $\overrightarrow{c_i}$ (i = 0, 1,...., N).

Hence, the distribution function $f(\vec{x}, \vec{c}, t)$, is approached by a set of discrete functions:

$$f_i(\vec{x}, t) \approx f\left(\vec{x}, \overrightarrow{c_i}, t\right), \tag{11}$$

where each function is associated with a specific direction of propagation $\overrightarrow{c_i}$. This step enables us to move from a continuous system to a discrete formulation suitable for numerical computations on a lattice. This leads to Boltzmann's discrete equation:

$$\frac{\partial f_i}{\partial t} + \overrightarrow{c_i} \cdot \vec{\nabla} f_i = -\frac{1}{\tau}\left(f_i - f_i^{(eq)}\right), \tag{12}$$

Afterwards, the temporal and spatial derivatives are discretized according to steps of time δt and a spatial δx, which are selected so that the particles move exactly from one node to another in the lattice during δt. This condition involves:

$$\delta x = \left\| \overrightarrow{c_i} \right\| \delta t, \tag{13}$$

Then, we apply an explicit Euler scheme to the time derivative:

$$\frac{\partial f_i}{\partial t} \approx \frac{f_i(\vec{x}, t + \delta t) - f_i(\vec{x}, t)}{\delta t}, \tag{14}$$

and to the convection term:

$$\vec{c_i} \cdot \vec{\nabla} f_i \approx \frac{f_i\big(\vec{x} + \vec{c_i}\,\delta t, t + \delta t\big) - f_i(\vec{x}, t + \delta t)}{\delta x}, \tag{15}$$

If we substitute the approximations in the discrete equation, while also taking in account the relationship $\delta x = \| \vec{c_i} \| \delta t$, we eventually obtain the discrete Boltzmann equation on a lattice:

$$f_i\big(\vec{x} + \vec{c_i}\,\delta t, t + \delta t\big) = f_i(\vec{x}, t) - \frac{\delta t}{\tau}\Big(f_i - f_i^{(eq)}\Big), \tag{16}$$

This explicit formulation forms the core of the LBM method. It is interpreted as the combination of two distinct processes:

- **Streaming**: the particles are transported from one node to another according to their velocity vector $\vec{c_i}$.
- **Collision**: at each node, the distribution is relaxed to a state of local equilibrium $f_i^{(eq)}$, depending on macroscopic quantities (ρ, $\vec{u}$, T).

2.5 *Discrete Velocity Models*

The quality of the discretization ultimately depends on the choice of the velocity set $\vec{c_i}$, which must satisfy several fundamental requirements:

- Mesh isotropy to ensure rotation invariance.
- Conservation of mass, quantity of motion, and energy.
- Compatibility with the symmetries of the simulation domain.

Among the most commonly used models, the most recommended are:

- D1Q3 in one dimension (3 discrete velocities),
- D2Q9 in two dimensions (9 discrete velocities),
- D3Q19 in three dimensions (19 discrete velocities).

These models provide a good compromise between accuracy, stability, and numerical efficiency. In this study, we mainly use D2Q9 scheme for both fluid flow and thermal transport (Fig. 2).

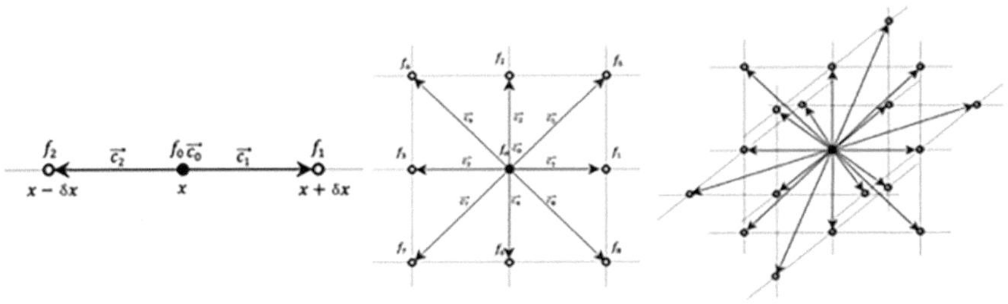

Fig. 2 Examples of discrete arrangements of the Boltzmann lattice

3 Thermal LBM Based on the Double-Population Approach

As part of the Boltzmann dual-population network method for thermal flows, the hydrodynamic and thermal fields are resolved via the distribution functions f_i and g_i, respectively. These functions are developed according to standard LBM equations, supplemented by source terms representing volume forces and heat sources [23–25].

$$f_i\left(\vec{x} + \vec{c_i}\,\delta t, t + \delta t\right) - f_i(\vec{x}, t) = -\frac{\delta t}{\tau_f}\left[f_i(\vec{x}, t) - f_i^{eq}(\vec{x}, t)\right] + \delta t\, S_{f_i}(\vec{x}, t), \tag{17}$$

$$g_i\left(\vec{x} + \vec{c_i}\,\delta t, t + \delta t\right) - g_i(\vec{x}, t) = -\frac{\delta t}{\tau_g}\left[g_i(\vec{x}, t) - g_i^{eq}(\vec{x}, t)\right] + \delta t\, S_{g_i}(\vec{x}, t), \tag{18}$$

where τ_f and τ_g respectively represent the relaxation times associated with the density and temperature distribution functions, which times are related to the cinematic viscosity ν and thermal diffusivity α through the following formulae:

$$\nu = \left(\tau_f - \frac{1}{2}\right)c_s^2\delta t, \quad \alpha = \left(\tau_g - \frac{1}{2}\right)c_s^2\delta t, \tag{19}$$

Here, c_s denotes the lattice sound speed. The source terms S_{f_i} and S_{g_i} take into account, respectively, the external force $\vec{F}(\vec{x}, t)$ and the surface heat source q''. They are written as follows:

$$S_{f_i}(\vec{x}, t) = \left(1 - \frac{\delta t}{\tau_f}\right)w_i\left[\frac{\vec{c_i} - \vec{u}}{c_s^2} + \frac{\left(\vec{c_i}\cdot\vec{u}\right)\vec{c_i}}{c_s^4}\right]\cdot\vec{F}(\vec{x}, t), \tag{20}$$

$$S_{g_i}(\vec{x}, t) = \left(1 - \frac{\delta t}{2\tau_g}\right)w_i\frac{q''}{\rho c_p}, \tag{21}$$

where c_p represents the specific heat at constant pressure, and w_i denotes the weight associated with the discrete velocity direction c_i. The equilibrium functions f_i^{eq} and g_i^{eq} describe the system at the equilibrium state. After discretization, they take the following forms:

$$f_i^{eq}(\vec{x}, t) = w_i\,\rho(\vec{x}, t)\left[1 + \frac{\vec{c_i}\cdot\vec{u}}{c_s^2} + \frac{\left(\vec{c_i}\cdot\vec{u}\right)^2}{2c_s^4} - \frac{\vec{u}\cdot\vec{u}}{2c_s^2}\right], \tag{22}$$

$$g_i^{eq}(\vec{x}, t) = w_i\,T(\vec{x}, t)\left[1 + \frac{\vec{c_i}\cdot\vec{u}}{c_s^2}\right], \tag{23}$$

Finally, macroscopic variables (density ρ, velocity $\vec{u}$, and temperature T) are calculated from the distribution function moments:

$$\rho(\vec{x}, t) = \sum_{i=0}^{8} f_i(\vec{x}, t), \tag{24}$$

$$\rho\,\vec{u}(\vec{x}, t) = \sum_{i=0}^{8} \vec{c_i} f_i(\vec{x}, t) + \frac{\delta t}{2}\vec{F}(\vec{x}, t), \tag{25}$$

$$T(\vec{x}, t) = \sum_{i=0}^{8} g_i(\vec{x}, t) + \frac{\delta t}{2}\frac{q''}{\rho c_p}, \tag{26}$$

4 Implementation of Boundary Conditions in LBM

In contrast to the traditional approaches based on the direct resolution of Navier–Stokes equations, the Boltzmann lattice method (LBM) is an approach based on the discrete evolution of fictive particles distribution functions on a meshed domain. In this context, the accurate and suitable treatment of boundary conditions is a major challenge to ensure numerical stability, accurate results and physical consistency. With this in mind, we present the main boundary conditions commonly used in LBM.

4.1 Wall Boundary Conditions: Solid Surface Treatment

a. **Bounce-back scheme**

This is the most common approach used to model a no-slip condition. This condition consists in reversing the direction of the particles when they collide with a wall,

Fig. 3 Bounce-back boundary conditions

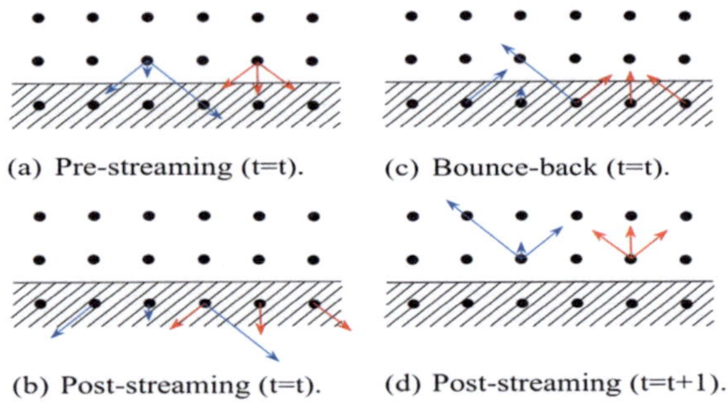

which cancels out the tangential velocity of the fluid on the surface [26, 27]. This scheme guarantees zero velocity at the half-way from the wall (Fig. 3).

$$f_i(\vec{x}_b, t + \delta t) = f_{\bar{i}}(\vec{x}_b, t), \tag{27}$$

where $\bar{i}$ is the velocity direction opposite to i post-collision one.

b. **Moving boundaries**

To model boundary walls moving at a speed $\vec{u}_w$, the bounce-back scheme is modified by a dynamic correction [28]:

$$f_i(\vec{x}_b, t + \delta t) = f_{\bar{i}}(\vec{x}_b, t) - 2w_i\rho\frac{\overrightarrow{c_i} \cdot \vec{u}_w}{c_s^2}, \tag{28}$$

4.2 Input and Output Conditions

a. **Method of Zou and He** [26]

The method of Zou and He allows a macroscopic quantity (density ρ or velocity $\vec{u}$) to be imposed on a boundary by reconstructing the missing distribution functions from the conservation equations of mass and momentum [26]:

$$\rho = \sum_i f_i, \tag{29}$$

$$\rho\vec{u} = \sum_i f_i \overrightarrow{c_i}, \tag{30}$$

At the frontier, some directions i are known (incoming populations) and others are unknown (outgoing populations). Solving this partial system, we express the unknown f_i as a function of the known $f_{\bar{i}}$, ρ and $\vec{u}$, according to:

$$f_i = f_{\bar{i}} + 2w_i\rho\frac{\overrightarrow{c_i}\cdot\vec{u}}{c_s^2}, \tag{31}$$

where $\bar{i}$ is the opposite direction to i, w_i is the weight associated with the direction $\overrightarrow{c_i}$, and c_s is the speed of sound on the network.

b. **Equilibrium extrapolation method**

This approach assumes that the unknown distributions at the frontier can be evaluated using the equilibrium distribution:

$$f_i = f_i^{eq}(\rho, \vec{u}), \tag{32}$$

The macroscopic quantities ϕ_b, ρ and$\vec{u}$, at the frontier are determined by extrapolation from adjacent internal cells. A typical linear extrapolation can be written as,

$$\phi_{\text{b}} = 2\phi_{int} - \phi_{int-1}, \tag{33}$$

This method is easy for implementation and suitable for open frontiers, although it is less accurate in the presence of strong gradients.

5 Thermal Boundary Conditions

In the Boltzmann lattice method (LBM), the treatment of thermal boundary conditions is essential to ensure the numerical stability and physical validity of heat transfer simulations. Three main categories are generally considered:

5.1 Dirichlet Condition (Imposed Temperature)

When we impose a constant temperature $T = T_0$ at the frontier, the unknown thermal distribution functions g_i are reconstructed using the equilibrium distribution [29]:

$$g_i = g_i^{eq}(\rho, T_0), \tag{34}$$

where g_i^{eq} is calculated from thermal equilibrium equations. This approach is simple, stable and suitable for domains with defined temperatures.

5.2 Adiabatic Boundaries (Zero Flow)

For an adiabatic boundary, the normal heat flow is zero [30]:

$$\left.\frac{\partial T}{\partial n}\right|_{\mathrm{b}} = 0, \tag{35}$$

This is often implemented by a rebound scheme applied to thermal distributions:

$$g_i = g_j, \tag{36}$$

which eliminates the net flow through the wall and preserves the physical properties of the system.

5.3 Dirichlet Neumann Boundary Conditions

a. **Traditional Neumann boundary conditions**

In the context of the Boltzmann lattice method (LBM), several classical approaches have been developed for implementing Neumann-type boundary conditions, in which we impose a heat flux q_w rather than a temperature value. A first method, proposed by Li et al. [31] for the D2Q9 model, relies on a modified thermal rebound scheme based on the direction of discrete velocities. It is written as follows:

$$g_i(\vec{x}_b, t + \delta t) = \widetilde{g}_i(\vec{x}_b, t) + \lambda_i \left(\frac{\delta t}{\delta x}\right)\left(\frac{q_w}{\rho c_p}\right), \tag{37}$$

where the factors λ_i are related to the nodal weights w_i and the speed of sound c_s by the relation $\lambda_i = 2\, w_i / c_s^2$.

A second method transforms the Neumann condition into an equivalent Dirichlet one, via a first-order finite-difference approximation of the Fourier law:

$$T_w(t) = T(\vec{x}_b, t) + \frac{3}{2}\delta x \frac{q_w}{k}, \tag{38}$$

This reformulation is beneficial for its simple implementation, but it may introduce a first-order error. For better accuracy, a third approach adopts a second-order finite volume formulation, taking into account the temporal and spatial evolution of the thermal field:

$$\begin{aligned} T_w(t) = {} & \frac{q_w \delta t \alpha}{k \delta x} + T_w(t - \delta t) \\ & + \frac{\alpha \delta t}{\delta x^2}\left[T(\vec{x}_b + \delta x \vec{n}, t - \delta t) - T(\vec{x}_b, t - \delta t)\right], \end{aligned} \tag{39}$$

In the presence of a volumetric heat source q‴, this equation is corrected using the following affectation:

$$T_w(t) = T_w(t) + \frac{q'''\delta t\, \alpha}{k}, \tag{40}$$

These approaches, even their effectiveness in standard configurations, sometimes suffer from limitations in terms of stability or accuracy in transient or highly non-linear regimes. This justifies the development of more advanced or more flexible schemes.

b. **Enhanced Neumann condition**

A new method for implementing the Neumann condition, proposed by Martins et al., improves the thermal accuracy of the LBM method, while maintaining a simple numerical structure. The approach consists in transforming the wall heat flux q_w into a volume flux ($q_v = q_w/\delta x$), incorporated as a source term in the Boltzmann equation for temperature. The rebound condition is applied to the boundary nodes, assuming an adiabatic wall:

$$g_i(\vec{x}_b, t + \delta t) = \widetilde{g}_i(\vec{x}_b, t), \tag{41}$$

The macroscopic temperature is then adjusted as follows:

$$T(\vec{x}, t) = \sum_i g_i(\vec{x}, t) + \frac{\delta t}{2}\frac{q_v}{\rho c_p}, \tag{42}$$

An additional correction is, therefore, applied to the boundary nodes during the post-treatment process:

$$T(\vec{x}_b, t) = T(\vec{x}_b + 3\,\delta x\vec{n}, t) - 3\,T(\vec{x}_b + 2\,\delta x\vec{n}, t) + 3\,T(\vec{x}_b + \delta x\vec{n}, t), \tag{43}$$

This method ensures better management of the flows imposed at the boundary and improves the stability and fidelity of thermal results.

6 Case Study and Numerical Validation

The physical configuration studied consists of a closed cavity with a square-section geometry, filled with air considered as an incompressible and Newtonian fluid, with a fixed Prandtl number of 0.71. This cavity is exposed to lateral differential heating under a uniform downward gravitational field (Fig. 4). The left and right vertical boundaries are respectively heated and cooled with a uniform heat flux, $q_{in} = q_{out}$, while the upper and lower horizontal boundaries are assumed to be adiabatic, allowing

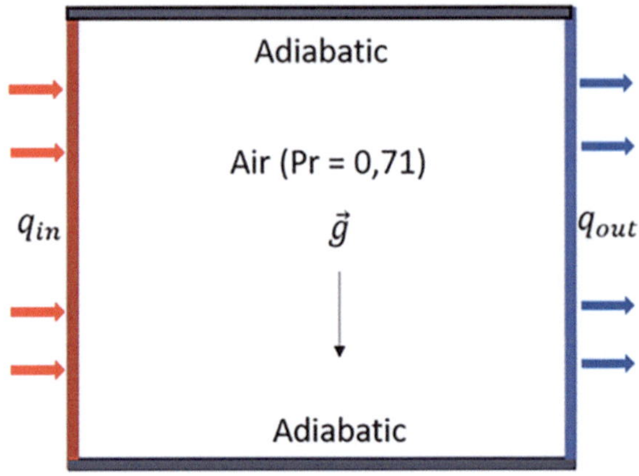

Fig. 4 Schematic diagram of the physical system

no heat exchange with the outside. This coupling between a horizontal thermal gradient and a vertical gravitational field induces a natural convection mechanism inside the cavity, giving rise to convective cells characteristic of the regime under study.

6.1 *Boundary Condition Implementation*

The accuracy and stability of lattice Boltzmann method (LBM) simulations are largely dependent on the rigorous treatment of boundary conditions. In the present study, we outline the numerical strategies applied to implement hydrodynamic and thermal boundary conditions in the considered configuration, using a dual-population LBM model based on the D2Q9 scheme. Particular focus is given to the specific application of Neumann conditions (imposed heat flux), based on the improved method proposed by Martins et al. [16].

(i) *Hydrodynamic boundary conditions*:

All cavity boundary walls have treated as rigid (no-slip), with zero velocity imposed. This condition is implemented using the standard bounce-back scheme applied to relate the hydrodynamic distribution functions f_i and their post-collision values f_{pos} at the boundaries. This approach ensures non-penetration of the fluid and the zero tangential velocity on the boundary walls. For a $[nx] \times [ny]$ meshing size, this condition is given, respectively, for vertical (horizontal) walls by:

$$\text{Left} - \text{wall} : f[j][0][1] = f_{pos}[j][0][3];$$

$$\text{Right} - \text{wall} : f[j][nx][3] = f_{pos}[j][nx][1];$$

$$\text{Bottom} - \text{wall} : f[0][i][2] = f_{pos}[0][i][4];$$

$$\text{Top} - \text{wall} : f[ny][i][4] = f_{pos}[ny][i][2];$$

(b) *Thermal boundary conditions*:

The temperature field is modeled using a second population distribution g_i, according to a thermal LBM scheme with BGK collision. The upper and lower boundary conditions are assumed to be adiabatic ($\partial T/\partial y = 0$). This condition is implemented via a specular symmetry (anti-bounce-back) applied to the thermal populations.

$$\text{Lower} - \text{wall} : g[0][i][2] = g_{pos}[0][i][4];$$

$$\text{Upper} - \text{wall} : g[ny][i][4] = g_{pos}[ny][i][2];$$

The left ($x = 0$) and right ($x = n_x$) boundaries are exposed to uniform and opposite heat flux ($q_{in} = -q_{out}$). To ensure precise imposition of these fluxes, Martins et al. [16] consists on several steps, summarized as follows. After applying the bounce-back condition on the hydrodynamic boundary nodes, assuming that both horizontal sides are adiabatic, a source term is added in the thermal collision phase to model the flows imposed on the vertical sides. At the left side, for example, the following correction is added in the collision step:

$$\begin{aligned} g_{pos}[j][0][k] = {} & g[j][0][k] + (\delta t/\tau_g) \\ & * \left(g^{eq}[j][0][k] - g[j][0][k]\right) \\ & + \delta t * \left(1 - \frac{\delta t}{(2 * \tau_g)}\right) * w[k] * \frac{q_w}{\rho * c_p}; \end{aligned}$$

Next, we adjust the macroscopic temperature at the boundary, in this way accounting for the contribution of heat flow in the reconstruction of the scalar field:

$$T[j][i] = \sum_{k=0}^{8} g[j][i][k];$$

Then, the correction is added only to the vertical walls as

$$T[j][0] = \sum_{k=0}^{8} g[j][i][k] + 0.5 * \delta t * \frac{q_w}{\rho * c_p};$$

$$T[j][nx] = \sum_{k=0}^{8} g[j][i][k] - 0.5 * \delta t * \frac{q_w}{\rho * c_p};$$

A second-order Lagrangian extrapolation correction is applied to the edge nodes to improve the accuracy of the output temperatures without altering the simulated thermal dynamics. According to the Eq. (43), the temperatures of the left and right walls become, respectively:

$$T[j][0] = T[j][3] - 3 * T[j][2] + 3 * T[j][1];$$

$$T[j][nx] = T[j][nx-3] - 3 * T[j][nx-2] + 3 * T[j][nx-1];$$

The implementation of boundary conditions is based on rigorous methods. The use of Ivan Martin's method ensures accurate simulation of heat flows at the boundary, without recourse to approximate extrapolations. Adjusted reconstruction of the scalar field guarantees local energy conservation and numerical stability of the solution. This strategy enables reliable modeling of natural convection in closed configurations under more realistic boundary condition of Neumann.

6.2 Validation of Simulation Code

To assess the accuracy of the Neumann-type boundary condition scheme implemented within the framework of the Lattice Boltzmann Method (LBM), a comparative study has carried out using benchmark results provided by Sharif and Mohammad [32], obtained via the finite volume method. The case selected for comparison with their work involves natural convection in a square cavity (aspect ratio A = 1), heated from below by a localized constant heat flux ($\varepsilon = 0.6$), with isothermal vertical walls and an adiabatic top wall. The flow is assumed to be laminar, and the natural convection regime is characterized by a variable Grashof number: $Gr = 10^3$, 10^4, 10^5, and 10^6, for a horizontal cavity ($\beta = 0°$).

The numerical results obtained using the LBM show excellent agreement with the literature. Figure 5 illustrates the variation of the average Nusselt number as a function of the Grashof number, comparing the simulated values with the reference values. The relative deviations between the two approaches remain consistently below 0.6%, confirming the reliability of the developed model in both conduction-dominated regimes $\left(Gr \leq 10^4\right)$ and convection dominated regimes $\left(Gr \geq 10^5\right)$.

7 Discussion of Results

In this study, the obtained results for different Rayleigh number ($Ra = 10^3$, 10^4, 10^5, 10^6) are influenced by the implementation of Neumann-type boundary conditions using Ivan's method. This approach guarantees the accuracy and stability of the model by ensuring adequate boundary conditions, in particular with the absence of heat flow

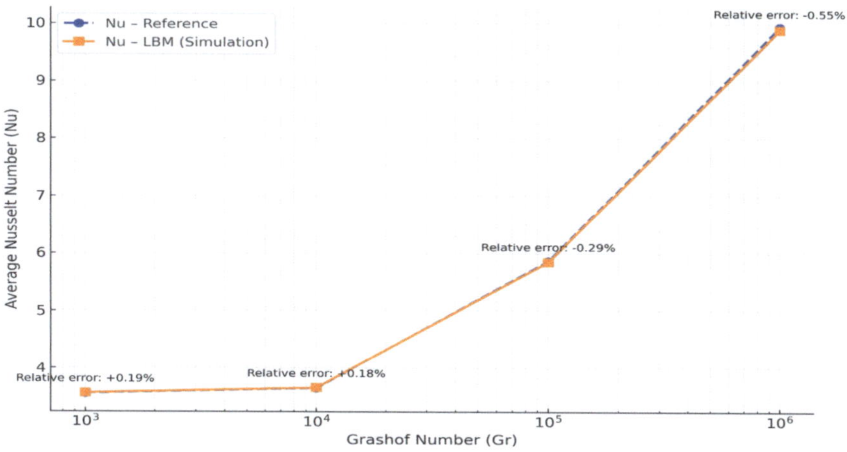

Fig. 5 Comparison of average Nusselt Numbers: LBM versus reference [32]

through horizontal boundary walls. The method used by Ivan is based on a numerical scheme using the Lattice Boltzmann Method (LBM), combined with Neumann conditions, where the temperature gradient is zero on adiabatic boundary walls. This technique effectively simulates natural convection in a confined geometry, while maintaining thermal equilibrium and minimizing numerical errors at the boundary. The main innovation of this method lies in the reconstruction of thermal gradients at boundaries (notably adiabatic walls) using second-order finite differences, ensuring a fluid transition between fluid and boundary without thermal discontinuity.

7.1 *Isotherms and Streamlines Distributions*

The application of Neumann conditions via Ivan's method has enabled us to improve our modeling of natural convection effects, particularly for high Rayleigh configurations ($Ra = 10^5$ and $Ra = 10^6$). These adiabatic boundary conditions, although theoretically simple, have a major impact on the development of thermal boundary layers and isotherms, as they prevent heat escaping through the horizontal boundaries, thus enhancing convection in the domain. This was particularly apparent in the way the isotherms closed in on the heated and cooled boundary walls at higher Ra, which was less clear with a lower-precision approximation of the boundary conditions. In addition, the current lines, illustrating fluid dynamics under the effect of the gravitational field and the thermal gradient, have greatly influenced by this implementation. At $Ra = 10^6$, the increased precision of the boundary conditions has enabled the formation of well-defined convective cells to be more accurately captured. These cells, characterized by fluid motion following the isotherms and generating secondary vortices near the corners of the cavity, demonstrated the impact of this approach. The use of

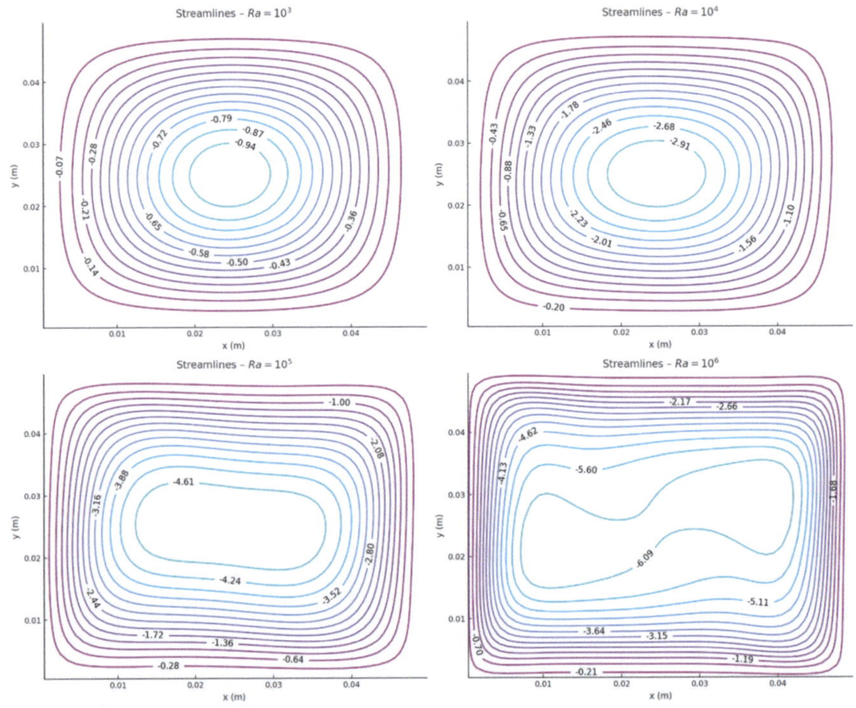

Fig. 6 Streamlines for different Rayleigh numbers

Neumann conditions not only ensured better energy conservation in the system, but also promoted greater stability in the evolution of these cells as the Rayleigh number increased (Figs. 6 and 7).

7.2 Variation of Local and Average Nusselt Number

Both the local Nusselt number Nu_{local} and the average Nusselt number Nu_{avg} showed a marked improvement in terms of consistency and accuracy, as a result of the implementation of the Neumann conditions by Ivan's method.

In particular, the profile of the local Nusselt number along the heated boundary showed a more pronounced increase at higher values of Ra, due to the increased efficiency of convective heat transfer permitted by more precise boundary conditions. The values of Nu_{local} and Nu_{avg} showed a better correspondence with theoretical expectations, particularly for $\text{Ra} = 10^6$, where improvements in the thermal gradient near the boundary allowed us to better model the intensification of convection. The increase in the average Nusselt number with Ivan's method is particularly evident when moving from $\text{Ra} = 10^3$ to $\text{Ra} = 10^6$. This increase reflects the

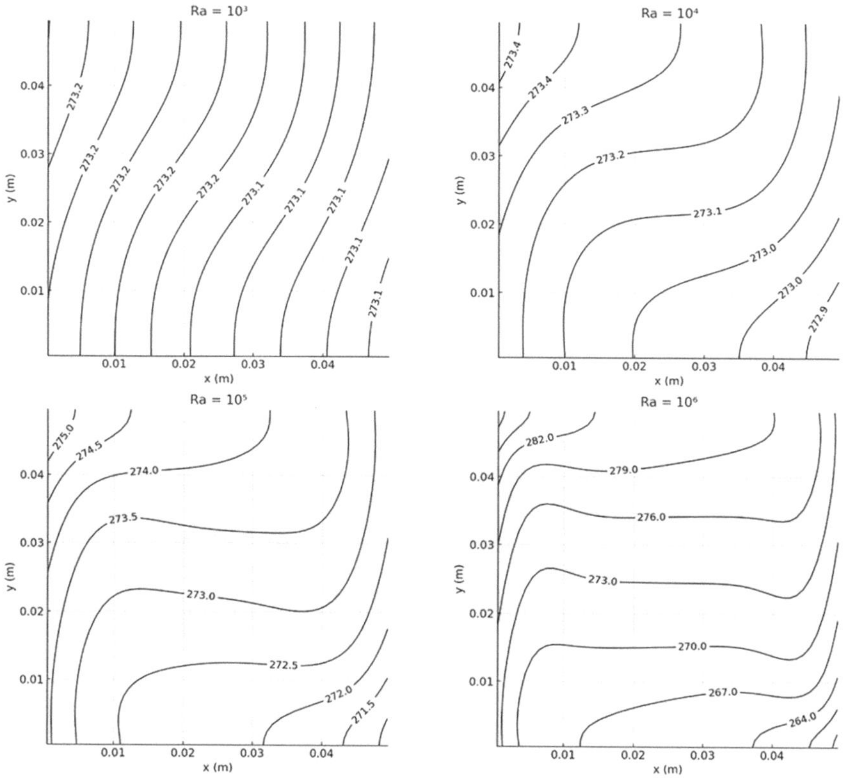

Fig. 7 Isotherms for different Rayleigh numbers

impact of more active convection and better coupling between heat, fluids and boundaries, through more precise management of boundary conditions and thermal gradients. This method significantly improved the accuracy of the results, enabling a better representation of natural convection phenomena, particularly at high Rayleigh numbers (Figs. 8 and 9).

8 Summary and perspectives

By guaranteeing more stable and accurate simulations, Ivan's method opens up new perspectives for the study of thermal flows in confined geometries. It is particularly crucial for practical applications such as the design of passive cooling systems, where natural convection plays an essential role in heat transfer efficiency. These improvements not only provide a better understanding of convective phenomena, but also enable thermal systems to be optimized for higher performance, opening

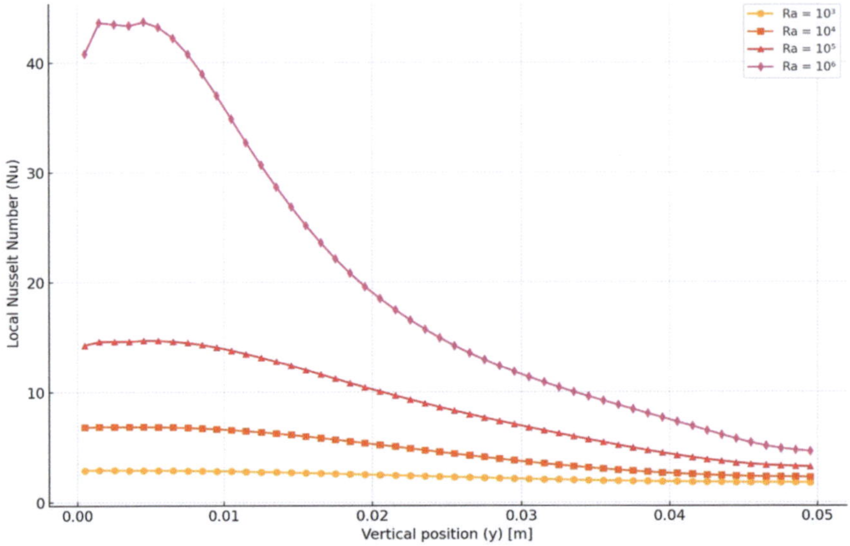

Fig. 8 Local Nusselt number profiles for different Rayleigh numbers

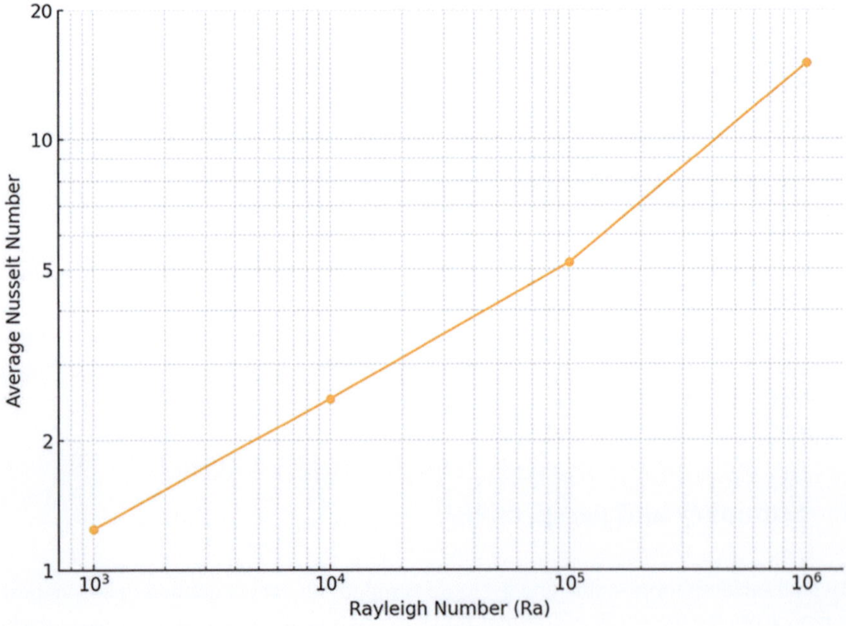

Fig. 9 Variation of average Nusselt number with Rayleigh number

up opportunities for future research into the optimization of natural convection in industrial or advanced thermal management contexts.

References

1. Ferreira, M., Carvalho, V., Ribeiro, J., et al.: Advances in microfluidic systems and numerical modeling in biomedical applications: a review. Micromachines **15**(7), 873 (2024)
2. Abidi, A., Ahmadi, A., Enayati, M., et al.: A review of the methods of modeling multi-phase flows within different microchannels shapes and their applications. Micromachines **12**(9), 1113 (2021)
3. Lach, Ł., Svyetlichnyy, D.: Advances in numerical modeling for heat transfer and thermal management: a review of computational approaches and environmental impacts. Energies **18**(5), 1302 (2025)
4. Martins, R.C., Lopes, V.V., Vicente, A.A. et al.: Numerical solutions: finite elements and finite volume methods. Optim. Food Eng. 49–107 (2009)
5. Dick, E.: Introduction to finite volume methods in computational fluid dynamics. In: Computational Fluid Dynamics, pp. 275–301. Springer, Berlin, Heidelberg (2009)
6. Moukalled, F., Mangani, L., Darwish, M.: Implementation of boundary conditions in the finite-volume pressure-based method—part I: segregated solvers. Numer. Heat Transf. Part B: Fundam. **69**(6), 534–562 (2016)
7. Ham, D.A., Cotter, C.J.: Finite Elements: Analysis and Implementation. Accessed 22 May 2024
8. Mohamad, A.A.: Lattice Boltzmann Method: Fundamentals and Engineering Applications with Computer Codes, 2nd edn. Springer-Verlag London Ltd., Londres. ISBN 978-1-4471-7422-6 (2019)
9. Agoujil, F., Elguennouni, Y., Ighris, Y., et al.: Modeling of natural convection by lattice Boltzmann in a partially heated cavity. In: E3S Web of Conferences, p. 00071. EDP Sciences (2023)
10. Hssikou, M., Elguennouni, Y., Baliti, J., et al.: Heat transfer of gas flow within a partially heated or cooled square cavity. Model. Simul. Eng. **2020**(1), 8886682 (2020)
11. Elguennouni, Y., Hssikou, M., Baliti, J., Alaoui, M.: Cavity inclination effect on convection flow in a triangular geometry. Heat Transf. **52**(4), 3253–3279
12. Oulahou, Y., Elguennouni, Y., Hssikou, M., Baliti, J., Alaoui, M.: Numerical investigation of natural convection heat transfer using TiO_2/Al_2O_3-water nanofluids. FME Trans. **52**(1), 158 (2024)
13. Ighris, Y., Qaffou, M., Baliti, J., Elguennouni, Y., Hssikou, M.: Thermal management optimization of natural convection in a triangular chamber: role of heating positions and ternary hybrid nanofluid. Phys. Fluid. **36**(9) (2024)
14. Ighris, Y., Bouhouchi, Y., Elguennouni, Y. et al.: Numerical study of natural convection in an inclined cavity filled with Al_2O_3/Cu-H_2O nanofluids. Numer. Heat Transf. Part A: Appl. 1–25 (2024)
15. Suzuki, K., Kawasaki, T., Furumachi N. et al.: A thermal immersed boundary–lattice Boltzmann method for moving-boundary flows with Dirichlet and Neumann conditions. Int. J. Heat Mass Transf. **121**, 1099–1117 (2018)
16. Martins, I.T., Matsuda, V.A., Cabezas-Gómez, L.: A new Neumann boundary condition scheme for the thermal lattice Boltzmann method. Int. Commun. Heat Mass Transf. **156**, 107653 (2024)
17. Succi, S.: The Lattice Boltzmann Equation for Fluid Dynamics and Beyond. Oxford University Press (2001)
18. Wolf-Gladrow, D.A.: Lattice-Gas Cellular Automata and Lattice Boltzmann Models: An Introduction. Springer (2000)

19. Bhatnagar, P.L., Gross, E.P., Krook, M.: A model for collision processes in gases. I. Small amplitude processes in charged and neutral one-component systems. Phys. Rev. **94**(3), 511–525 (1954)
20. Noriega, H.N.: Convection naturelle de nanofluides par la méthode de Boltzmann sur réseau. École Polytechnique de Montréal. Mémoire de maîtrise en génie mécanique (2011)
21. Mahfoudi, A.: Simulation numérique de la convection naturelle en géométrie complexe par la méthode de Boltzmann sur réseau. Université Mohammed Premier, Faculté des Sciences Oujda. Mémoire de Master en physique numérique (2021)
22. Aboubakr, K.: Simulation numérique de la convection naturelle dans une cavité annulaire remplie d'un nanofluide par la méthode de Boltzmann sur réseau. Université Ibn Tofail, Faculté des Sciences Kénitra. Mémoire de Master en physique et modélisation (2021)
23. Guo, Z., Zheng, C., Shi, B.: Discrete lattice effects on the forcing term in the lattice Boltzmann method. Phys. Rev. E **65**(4), 046308 (2002)
24. Shan, X., Yuan, X.-F., Chen, H.: Kinetic theory representation of hydrodynamics: a way beyond the Navier–Stokes equation. J. Fluid Mech. **550**, 413–441 (2006)
25. Mezrhab, A., Bouzidi, M.: Double-population lattice Boltzmann model for thermal flows. Int. J. Therm. Sci. 45(12), 1136–1146
26. Zou, Q., He, X.: On pressure and velocity boundary conditions for the lattice Boltzmann BGK model. Phys. Fluid. **9**(6), 1591–1598 (1997)
27. Latt, J., Chopard, B.: Lattice Boltzmann method with regularized pre-collision distribution functions. Math. Comput. Simul. **72**(2–6), 165–168 (2006)
28. Rüger, T., Kusumaningrum, R.D., Varnicka, L., et al.: The Lattice Boltzmann Method: Principles and Practice. Chapitre 6: Moving Boundaries. Springer (2017)
29. Liu, H., Zhang, Y., Valocchi, A.J.: Lattice Boltzmann method for simulating thermal multiphase flows. Comput. Math. Appl. **65**(2), 262–273 (2012)
30. Yang, J., Shu, C.-W.: Boundary condition treatment in LBM: a review. Int. J. Mod. Phys. C **16**(8), 1373–1393 (2005)
31. Li, Q., Mei, R., Klausner, J.F.: Thermal boundary conditions for the lattice Boltzmann equation method. Int. J. Heat Mass Transf. **47**(10–11), 2219–2230 (2004)
32. Sharif, M.A.R., Mohammad, T.R.: Natural convection in cavities with constant flux heating at the bottom wall and isothermal cooling from the sidewalls. Int. J. Therm. Sci. **44**(9), 865–878 (2005)

Influence of Process Parameters on Tensile Strength of FDM Printed ABS-Carbon Fiber Composites: Statistical and Experimental Analysis

F. Ait Hmazi, T. Hachimi, H. Rejdali, H. El Assad, Y. Riyad, F. Arhouni, H. Doghmi, A. Madani, and I. Mrani

Abstract This study investigates the influence of key process parameters on the tensile strength of ABS-carbon fiber composite parts fabricated using the Fused Deposition Modeling (FDM) process. Four parameters were considered, namely layer thickness, layer orientation, extrusion temperature, and printing speed. The Taguchi design of experiments was applied to efficiently reduce the number of tests while systematically evaluating the effect of each parameter. Tensile specimens were produced according to the ASTM D638 Type I standard and tested using a universal testing machine. The results showed that layer orientation and layer thickness have the most significant influence on tensile strength, while extrusion temperature and printing speed also have a notable but less dominant effect. Aligning the printed filaments with the load direction and reducing the layer thickness enhance interlayer bonding and lead to a significant improvement in mechanical performance. These findings provide valuable insights into the relationships between process parameters, internal structure, and mechanical properties, and form a solid basis for optimizing the FDM process to improve the quality of printed parts.

Keywords Additive manufacturing · Fused deposition modeling (FDM) · ABS-carbon fiber composite · Tensile strength · Process parameters · Taguchi

---

F. Ait Hmazi (✉) · T. Hachimi · H. El Assad · F. Arhouni · H. Doghmi · I. Mrani
Laboratory of Nuclear, Atomic, Molecular, Mechanical and Energetic Physics, Physics Department, Faculty of Sciences, Chouaib Doukkali University, El Jadida, Morocco
e-mail: aithmazi.f@ucd.ac.ma

H. Rejdali · Y. Riyad
Laboratory of Engineering Sciences for Energy (LabSIPE), National School of Applied Sciences, Chouaïb Doukkali University, El Jadida, Morocco

H. Rejdali
Centre for Energy and Thermal Sciences of Lyon (CETHIL), National Institute of Applied Sciences of Lyon, Villeurbanne, France

A. Madani
Laboratory LAROSERI, Department of Computer Science, Faculty of Sciences, Chouaib Doukkali University, El Jadida, Morocco

E. El Mehdi and M. Boutahir (eds.), *Advanced Materials for Sustainable Energy and Engineering*, Engineering Materials, https://doi.org/10.1007/978-3-032-10069-6_14

method · Analysis of variance (ANOVA) · Mechanical properties · Layer orientation · 3D printing optimization

1 Introduction

Additive Manufacturing (AM), commonly referred to as 3D printing, has emerged as a disruptive technology across various engineering and industrial sectors, offering numerous advantages over traditional manufacturing processes. These advantages include the ability to fabricate complex geometries, enable rapid prototyping, minimize material waste, and provide substantial design freedom [1, 2]. AM technologies have found applications in a wide range of fields such as aerospace, automotive, biomedical engineering, and consumer goods, where lightweight structures and customized components are in increasing demand [3].

Among the various AM processes, Fused Deposition Modeling (FDM) (Fig. 1) has become one of the most widely adopted methods, particularly for thermoplastic based materials, due to its simplicity, low cost, and versatility [4]. In FDM, a thermoplastic filament is heated to a semi molten state and extruded through a nozzle to deposit material layer by layer onto a build platform, gradually forming the final geometry of the part. However, this layer-by-layer deposition results in anisotropic properties, leading to variations in mechanical behavior depending on the orientation and bonding quality between successive layers.

Despite its practical advantages, the mechanical performance of FDM printed parts is generally inferior to that of components produced using conventional manufacturing techniques such as injection molding or subtractive machining [5]. This limitation is primarily due to weak interlayer adhesion and the presence of internal defects, such as voids or gaps, formed during material deposition. The anisotropic nature of the process, influenced by filament orientation relative to the applied load, makes printed parts highly sensitive to changes in processing parameters. Therefore, optimizing these parameters is crucial to improving the strength, durability, and reliability of FDM fabricated components [6].

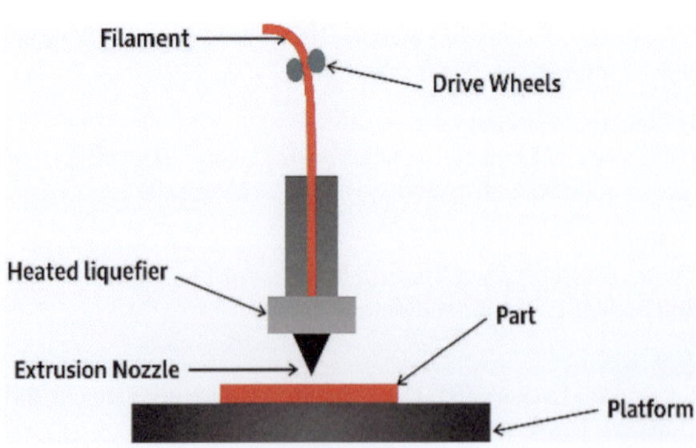

Fig. 1 Schematic representation of the fused deposition modeling (FDM) 3D printing process

Among thermoplastics commonly used in FDM, Acrylonitrile Butadiene Styrene (ABS) remains a widely employed material due to its favorable characteristics, such as impact resistance, thermal stability, dimensional accuracy, and ease of processing [7]. However, to meet the increasing demands for higher strength to weight ratios, particularly in structural applications, researchers have turned to carbon fiber (CF) reinforcement. The incorporation of short carbon fibers into the polymer matrix has been shown to significantly improve mechanical properties such as tensile strength, stiffness, and wear resistance [8]. These ABS carbon fiber composites thus present strong potential for lightweight structural components in demanding sectors like aerospace, transportation, and mechanical systems [9].

Nevertheless, the mechanical performance of fiber reinforced FDM composites is not determined solely by their composition. Instead, it is highly dependent on the processing parameters used during fabrication. Interactions between material flow, cooling rates, interlayer bonding quality, and fiber orientation play a crucial role in determining the mechanical properties of the final printed part. As such, parameters such as layer thickness, layer orientation, printing speed, and extrusion temperature must be carefully controlled to achieve optimal performance.

Given the multivariate nature of the FDM process, conventional trial and error approaches to parameter tuning are inefficient and time consuming. Therefore, Design of Experiments (DoE) methodologies are widely employed to systematically investigate the influence of multiple parameters simultaneously while minimizing the number of required tests. Among these techniques, the Taguchi method is recognized as a powerful and efficient statistical tool. By using orthogonal array designs, the Taguchi method allows for the identification of key factors and optimal parameter settings with a reduced number of experimental runs [10].

Numerous studies have successfully applied the Taguchi method to optimize FDM process parameters for both unreinforced and reinforced polymer systems. For instance, Raju et al. [11]. used a Taguchi approach combined with grey relational analysis to optimize printing parameters for PLA composites reinforced with carbon nanoparticles. Their work examined the effects of printing speed, carbon deposition, and layer count on tensile strength, toughness, and surface roughness. Similarly, Maideen et al. [12]. employed univariate Analysis of Variance (ANOVA) to evaluate the effect of layer thickness on the dimensional repeatability of PLA parts printed using an Openware Espresso F220 printer.

In addition to identifying the main effects of each parameter, statistical tools such as ANOVA provide a quantitative understanding of the contribution of each factor to the variability in mechanical properties. ANOVA can be used to determine the statistical significance and relative weight of each parameter, offering valuable insights into process sensitivity. This statistical framework is especially useful in FDM, where complex, nonlinear interactions often exist among parameters and are difficult to capture through single variable analysis.

Although substantial research has been conducted on parameter optimization for standard thermoplastics and some composite systems [13, 14], fewer studies have focused specifically on ABS carbon fiber composites fabricated by FDM. Given the growing industrial interest in fiber reinforced thermoplastics, especially for structural applications, a comprehensive investigation of how FDM process parameters influence the mechanical behavior of ABS CF composites is timely and necessary.

The present study aims to fill this gap by investigating the tensile strength of 3D-printed ABS carbon fiber composites. A Taguchi L9 orthogonal array is used to systematically evaluate the effects of four key process parameters: layer thickness, printing orientation, extrusion temperature, and printing speed, each studied at three levels. Tensile tests were conducted in accordance with the ASTM D638 Type I standard. The resulting data were analyzed using ANOVA and main effect plots to identify the most influential parameters. The outcomes of this study are expected to contribute to the existing literature on the additive manufacturing of fiber reinforced composites and to provide practical guidance for optimizing process parameters to enhance mechanical performance.

2 Method and Process

The selection of the material represents a critical step in the 3D printing process, as it directly determines the mechanical and functional performance of the fabricated parts. Depending on the specific requirements of each application, certain materials are favored for their mechanical strength, stiffness, flexibility, thermal resistance, or lightness.

In addition to the material itself, various printing process parameters significantly affect the quality of the printed parts and their mechanical properties. The interaction between the intrinsic properties of the material and the printing conditions largely governs the final performance of the 3D printed components.

2.1 Material

In this study, a composite based on Acrylonitrile Butadiene Styrene (ABS) reinforced with carbon fibers was selected. This material was chosen due to its enhanced mechanical properties and its good compatibility with the Fused Deposition Modeling (FDM) process, which is widely used in 3D printing.

ABS is well known for its impact resistance, dimensional stability, and ease of processing. The addition of carbon fibers significantly increases the material's stiffness, reduces its density, and improves its overall mechanical performance. Owing to this combination of properties, the ABS carbon composite is particularly suitable for applications requiring both light weight and high strength, such as in the automotive, aerospace, and functional mechanical component manufacturing sectors.

Table 1 FDM process parameters at three distinct levels

Process variables	Levels		
	1	2	3
Layer thickness (mm)	0.2	0.32	0.4
Orientation (°)	0	45	60
Printing speed (mm/s)	40	60	80
Printing temperature (°C)	220	240	260

2.2 *Process Parameters*

The objective of this study is to evaluate the influence of 3D printing parameters on the tensile strength of carbon fiber reinforced ABS composite parts. Special attention is given to the analysis of key parameters such as layer thickness, orientation, extrusion temperature, and printing speed, all of which play a decisive role in the development of the final mechanical properties of printed components.

Indeed, 3D printing process parameters exert a significant impact on the geometrical quality, microstructure, and mechanical performance of the produced parts. In the case of thermoplastic composites such as carbon fiber reinforced ABS, precise control of these parameters is particularly critical, as they directly affect interlayer adhesion, porosity formation, and the development of residual stresses that may compromise the overall mechanical performance.

The process variables and their respective levels considered in this study are summarized in Table 1.

2.3 *Experiment Design*

The study of the influence of process parameters on the mechanical properties of printed parts was conducted using the Taguchi method. This experimental approach allows for the optimization of the number of tests required while efficiently evaluating the impact of each factor under investigation.

In this study, four factors were selected: layer thickness, layer orientation, printing speed, and extrusion temperature. Each of these parameters was examined at three levels, defined according to the capabilities of the 3D printer used (FlashForge Creator 3 Pro) and the specific characteristics of the ABS carbon composite material.

To reduce the total number of experiments while ensuring sufficient statistical coverage, an L9 (34) orthogonal array was employed (Table 2). This design allowed for only nine experimental runs to analyze the combined influence of the four parameters. This approach provides an optimal compromise between experimental workload and the quality of the statistical conclusions obtained.

Table 2 Taguchi orthogonal array design (L9) with experimental data

	Layer thickness (mm)	Orientation (°)	Printing speed (mm/s)	Printing temperature (°C)
1	0.20	0	40	220
2	0.20	45	60	240
3	0.20	60	80	260
4	0.32	0	60	260
5	0.32	45	80	220
6	0.32	60	40	240
7	0.40	0	80	240
8	0.40	45	40	260
9	0.40	60	60	220

The objective of the experimental design is to identify the most favorable process configurations for improving the tensile strength of the printed parts, while also assessing the relative influence of each factor.

2.4 Specimens Fabrication

The tensile test specimens were designed in accordance with the ASTM D638 Type I standard, which specifies the geometry of samples intended for tensile testing of plastic materials. This standard ensures consistency across tests and allows for reliable comparison of mechanical properties. The exact dimensions of the specimens are shown in Fig. 2.

The digital models of the specimens were prepared in STL format and imported into FlashPrint software, which is dedicated to preparing print jobs for the FlashForge

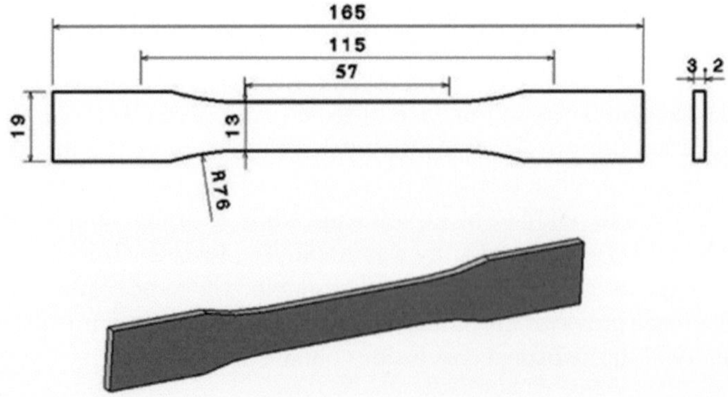

Fig. 2 ASTM D638-14 Type 1 specimen

Creator 3 Pro printer. The slicing process was performed using this software to generate the corresponding G-code files required for printer operation.

The 3D printing process was carried out using a 0.4 mm diameter nozzle. Each specimen was printed following the parameter configurations established in the Taguchi experimental matrix. Upon completion of printing, the specimens were carefully removed from the build plate to avoid any damage, and then allowed to rest at room temperature for 24 h to enable the stabilization of residual stresses generated during the cooling phase. This conditioning step ensures reliable and reproducible mechanical testing results.

2.5 Tensile Testing

Tensile tests were conducted to characterize the mechanical behavior of the printed specimens and to assess the influence of the selected process parameters on their tensile strength. All tests were performed using an MTS universal testing machine equipped with a 50 kN load cell, ensuring high precision in force and displacement measurements.

Before testing, each specimen was carefully positioned between the machine's grips according to the ASTM D638 Type I standard, ensuring proper alignment and consistent load application (Fig. 3). The tests were performed at a constant crosshead speed of 1 mm/min, as recommended for polymer materials to minimize strain rate effects and to simulate quasistatic loading conditions.

During testing, load and displacement were continuously recorded, enabling the generation of engineering stress–strain curves for each specimen. The ultimate tensile strength (UTS) was determined directly from the maximum stress value reached on these curves, representing the highest tensile load the material can sustain before failure. As a critical indicator of the overall strength of the printed parts, UTS reflects both the intrinsic material properties and the quality of the bonding between printed layers.

The obtained tensile strength values were subsequently analyzed in relation to the process parameters defined in the experimental design, allowing for the identification of significant factors and their contributions to the mechanical performance of the ABS-carbon composite specimens.

3 Results and Discussion

The analysis of the results obtained from the tensile tests highlighted the significant influence of the 3D printing parameters on the ultimate tensile strength (UTS) of the carbon fiber reinforced ABS composite specimens. The tensile strength values measured for each experimental combination derived from the Taguchi L9 (34) orthogonal array are presented in Fig. 4.

Fig. 3 Experimental setup for tensile testing using the MTS universal testing machine

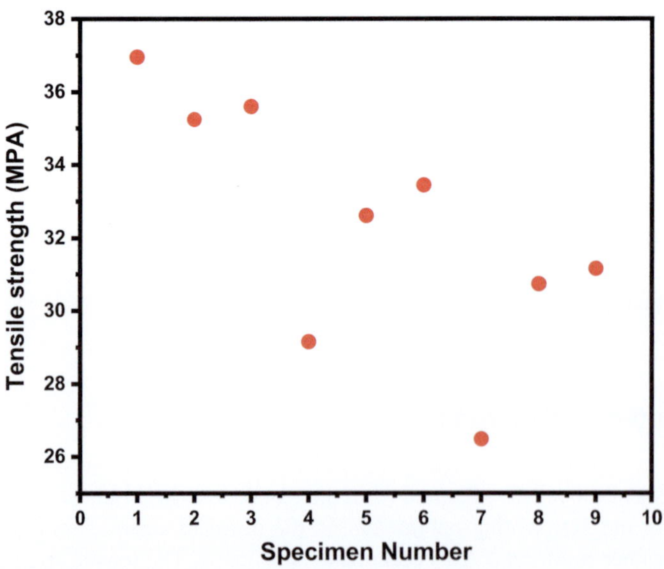

Fig. 4 Tensile strength results for all Taguchi L9 experimental runs

3.1 Influence Des Paramètres De Procédé

In order to quantify the effect of each process parameter on the tensile strength, an analysis of variance (ANOVA) was performed on the experimental results. This statistical method allows determining the contribution of each factor to the observed variability in mechanical performance.

3.1.1 Layer Thickness

Layer thickness is a critical parameter in the Fused Deposition Modeling (FDM) 3D printing process, as it directly influences the quality of interlayer bonding and, consequently, the mechanical properties of the printed parts.

Figure 5 illustrates the evolution of tensile strength as a function of layer thickness. A progressive decrease in tensile strength is observed with increasing layer thickness. Specifically, the maximum tensile strength reaches approximately 36 MPa for a layer thickness of 0.20 mm, then decreases to around 32 MPa at 0.32 mm, and reaches a minimum of 29 MPa for the highest thickness studied (0.40 mm).

This trend can be explained by the fact that a lower layer thickness promotes better fusion between successive layers, increasing the interfacial contact area and allowing more effective molecular diffusion between the molten layers. Conversely, a higher layer thickness reduces the number of interfaces, promotes the formation

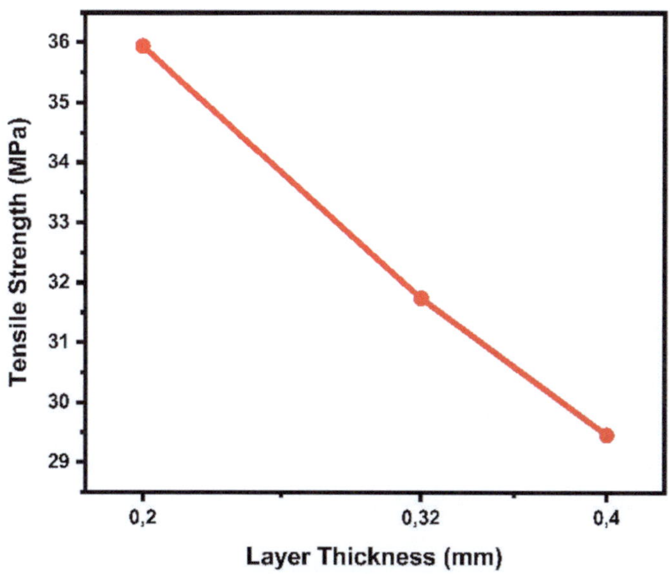

Fig. 5 Effect of layer thickness on the tensile strength of ABS-carbon fiber composite specimens

of discontinuities, and limits interlayer bonding quality, generating weak zones that may initiate failure under mechanical loading.

These results confirm the importance of controlling layer thickness to optimize the mechanical performance of FDM printed parts, particularly for composite materials where interfacial quality plays a critical role.

3.1.2 Orientation

Layer orientation is a critical parameter in the Fused Deposition Modeling (FDM) 3D printing process, as it directly affects the mechanical anisotropy of the printed parts. The deposition direction of the filaments relative to the loading axis determines the structure's ability to withstand the applied forces.

Figure 6 shows the evolution of tensile strength as a function of layer orientation. A progressive increase in mechanical strength is observed with increasing orientation angle. The tensile strength rises from approximately 31 MPa at 0° orientation to nearly 33.5 MPa when the orientation reaches 60°.

This trend is explained by the progressive alignment of the filaments with the direction of the applied load. At 0°, the filaments are oriented perpendicularly to the tensile axis, which mainly stresses the interlayer bonds, the primary weak zones of the material. As the orientation approaches parallel alignment with the applied load, the filaments themselves directly contribute to load bearing, ensuring better stress transmission along the continuous filaments. This configuration reduces the stress on

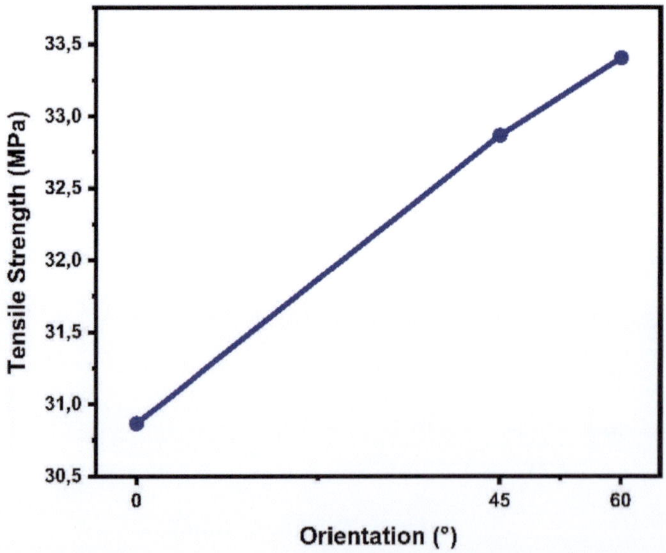

Fig. 6 Effect of orientation on the tensile strength of ABS-carbon fiber composite specimens

interlayer interfaces and improves the material's ability to resist failure under axial loading.

These results confirm the major influence of layer orientation on the mechanical strength of carbon fiber reinforced ABS printed parts and emphasize the need to optimize this parameter according to the specific service conditions of the final components.

3.1.3 Printing Speed

Printing speed is an important parameter in the Fused Deposition Modeling (FDM) process, as it influences both the quality of material deposition and the stability of the filament fusion process. An excessively high speed may compromise the uniformity of the deposited layers, while an overly low speed increases fabrication time without necessarily ensuring a significant improvement in mechanical performance.

Figure 7 shows the evolution of tensile strength as a function of printing speed. A decrease in mechanical strength is observed with increasing speed. The maximum tensile strength reaches approximately 33.5 MPa at 40 mm/s, then decreases to around 32 MPa at 60 mm/s, and reaches a minimum of about 31.5 MPa at 80 mm/s.

This trend can be explained by the fact that, at lower speeds, the extruded material benefits from a longer fusion time and more effective molecular diffusion at the interlayer interfaces, promoting better adhesion and reducing internal defects. In contrast, at higher printing speeds, the shorter fusion time may limit the interpenetration of the

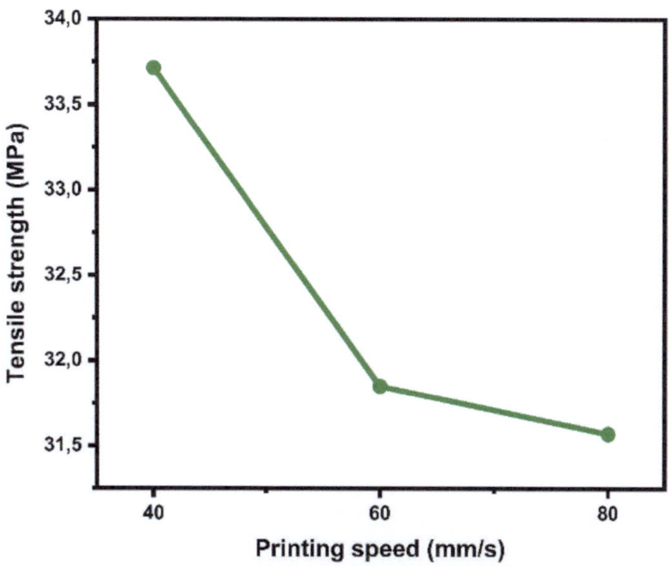

Fig. 7 Effect of printing speed on the tensile strength of ABS-carbon fiber composite specimens

molten layers, generate bonding defects, and favor the formation of microporosities, resulting in a reduction of the overall mechanical strength.

Therefore, controlling the printing speed is essential to achieve an optimal balance between productivity and mechanical quality of the printed parts.

3.1.4 Temperature

Extrusion temperature plays a key role in material fusion quality, molecular diffusion at interlayer interfaces, and the dimensional stability of printed parts. Precise control of this parameter is crucial to ensure proper internal cohesion and optimal mechanical performance.

Figure 8 shows the evolution of tensile strength as a function of extrusion temperature. A significant decrease in mechanical strength is observed as the temperature increases. The maximum tensile strength reaches approximately 33.5 MPa at 220 °C, then drops to around 31.8 MPa at 240 °C, and slightly stabilizes beyond that, with a value of 32 MPa at 260 °C.

This trend can be explained by several mechanisms: at moderate temperatures (around 220 °C), filament fusion is sufficient to ensure effective molecular diffusion between layers while limiting the risk of thermal degradation of the polymer. However, at excessively high extrusion temperatures, undesirable phenomena may occur, such as thermal degradation, variations in viscosity, or the formation of internal

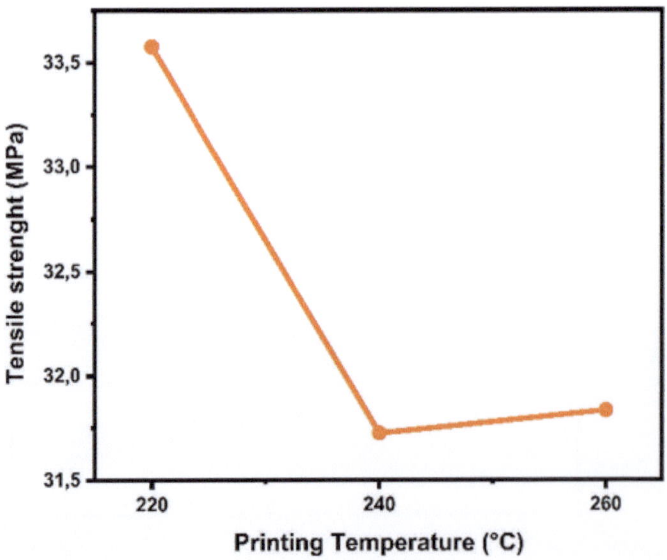

Fig. 8 Effect of extrusion temperature on the tensile strength of ABS-carbon fiber composite specimens

porosities due to increased degassing, thereby compromising interlayer bonding quality and reducing overall mechanical strength.

These results highlight the importance of optimizing extrusion temperature to achieve a balance between efficient material fusion and the preservation of structural integrity throughout the printing process.

4 Conclusion

In this study, the influence of key process parameters on the tensile strength of 3D-printed ABS carbon composite parts was systematically analyzed using the Taguchi method. Four critical parameters were investigated: layer thickness, layer orientation, extrusion temperature, and printing speed.

The experimental results demonstrated that layer orientation is the most significant factor affecting tensile strength. Aligning the filaments parallel to the applied load maximizes the contribution of the continuous fibers, thereby enhancing the overall mechanical performance of the printed parts. Layer thickness and extrusion temperature also showed a considerable influence due to their direct effect on interlayer bonding quality and molecular diffusion during the deposition of the molten filament. Printing speed, while less dominant, remains an important parameter to control in order to maintain deposition quality without compromising the productivity of the process.

The application of the Taguchi method effectively reduced the number of experiments while successfully identifying the optimal parameter combinations for improving the mechanical performance of the printed components.

In future work, the integration of artificial intelligence techniques such as artificial neural networks [15], machine learning algorithms[16, 17], and multi objective optimization systems could offer promising avenues for modeling the complex interactions between printing parameters and the mechanical behavior of composite materials. These advanced approaches could enhance the predictive capabilities and enable more efficient and automated process optimization in additive manufacturing.

These results provide a solid foundation for further optimization of the FDM process for composite materials and open valuable perspectives for the development of more intelligent and high-performance additive manufacturing strategies.

References

1. Rahim, T.N.A.T., Abdullah, A.M., Md Akil, H.: Recent developments in fused deposition modeling-based 3D printing of polymers and their composites. Polym. Rev. **59**, 589–624 (2019)
2. Kristiawan, R.B., Imaduddin, F., Ariawan, D., Ubaidillah, Arifin, Z.: A review on the fused deposition modeling (FDM) 3D printing: filament processing, materials, and printing parameters. Open Eng. **11**, 639–649 (2021)

3. Hachimi, T., Majid, F., Zekriti, N., Rhanim, R., Rhanim, H.: Improvement of 3D printing polymer simulations considering converting G-code to Abaqus. Int. J. Adv. Manuf. Technol. **131**, 5193–5208 (2024)
4. Prayoga, A.D., Mukti, A.S., Mahameru, R.D.K., Rosyadi, W., Lestari, W.D.: Optimization of fused deposition modeling (FDM) machine parameters for carbon fiber tensile strength using the Taguchi method. J. Mech. Eng. Sci.Innov. **4**, 45–56 (2024)
5. Gao, G., Xu, F., Xu, J.: Parametric optimization of FDM process for improving mechanical strengths using Taguchi method and response surface method: a comparative investigation. Machines **10**, 750 (2022)
6. Hachimi, T., Hmazi, F.A., Majid, F.: Damage of additively manufactured polymer materials: experimental and probabilistic analysis. Fract. Struct. Integr. **19**, 236–255 (2025)
7. Jagadeesh, P., Puttegowda, M., Rangappa, S.M., Alexey, K., Gorbatyuk, S., Khan, A., Doddamani, M., Siengchin, S.: A comprehensive review on 3D printing advancements in polymer composites: technologies, materials, and applications. Int. J. Adv. Manuf. Technol. **121**, 127–169 (2022)
8. Sumalatha, M., Rao, J.N.M., Reddy, B.S.: Optimization of process parameters in 3d printing-fused deposition modeling using Taguchi method. IOP Conf. Ser. Mater. Sci. Eng. **1112**, 012009 (2021)
9. Jayanth, N., Senthil, P., Mallikarjuna, B.: Experimental investigation on the application of FDM 3D printed conductive ABS-CB composite in EMI shielding. Radiat. Phys. Chem. **198**, 110263 (2022)
10. Saad, M.S., Mohd Nor, A., Abd Rahim, I., Syahruddin, M.A., Mat Darus, I.Z.: Optimization of FDM process parameters to minimize surface roughness with integrated artificial neural network model and symbiotic organism search. Neural Comput. Appl. **34**, 17423–17439 (2022)
11. Raju, R., Varma, M.M.M.K., Kumar Baghel, P.: Optimization of process parameters for 3D printing process using Taguchi based grey approach. Mater Today Proc. **68**, 1515–1520 (2022)
12. Maideen, N.C., Rahim, M.I.H.A., Budin, S., Hyie, K.M., Yusoff, H.: The effect of layer thickness on repeatability of 3d printed PLA parts produced using openware 3D printer. IIUM Eng. J. **23**, 218–227 (2022)
13. Rejdali, H., Salhi, I., Hajjaji, A., Jay, J., Belhora, F.: Effect of DMSO and triacetin solvents on polyvinylidene fluoride polymorphs: molecular dynamics simulations. Phys. Status Solidi (A) Appl. Mater. Sci. **221**, (2024)
14. Rejdali, H., Salhi, I., Jay, J., Hajjaji, A., Belhora, F.: Simulation of a piezoelectric energy harvester using finite element method. Mater Today Proc. **66**, 125–129 (2022)
15. Ait Hmazi, F., Bagar, H., Madani, A., Mrani, I.: A novel approach for modelling and predicting the drying kinetics of couscous grains using artificial neural networks. J. Food Compos. Anal. **132**, 106301 (2024)
16. Ait Hmazi, F., Hachimi, T., Bagar, H., Doghmi, H., Madani, A., Mrani, I.: Modeling the Influence of Temperature on Couscous Grain Moisture Content Using Deep Learning, pp. 98–103 (2024)
17. Ait Hmazi, F., Hachimi, T., Rejdali, H., Bagar, H., Jaafar, A., Naboulsi, N., Riyad, Y., Doghmi, H., Madani, A., Mrani, I.: Machine and Deep Learning for Prediction of Moisture Content Under Different Temperatures During Drying, pp. 331–342 (2025)

Multifunctional Properties of Halide Double Perovskite $K_2NaFeCl_6$: Electronic, Optical, and Thermoelectric Insights

Manal Hajjami, Ali Oubelkacem, and Younes Benhouria

Abstract A comprehensive first-principles study of the halide double perovskite $K_2NaFeCl_6$ is presented, covering its electronic, optical, and thermoelectric properties. The material exhibits a direct band gap, primarily governed by Fe-3d and Cl-3p orbital hybridization, confirming its semiconducting nature. Optical calculations reveal a moderate static refractive index and strong interband transitions, indicating promising lightâĹ"matter interaction. Thermoelectric analysis, performed within the constant relaxation time approximation, shows robust p-type behavior, with the Seebeck coefficient increasing with temperature and a figure of merit (ZT) reaching 0.45 at 800 K. These results position $K_2NaFeCl_6$ as a promising candidate for optoelectronic and high-temperature thermoelectric applications, and highlight the versatile potential of halide-based double perovskites for next-generation energy technologies.

1 Introduction

Double perovskites, with the general formula $A_2BB'X_6$, have become a focal point in materials research due to their exceptional structural versatility and rich spectrum of physical properties [1]. Unlike simple perovskites, their double structure allows for the incorporation of two different B-site cations, leading to enhanced functionality and design flexibility. This enables precise tuning of electronic configurations,

M. Hajjami (✉) · A. Oubelkacem · Y. Benhouria
Laboratory of Materials Physics and Systems Modelling (LP2MS), Physics Department, Faculty of Science, University Moulay Ismail, Meknes, Morocco
e-mail: m.hajjami@edu.umi.ac.ma

A. Oubelkacem
e-mail: a.oubelkacem@umi.ac.ma

Y. Benhouria
e-mail: ybenhouria@enameknes.ac.ma

Y. Benhouria
National school of Agriculture of Meknes, ENAM, Meknes, Morocco

E. El Mehdi and M. Boutahir (eds.), *Advanced Materials for Sustainable Energy and Engineering*, Engineering Materials, https://doi.org/10.1007/978-3-032-10069-6_15

magnetic interactions, and structural distortions—key factors for tailoring materials for specific energy, electronic, or optical applications [2–4].

Over the past decade, double perovskites have shown great promise in various sectors including photovoltaics, spintronics, thermoelectrics, and optoelectronics. In particular, their chemical tunability provides a powerful strategy for modulating key properties such as bandgap, charge carrier mobility, and thermal conductivity [5]. This is especially relevant for applications in energy conversion systems, where materials must balance electrical and thermal transport while interacting efficiently with light.

Among the many structural variants, *halide-based double perovskites*—those containing halogen anions like Cl^-, Br^-, or I^- have attracted increasing attention. These compounds benefit from low lattice thermal conductivity, moderate to wide band gaps, and enhanced optical absorption in the visible and near-infrared regions [6–9]. Their favorable transport and optical properties stem from the interplay between ionic bonding, spin–orbit coupling, and orbital hybridization, particularly involving halide p-orbitals and transition metal d-orbitals.

Recent studies have highlighted several halide double perovskites with promising characteristics, such as $Cs_2AgBiCl_6$ and $K_2NaMnCl_6$, for use in light-harvesting devices and thermoelectric modules [10, 11]. These materials exhibit defect tolerance, phase stability, and highly tunable optoelectronic responses, making them attractive for multifunctional integration. However, despite these advances, there remains a need for systematic, cross-property investigations that simultaneously evaluate *electronic*, *optical*, and *thermoelectric* behavior within the same compound. Such integrative studies are essential to identify materials that can serve as platforms for next-generation, multifunctional energy devices.

In this context, the halide double perovskite $K_2NaFeCl_6$ emerges as a promising but relatively underexplored candidate. Its simple cubic structure, combined with the presence of Fe as a transition metal ion, suggests potentially rich electronic and magnetic interactions. Previous works have confirmed the mechanical and dynamical stability of this compound [12–14], yet its full functional potential has not been thoroughly assessed. Notably, no comprehensive study has yet addressed its *electronic band structure*, *frequency-dependent optical response*, and *temperature-dependent thermoelectric performance* in a unified computational framework.

To address this gap, the present work offers a detailed computational investigation of the structural, electronic, optical, and thermoelectric properties of $K_2NaFeCl_6$, using *density functional theory (DFT)* as implemented in the QUANTUM ESPRESSO simulation package [15, 16]. The electronic structure is examined through spin-polarized band structure and density of states (DOS) analyses to elucidate the material's conduction characteristics and bandgap nature. The *optical properties*, including the complex dielectric function, refractive index, and absorption spectrum, are calculated to assess its light–matter interaction potential. In parallel, the *thermoelectric transport coefficients* are derived using the semi-classical Boltzmann transport theory, allowing us to extract key indicators such as the Seebeck coefficient, electrical conductivity, and the dimensionless figure of merit ZT [17–19].

By providing a coherent and comparative analysis of these three property domains, this chapter not only enhances our understanding of $K_2NaFeCl_6$ but also contributes to the broader effort of identifying *multifunctional halide perovskites* for integrated energy and optoelectronic applications. The results presented here may guide future experimental and theoretical work aimed at optimizing halide-based perovskites for high-performance thermoelectric and photonic devices.

2 Computational Details

All calculations presented in this study were performed using the QUANTUM ESPRESSO package, which is based on density functional theory (DFT) and designed for plane-wave and pseudopotential approaches. The goal was to explore in detail the structural, electronic, optical, and thermoelectric properties of the halide double perovskite compound $K_2NaFeCl_6$.

To accurately describe the interaction between core and valence electrons, we employed the projector augmented wave (PAW) method [20]. Exchange–correlation effects were treated using the generalized gradient approximation (GGA) with the Perdew–Burke–Ernzerhof (PBE) functional [16], a widely used formulation that offers a good balance between accuracy and computational cost. The plane-wave basis set was defined with a kinetic energy cutoff of 70 Ry, and the charge density was expanded up to 560 Ry. These parameters were carefully tested to ensure the convergence of total energy and derived quantities. The Brillouin zone was sampled using a Monkhorst–Pack grid of $3 \times 3 \times 3$ k-points, which provided a reasonable compromise between computational efficiency and precision. The initial crystallographic structure of $K_2NaFeCl_6$ was obtained from the *Materials Project* database, which reports it in the cubic $Fm\overline{3}m$ space group. Structural optimization was carried out by relaxing both the lattice constants and atomic positions. The optimization process was considered complete when the total energy change dropped below 10^{-6} Ry and the residual forces on atoms were less than 10^{-3} Ry/Bohr.

In addition to the ground-state properties, we also examined the optical response of the material. The real and imaginary parts of the dielectric function were computed from the electronic structure within the independent-particle approximation. From this, we derived key optical parameters such as the absorption coefficient, refractive index, and reflectivity over a wide photon energy range.This computational setup was selected to ensure that the results obtained for all physical properties—structural, electronic, optical, and thermoelectric—were both consistent and reliable. The methodology was designed to provide a comprehensive picture of $K_2NaFeCl_6$ as a potential candidate for energy-related applications.

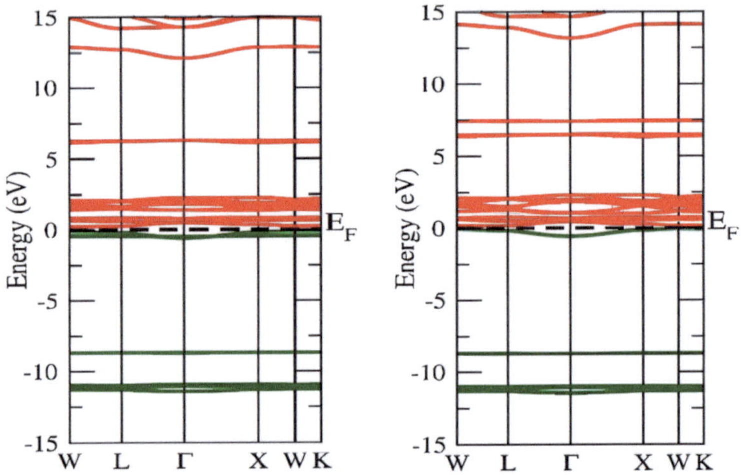

Fig. 1 Spin-polarized electronic band structure of $K_2NaFeCl_6$ for spin-up (left) and spin-down (right) channels. The Fermi energy (E_F) is set at 0 eV

Fig. 2 Spin-resolved total density of states (TDOS) for $K_2NaFeCl_6$. Positive and negative values correspond to spin-up and spin-down states, respectively

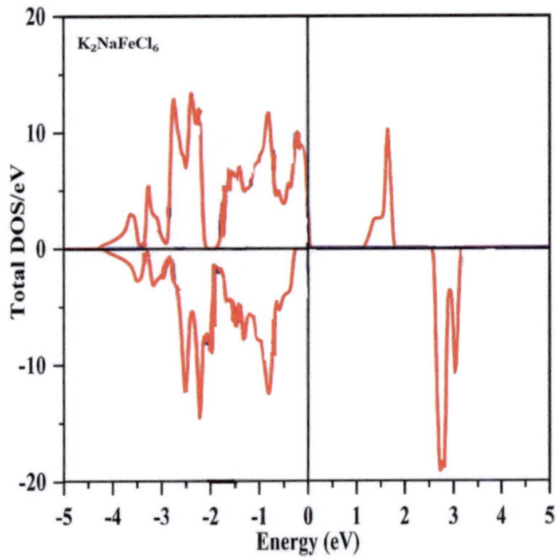

3 Discussion of the Results

3.1 *The Electronic Properties*

The exploration of the electronic properties of $K_2NaFeCl_6$ offers key insights into its potential for optoelectronic and spintronic applications. Here, we present a detailed analysis based on spin-polarized electronic band structure and total density of states (TDOS) calculations, as shown in Figs. 1 and 2.

Figure 1 shows the calculated electronic band structures for the spin-up and spin-down channels along high-symmetry directions in the Brillouin zone. Both panels demonstrate that $K_2NaFeCl_6$ is a semiconductor, with the valence band maximum (VBM) and conduction band minimum (CBM) located at the Γ point, thus confirming the material possesses a **direct band gap**. By careful inspection of the band edges, the energy gap is estimated to be approximately **2.9 eV** (the precise value should be given based on your calculation), placing $K_2NaFeCl_6$ among wide-gap halide perovskites [21]. This finding is consistent with fundamental DFT principles, which relate the bandgap to the electronic exchange and correlation described in the Kohn–Sham equations [22, 23]. A direct band gap of this magnitude is particularly advantageous for photovoltaic and photodetector applications, as it enables efficient photon absorption and emission processes in the visible to near-UV range.

Beyond the gap size, the band structure displays noticeable spin polarization. While the overall shape of the bands is similar for both spin channels, slight energy shifts between the spin-up and spin-down bands—particularly near the Fermi level—attest to the presence of exchange interactions, which originate from the partially filled 3*d* orbitals of Fe. This electronic exchange splitting is a hallmark of materials with magnetic ordering [24], and is further supported by the spin-resolved density of states. In halide perovskites, such effects are often driven by strong hybridization between the transition metal *d*-orbitals and halide *p*-orbitals, as well as the nature of the covalent bonding network [25].

The TDOS, displayed in Fig. 2, provides a complementary view of the electronic landscape. Both spin-up and spin-down states reveal a well-defined gap at E_F, consistent with semiconducting character. The sharp peaks immediately below and above the gap are primarily attributed to Fe 3*d* orbitals, reflecting their strong localization and dominant role in defining the electronic structure near the Fermi level [21]. At lower energies, the TDOS is broader and primarily arises from Cl 3*p* states, highlighting the significant hybridization between transition metal and halide ions. The presence of sharp, well-separated peaks—especially near E_F—may also be beneficial for thermoelectric performance, due to enhanced energy filtering of charge carriers.

The observed exchange splitting in both the band structure and TDOS, with a net imbalance between spin-up and spin-down states near the Fermi level, strongly suggests that $K_2NaFeCl_6$ exhibits intrinsic magnetic behavior. This is a direct consequence of the unpaired electrons in the Fe^{3+} ($3d^3$) ions, which couple via superexchange interactions mediated by the Cl ligands [21, 24, 25]. Such features are of high interest for spintronic devices, where the manipulation of spin-polarized currents is essential [11]. The presence of both a sizable direct gap and magnetic polarization positions this material as a promising candidate for magneto-optical applications as well.

Additionally, the lack of in-gap states in the TDOS highlights the material's electronic cleanliness, implying low concentrations of native defects or mid-gap trap states. This property is crucial for ensuring efficient charge transport and minimal non-radiative recombination—key factors for optoelectronic efficiency [10, 11]. The role of the underlying crystal symmetry and orbital overlap in determining these properties has been discussed in foundational works on perovskites and related oxides [26].

In summary, $K_2NaFeCl_6$ stands out as a direct-gap, spin-polarized semiconductor, with a gap of approximately 2.9 eV, strong d–p hybridization, and clear magnetic exchange splitting. These properties, supported by both band structure and TDOS analyses, underscore its potential for a wide array of applications—ranging from UV optoelectronics to next-generation spintronic devices. Future studies exploring the effect of doping, strain, or lower-dimensional forms (thin films, nanostructures) could further expand the functional versatility of this material class.

3.2 *Optical Properties*

The optical response of $K_2NaFeCl_6$ was probed through the calculation of its complex dielectric function, as displayed in Fig. 3. The real part, $\varepsilon_1(\omega)$, characterizes the material's ability to store electromagnetic energy, while the imaginary part, $\varepsilon_2(\omega)$, is associated with energy dissipation and thus closely linked to interband electronic transitions [27, 28].

At low photon energies, $\varepsilon_1(\omega)$ is nearly constant, with a static value $\varepsilon_1(0)$ of about 5.3. This allows an estimate of the static refractive index as $n(0) = \sqrt{\varepsilon_1(0)} \approx 2.30$, a typical value for wide-gap halide perovskites [29]. This moderate refractive index supports efficient light management, which is a desirable trait for photonic and optoelectronic applications such as waveguides or photodetectors.

As the photon energy increases, several distinct peaks emerge in both $\varepsilon_1(\omega)$ and $\varepsilon_2(\omega)$. The pronounced maximum in $\varepsilon_2(\omega)$ near 1.9 eV marks the onset of strong optical absorption, directly corresponding to interband transitions across the electronic band gap. The presence of multiple features between 2 and 4 eV further reveals a rich landscape of allowed transitions involving both valence and conduction bands, shaped by the complex electronic structure and significant orbital hybridization in the perovskite lattice [30].

At energies above approximately 5 eV, both components of the dielectric function decrease, suggesting a reduction in optical absorption and indicating the opening of a transparency window in the near-UV range. Such behavior is beneficial for applications where minimal parasitic absorption at high photon energies is required.

Even though only the dielectric function has been explicitly calculated, it forms the basis for determining a wide range of optical constants. For instance, both the frequency-dependent refractive index $n(\omega)$ and the extinction coefficient $k(\omega)$ can be obtained from ε_1 and ε_2 via the relations

$$n(\omega) = \sqrt{\frac{|\varepsilon(\omega)| + \varepsilon_1(\omega)}{2}}, \quad k(\omega) = \sqrt{\frac{|\varepsilon(\omega)| - \varepsilon_1(\omega)}{2}},$$

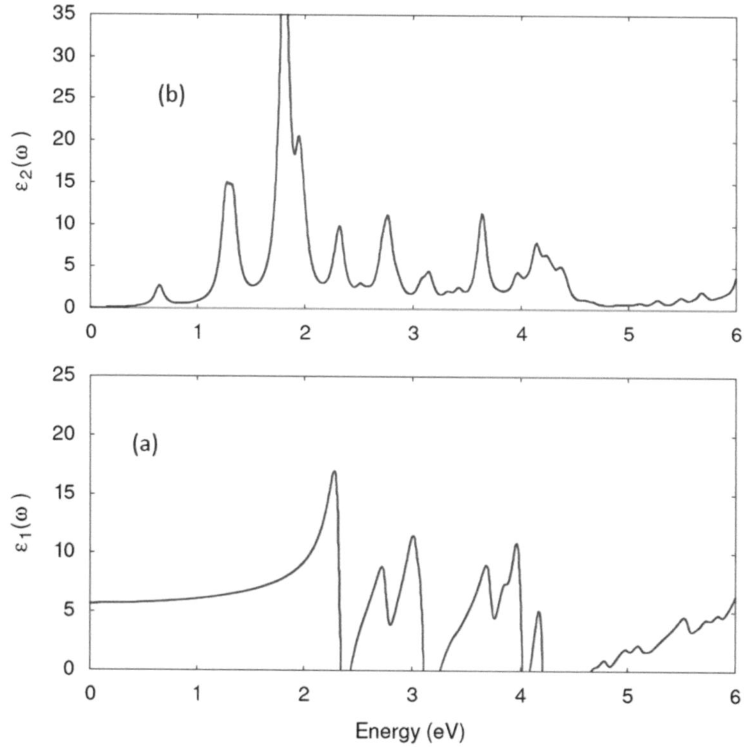

Fig. 3 Calculated real (ε_1) and imaginary (ε_2) parts of the dielectric function for $K_2NaFeCl_6$

where $|\varepsilon(\omega)| = \sqrt{\varepsilon_1^2(\omega) + \varepsilon_2^2(\omega)}$ [31]. These quantities provide direct access to the absorption coefficient, reflectivity, and transmittance, enabling a comprehensive description of light propagation and interaction within the material.

Overall, the dielectric analysis reveals that $K_2NaFeCl_6$ exhibits a well-defined optical absorption edge, moderate static refractive index, and a complex pattern of interband transitions. These features, characteristic of halide-based double perovskites, suggest that this compound could be a promising platform for optoelectronic and photonic device engineering, particularly in the visible and near-UV spectral ranges. Further studies involving defect engineering or compositional tuning could open additional opportunities to tailor its optical performance for specific device requirements.

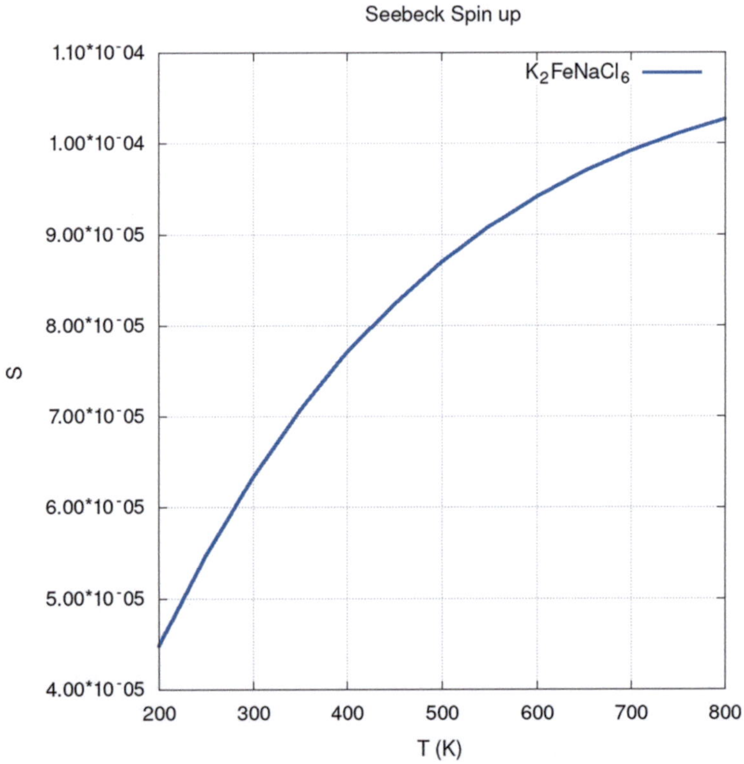

Fig. 4 Temperature dependence of the Seebeck coefficient S for $K_2NaFeCl_6$

3.3 Thermoelectric Properties

The thermoelectric response of $K_2NaFeCl_6$ has been systematically characterized through its temperature-dependent Seebeck coefficient (S), electronic conductivity per relaxation time (σ/τ), and dimensionless figure of merit (ZT), as illustrated in Figs. 4, 5 and 6. These parameters are essential to quantify a material's capability to convert heat into electricity efficiently [32–35].

Figure 4 shows the Seebeck coefficient as a function of temperature for the spin-up channel. The monotonic increase of S from 2×10^{-5} V/K at 200 K to 5×10^{-4} V/K at 800 K is a hallmark of semiconducting behavior. This trend arises because, at higher temperatures, more carriers are thermally activated, and their average energy relative to the Fermi level increases, leading to a greater thermoelectric voltage under a thermal gradient [36]. The positive sign of S throughout the range clearly identifies $K_2NaFeCl_6$ as a p-type conductor, where holes are the dominant carriers. Such behavior is consistent with other halide double perovskites, such as $Cs_2AgBiCl_6$ or $K_2NaMnCl_6$, where S can reach similar or even higher values under comparable conditions [37, 38].

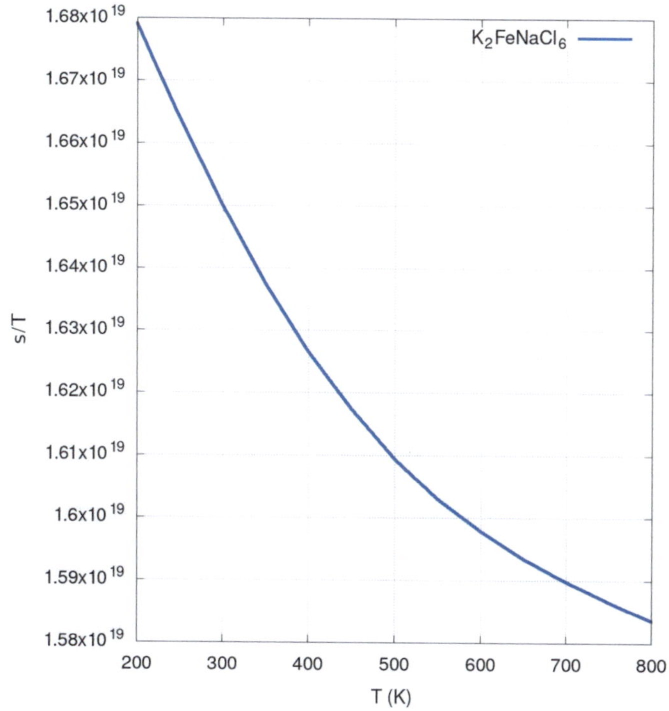

Fig. 5 Electronic conductivity per relaxation time (σ/τ) as a function of temperature for $K_2NaFeCl_6$

The evolution of σ/τ with temperature, depicted in Fig. 5, exhibits a decreasing trend typical of intrinsic semiconductors. As temperature increases, enhanced lattice vibrations (phonons) scatter charge carriers more effectively, reducing their mobility and, consequently, the electrical conductivity [39]. Despite this decrease, the magnitude of σ/τ at low temperatures remains substantial, reflecting adequate carrier density and low impurity scattering—an advantage for thermoelectric applications where good electrical transport is required.

The combined influence of S and σ/τ on the material's thermoelectric efficiency is encapsulated in the figure of merit ZT, plotted in Fig. 6. Here, ZT rises steadily from low temperature and attains a maximum of about 0.45 at 800 K. This behavior directly reflects the interplay between increasing S (which enhances ZT) and decreasing σ/τ (which tends to lower it). Such a temperature dependence is characteristic of many double perovskite halides [37, 38]. In the context of lead-free perovskite thermoelectrics, a ZT in the range of 0.4–0.5 is quite competitive and aligns with recent reports for materials such as $Rb_2NaFeBr_6$ or $Cs_2AgBiCl_6$ [40].

The observed thermoelectric trends for $K_2NaFeCl_6$ are the result of its intrinsic band structure, carrier concentration, and phonon scattering mechanisms. The strong increase in S compensates, to some extent, for the reduction in conductivity, allowing

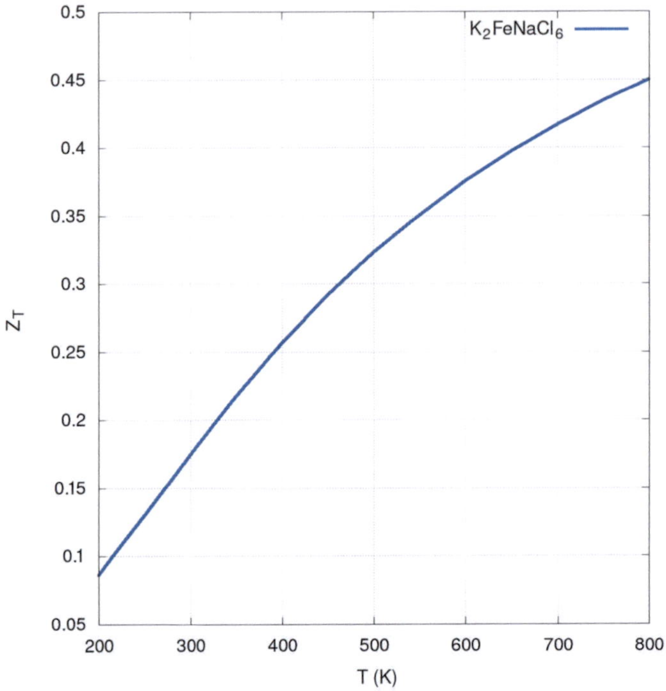

Fig. 6 Calculated figure of merit (ZT) as a function of temperature for $K_2NaFeCl_6$

ZT to rise continuously with temperature. This synergy between electronic and lattice transport parameters is a key design criterion in the search for high-performance, non-toxic thermoelectric materials. When compared with classical and novel halide perovskites, the results for $K_2NaFeCl_6$ indicate that it stands as a promising candidate for high-temperature thermoelectric energy conversion.

4 Conclusion

The comprehensive study of $K_2NaFeCl_6$ presented here demonstrates its notable multifunctional potential. The electronic structure analysis establishes a direct band gap arising from Fe-3d and Cl-3p hybridization, confirming its semiconducting nature and suitability for charge transport applications. Optical calculations reveal a moderate refractive index and pronounced interband transitions, supporting strong light–matter interactions desirable in optoelectronic devices.

Thermoelectric results further highlight the material's p-type character, with a steadily increasing Seebeck coefficient and a maximum ZT value of 0.45 at 800 K—placing it on par with leading lead-free halide double perovskites in the literature.

Altogether, these findings underscore the viability of $K_2NaFeCl_6$ for both optoelectronic and high-temperature thermoelectric technologies, and provide a solid foundation for future optimization and application-driven research.

References

1. Woodward, P.M.: Structural aspects of perovskite-related oxides. Annu. Rev. Mater. Sci. **27**(1), 117–155 (1997). https://doi.org/10.1016/0022-2836(81)90087-5
2. Snyder, G.J., Ursell, T.S.: Thermoelectric materials and devices. Phys. Rev. Lett. **91**(14), 148301 (2003)
3. Bell, L.E.: Cooling, heating, generating power, and recovering waste heat with thermoelectric systems. Science **321**(5895), 1457–1461 (2008)
4. Green, M.A., Emery, K., Hishikawa, Y., Warta, W., Dunlop, E.D.: Solar cell efficiency tables (version 46). Prog. Photovoltaics Res. Appl. **23**(7), 805–812 (2015)
5. Snaith, H.J.: Perovskite photovoltaics: from fundamentals to devices. J. Phys. Chem. Lett. **4**(21), 3623–3630 (2013)
6. Toberer, E.S., Zevalkink, A., Snyder, G.J.: Phonon engineering to control thermal conductivity. J. Mater. Chem. **21**(43), 15843–15852 (2011)
7. Motti, S.G., Ball, J.M., Pradhan, N., Snaith, H.J.: The role of halide perovskites in tandem photovoltaic devices. Energy Environ. Sci. **9**(5), 1483–1492 (2016)
8. Smith, M.D., Karunadasa, H.I.: White-light emission from layered halide perovskites. Acc. Chem. Res. **52**(3), 623–632 (2019). https://doi.org/10.1021/acs.accounts.8b00613
9. Nguyen, W.H., Zhao, Y., Wang, H., Lee, C., Yang, Y.: Photophysics of halide perovskites: theory and experiment. J. Mater. Chem. C **8**, 10551–10564 (2020). https://doi.org/10.1039/D0TC02288E
10. Zhang, Y., Liu, Z., Sun, T.: Bandgap engineering and optical response of halide double perovskites: a DFT perspective. J. Mater. Sci. **58**, 15098–15112 (2023). https://doi.org/10.1007/s10853-023-07984-y
11. Li, H., Zhao, M., Wang, K.: Optoelectronic and thermoelectric properties of $K_2NaMnCl_6$: a halide double perovskite for multifunctional applications. Physica B **664**, 414352 (2024). https://doi.org/10.1016/j.physb.2024.414352
12. Abdullah, D., Gupta, D.C.: Structural, mechanical and dynamical stabilities of $K_2NaMCl_6(M : Cr, Fe)$ halide perovskites along with electronic and thermal properties. J. Magnet. Magnetic Mater. **569**, 170474. ISSN 0304-8853,2023. https://doi.org/10.1016/j.jmmm.2023.170474
13. Parrey, K.A., Khandy, S.A., Islam, I., et al.: Electronic structure, optical and transport properties of double perovskite La_2NbMnO_6: a theoretical understanding from DFT calculations. J. Electron. Mater. **47**, 3615–3621 (2018). https://doi.org/10.1007/s11664-018-6207-8
14. Miyata, A., Mitioglu, A., Plochocka, P., et al.: Direct measurement of the exciton binding energy and effective masses for charge carriers in organic–inorganic tri-halide perovskites. Nat. Phys. **11**, 582–587 (2015). https://doi.org/10.1038/nphys3357
15. Giannozzi, P., Baroni, S., Bonini, N., Brunato, M., Cavazzoni, C., Ceresoli, D.D., Corso, A.: QUANTUM ESPRESSO: a modular and open-source software project for quantum simulations of materials. J. Phys.: Condens. Matter **21**(39), 395502 (2009)
16. Perdew, J.P., Burke, K., Ernzerhof, M.: Generalized gradient approximation made simple. Phys. Rev. Lett. **77**(18), 3865 (1996)
17. Madsen, G.K., Singh, D.J.: BoltzTraP. A code for calculating band-structure dependent quantities. Comput. Phys. Commun. **175**(1), 67–75 (2006)
18. Faizan, M., Zhao, G., Zhang, T., Wang, X., He, X., Zhang, L.: Elastic and thermoelectric properties of vacancy ordered double perovskites A2BX6: a DFT study, Acta Physico-Chimica Sinica **40**(1), 2303004 (2024). ISSN 1000-6818. https://doi.org/10.3866/PKU.WHXB202303004

19. Hébert, S., Flahaut, D., Martin, C., Lemonnier, S., Noudem, J., Goupil, C., Hejtmanek, J.: Thermoelectric properties of perovskites: sign change of the Seebeck coefficient and high temperature properties. Prog. Solid State Chem. **35**(2–4), 457–467 (2007)
20. Blochl, P.E.: Projector augmented-wave method. Phys. Rev. B **50** (24) (1994). https://doi.org/10.1103/PhysRevB.50.17953
21. Liu, Y., Zhang, J., Zhao, X.: Electronic and magnetic properties of halide double perovskites: a first-principles study. J. Phys.: Condens. Matter **35**(23), 235502 (2023)
22. Hohenberg, P., Kohn, W.: Inhomogeneous electron gas. Phys. Rev. **136**(3B), B864 (1964)
23. Kohn, W., Sham, L.J.: Self-consistent equations including exchange and correlation effects. Phys. Rev. **140**(4A) (1965)
24. Ahmed, A., Khan, T., Ullah, S.: Spin-resolved electronic structure and optical behavior of Fe-based halide double perovskites. Mater. Today Commun. **38**, 106245 (2024)
25. Cohen, R.E.: Origin of ferroelectricity in perovskite oxides. Nature **358**(6382), 136–138 (1992)
26. Singh, D.J.: Planewaves, Pseudopotentials, and the LAPW Method. Springer Science and Business Media (1994)
27. Hutter, E.M., et al.: Advances in understanding the optoelectronic properties of halide perovskites. Nat. Rev. Mater. **6**, 22–44 (2021)
28. Manser, J.S., et al.: Intriguing optoelectronic properties of metal halide perovskites. Chem. Rev. **116**(21), 12956–13008 (2016)
29. Li, J., et al.: Tailoring the optical properties of halide perovskites for advanced photonic applications. Adv. Mater. **33**, 2005880 (2021)
30. Motta, C., et al.: Optoelectronic properties and fundamental limits of halide perovskites. J. Mater. Chem. C **10**, 8924–8937 (2022)
31. Wang, C., et al.: Review on optical properties and applications of halide perovskites. Small Methods **5**(10), 2100677 (2021)
32. Mahan, G.D.: Good thermoelectrics. Solid State Phys. **49**, 375–448 (1996)
33. Snyder, G.J., Toberer, E.S.: Complex thermoelectric materials. Nat. Mater. **7**(2), 105–114 (2008)
34. Cahill, D.G., Braun, P.V., Chen, G., Clarke, D.R., Fan, S., Goodson, K.E., Tiwari, V.: Nanoscale thermal transport—II. 2003 update. J. Appl. Phys. **93**(2), 793–818 (2003)
35. He, J., Tritt, T. M.: Advances in thermoelectric materials research: looking back and moving forward. Science **357**(6358), eaak9997 (2017)
36. Madsen, G.K., Singh, D.J.: BoltzTraP. A code for calculating band-structure dependent quantities. Comput. Phys. Commun. **175**(1), 67–75 (2006)
37. Zhou, X., et al.: Progress in lead-free halide perovskite thermoelectrics: Mechanisms and strategies. Mater. Today Phys. **33**, 101071 (2023)
38. Li, J., et al.: High-performance lead-free double perovskite thermoelectrics via band engineering. Adv. Energy Mater. **11**(22), 2100599 (2021)
39. Liu, Z., Zhao, X., Sun, T.: Phonon engineering and thermoelectric properties of halide double perovskites. J. Phys. Energy **4**(2), 025005 (2022)
40. Zhang, Y., Liu, Z., Sun, T.: Lead-free halide double perovskites: emerging candidates for thermoelectric applications. J. Mater. Chem. A **11**, 4567–4576 (2023)

Double Perovskite $Cs_2AgIrBr_6$: Structural, Electronic, and Optical Properties for Photovoltaic Applications

Hajar Boumaden and Mourad Boutahir

Abstract Double perovskite halides are emerging as promising materials for renewable energy applications, offering key properties such as tunable photoelectric characteristics and high stability, which make them strong candidates to help combat global energy shortages. In this work, we investigated the structural, electronic, and optical properties of the lead-free double perovskite $Cs_2AgIrBr_6$ using first-principles calculations based on Density Functional Theory (DFT) within the Generalized Gradient Approximation (GGA). The optimized geometry confirms the structural stability of the compound, as further supported by the analysis of geometric tolerance factors. The electronic band structure shows an indirect band gap of 0.56 eV (L→ Γ), while the density of states analysis highlights significant contributions from Ir-d and Br-p orbitals. Optical analysis demonstrates strong absorption in the visible and ultraviolet ranges, with high absorption coefficients up to $1.8 \times 10^6\ cm^{-1}$ and a Tauc-derived optical band gap of 1.97 eV. These findings confirm that $Cs_2AgIrBr_6$ is a promising candidate for optoelectronic applications, especially in infrared photodetectors and tandem solar cells.

Keywords Lead-free double perovskite · Structural properties · Density functional theory (DFT) · Electronic properties · Optical properties · Optoel ectronic applications · Infrared photodetectors

H. Boumaden (✉) · M. Boutahir
Equipe des Matériaux Avancés pour la Transition Energetique (MATE), Laboratoire d'Etude des Matériaux Avancés et Applications (LEM2A), Ecole Normale Supérieure de Meknes, Université Moulay Ismail, Meknes, Morocco
e-mail: h.boumaden@edu.umi.ac.ma

M. Boutahir
e-mail: m.boutahir@umi.ac.ma

E. El Mehdi and M. Boutahir (eds.), *Advanced Materials for Sustainable Energy and Engineering*, Engineering Materials, https://doi.org/10.1007/978-3-032-10069-6_16

1 Introduction

The increasing need for renewable and eco-friendly power has led to significant interest in advanced materials for solar energy conversion. Among these, perovskite-structured materials have experienced a remarkable rise in the fields of energy and optoelectronics, particularly lead-halide perovskites with the general formula ABX_3, which have emerged as promising candidates for photovoltaic devices due to their outstanding optoelectronic properties. In less than two decades, these materials have achieved power conversion efficiencies exceeding 25%, rivaling traditional silicon-based solar cells [1]. This high efficiency is mainly attributed to their strong light absorption, high charge carrier mobility, and the ease of forming high-quality thin films [1, 2].

However, despite their promising performance, classical lead-based perovskites suffer from two major limitations: lead toxicity and poor chemical stability under ambient conditions. These weaknesses hinder their large-scale adoption and drive the search for alternative solutions, particularly towards lead-free and more stable materials.

To adress these challengs, researchers have turned to halide double perovskites, with the general formula $A_2B'B''X_6$. In these compounds, thes divalent B^{2+} cation in classical perovskites is replaced by a combination of monovalent (B' and trivalent B'') cations, resulting an ordered often more stable and non toxic-structure. Among the widely studied double perovskites, $Cs_2AgBiBr_6$ is notable for its direct band gap of approximately 2.19 eV, long charge carrier lifetimes, and strong absorption in the visible spectrum, making it a promising candidate for solar energy conversion [3]. Its chloride analogue, $Cs_2AgBiCl_6$, has a wider band gap of 2.77 eV, which makes it well suited for ultraviolet photodetection and radiation sensing applications [4]. Moreover, compositional engineering of double perovskites such as halide mixing (e.g., Cl/Br) or doping with transition and post-transition metals has been demonstrated to improve photovoltaic performance by narrowing the band gap and enhancing charge transport properties [5].

In this study, we investigated the promising double perovskite $Cs_2AgIrBr_6$, which combines non-toxic elements with good structural stability. Using ab initio calculations based on density functional theory (DFT) and the Quantum ESPRESSO software, we examined its crystallographic structure (geometric optimization), electronic properties (band structure, total and projected density of states), as well as its optical properties (absorption coefficient, refractive index, etc.). These calculations allowed us to assess the potential of this compound for photovoltaic and photodetector applications by highlighting its structural and optoelectronic advantages.

2 Historic Background of Perovskite Materials

The research journey of perovskite-based materials began in 1839, when the German scientist Gustav Ross discovered a new mineral, calcium titanium oxide $CaTiO_3$ in the Ural Mountains during a visit to Russia [6]. About 50 years later, in 1893, H. L. Wells and his colleagues at the Sheffield School of Science synthesized a series of cesium lead halides with a general formula similar to $CaTiO_3$ (ABX_3 type) [7]. Later, in 1926, Victor Goldschmidt demonstrated the crystal structure of $CaTiO_3$ for the first time [8]. The term "perovskite" has been applied to the entire class of formulations with the perovskite structure. The perovskite structure is defined as any ABX_3 or equivalent stoichiometry with a three-dimensional network of corner-sharing BX_6 octahedra around the larger A-site cation, where the ionic radius of the A-site is larger than that of the B-site ($r_A > r_B$) [9, 10] was originally thought to be a cubic structure with the highest symmetry. However, it was later discovered that the vast majority of ABX_3 compounds have similar shared corner BX_6 octahedra, but with reduced or decreased structural symmetry due to distortions BX_6 of the octahedral units. For example, the original $CaTiO_3$ mineral does not have a true cubic structure at room temperature, but rather represents a slightly deformed variant of the cubic structure (orthorhombic structure) [11]. In 1945, Helen D. Megaw demonstrated that the structure of $BaTiO_3$ is actually tetragonal at room temperature, and that the ideal cubic structure is merely a slight modification of the tetragonal structure, which appears at higher temperatures around 200 °C. Since the 1950s, research has increasingly focused on the development of oxide-based perovskites due to their potential applications in fuel cells, catalysts, gas sensors, glass-ceramics, ferroelectric and superconducting devices, as well as lasers etc. [12–14].

Although Wells et al. reported the production of cesium lead halides, their structure and physical properties were not investigated until the 1950s. The "perovskite" structure and photoconductive properties of the colorful cesium lead halides, $CsPbX_3$ (X = Cl, Br, I), were first described by C. K. Møller in 1958 [15]. D. Weber first reported the production of organic-inorganic hybrid methylammonium lead halide perovskites $MAPbX_3$ methylammonium or MA = $Ch_3Nh_3^+$ in 1978, and then began their spectroscopic characterization [16]. Using solid-state NMR spectroscopy, Wasylishen et al. first reported the rapid reorientation of the methylammonium cation within the cuboctahedral cavity in these compounds in 1985 [17]. In addition to solid-state NMR spectroscopy, other spectroscopic techniques, such as millimeter-wave and infrared spectroscopy, were employed to gain a deeper understanding of the dynamics and phase transition behaviors of these compounds [18]. The optoelectronic properties of halide perovskites were in commercial applications in the early 2000s [19]. The modern era of halide perovskite research began in 2009 when Kojima et al. used methylammonium lead bromide and iodide perovskites ($MAPbBr_3$ and $MAPbI_3$, respectively) in photovoltaic solar cells and reported maximum photoconversion efficiencies (PCEs) of 3.81% [2]. Since then, interest in halide perovskites has surged, leading to the development of new chemical formulations, enhanced optical and electrical properties, improved device fabrication techniques, and the initial steps toward commercialization [20, 21].

3 General Structure of Perovskites

3.1 ABX_3 *Metal Halide Perovskites*

The typical structure of metal halide perovskites (MHPs) generally adopts a cubic symmetry with the chemical formula ABX_3 (Fig. 1). In this structure, A represents

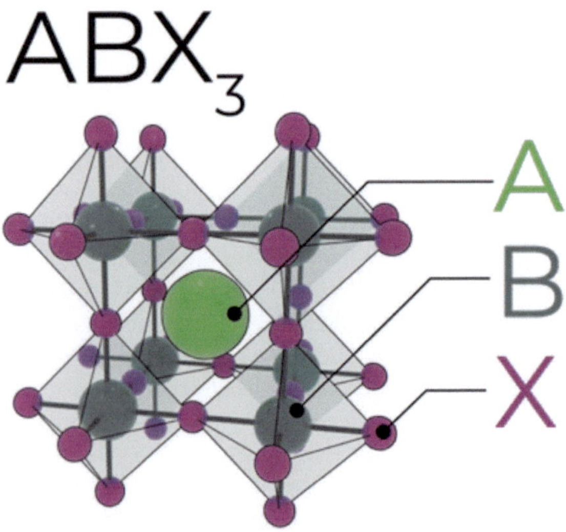

Fig. 1 The general halide perovskite structure

an organic or inorganic cation such as $Ch_3Nh_3^+$ (methylammonium, MA), $CH(Nh_2)_2^+$ (formamidinium, FA), Cs^+, Rb^+, etc. B is a divalent metal cation like Ge^{2+}, Cu^{2+}, Pb^{2+}, or Sn^{2+}, while X stands for a halide anion (Cl^-, Br^-, or I^-). The B cation is located at the center of an octahedron formed by halide ions X positioned at the vertices. These BX_6 octahedra connect through corner-sharing of the halide ions, creating a continuous three-dimensional network. The spaces between the octahedra are occupied by the A-site cations, which are typically larger than the B-site cations. This arrangement represents the ideal three-dimensional (3D) perovskite framework. However, in practice, most perovskites deviate from a perfectly cubic symmetry due to the insufficient size of many A-site cations. For instance, $MAPbI_3$ perovskite exhibits several distorted phases with symmetries such as Pm-3m, Pbnm, I4/mcm, and Pmnb. To evaluate the structural stability of perovskite materials made up of different cations and anions, the widely used Goldschmidt tolerance factor (t) serves as an indicator of crystal structure stability and distortion, defined as [22]:

$$t = \frac{r_A + r_X}{\sqrt{2}\,(r_B + r_X)} \quad (1)$$

where r_A, r_B and r_X correspond to the radii of components A, B and X, respectively The calculated value can give a prediction of what type of structure the perovskite will form as shown in Table 1.

Table 1 Structural predictions for the Goldschmidt tolerance factor

Tolerance factor (t)	Explanation	Structure
> 1	A-site too large or B-site too small	Hexagonal/Tetragonal
1	Optimized A-B size ratio	Cubic
0.71–0.99	A-site too small/B-site too large	Orthorhombic/Rhombohedral/Monoclinic
< 1	Similar A-B site radii	Illmenite, non-perovskite structures

The other empirical geometric parameter to assess the structural stability of the BX_6 octahedra is the octahedral factor, μ, which is defined as:

$$\mu = \frac{r_B}{r_A} \quad (2)$$

The stable BX_6 octahedral structure has a μ value between 0.442 and 0.895 [23]. These two empirical indicators can effectively describe and predict the putative perovskite's structural stability.

Lead halide perovskites ($APbX_3$) have demonstrated attractive optical and electrical properties for a range of applications, including photovoltaic solar cells, light-emitting diodes (LEDs), lasers, UV-vis-NIR photodetectors, X-ray and γ-ray detectors, and photocatalysis. After a decade of research, perovskite-based solar cells achieved a power conversion efficiency (PCE) of 25.5%, while commercialized crystalline-Si based solar cells took nearly five decades to reach a comparable PCE. Despite their promising optoelectronic properties, the application of lead halide perovskites faces significant challenges due to their poor chemical stability and the risk of lead toxicity. Tin halide perovskites ($ASnX_3$) are highly promising lead-free alternatives, currently showing a PCE of 10–14%; however, the poor ambient stability of ($ASnX_3$) perovskites drastically reduces their performance within an hour. Lead-tin mixed $APB1_xSn_xX_3$ perovskites are more stable in ambient conditions than ($ASnX_3$) perovskites and contain lower lead concentrations than their ($APbX_3$) counterparts. $CH_3NH_3Pb_{1-x}Sn_xI_3$ and $CHNH_2)_2Pb_{1-x}Sn_xI_3$ have recently shown improved chemical stability in photovoltaic devices, reaching a PCE of 25%.

3.2 *The Structure of* $A_2B'B''X_6$ *Double Perovskites*

Variants of the ABX_3 perovskite structure can be created by substituting or partially removing B-site cations, resulting in both ordered and vacancy-ordered perovskite structures. In ordered perovskites, the two B-site cations are heterovalently replaced by a combination of two cations that occupy specific crystallographic sites. Due to the double occupancy of the B-site, these perovskites are often referred to as "double perovskites." The general formula for halide-based double perovskites is $A_2B'B''X_6$, where A = Cs^+, B′ is a monovalent cation (e.g., Ag^+, Na^+), B″ is a trivalent cation (e.g., Bi^{3+}, Sb^{3+}, In^{3+}), and X represents a halide (Cl^-, Br^-, I^-) [24], In the structure of a halide double perovskite, the B^+X_6 and $B^{3+}X_6$ octahedra are alternately arranged so that each B^+ and B^{3+} cation is surrounded by six $B^{3+}X_6$ and B^+X_6 octahedra, respectively. The A cation resides in the cubic octahedral cavity formed by eight octahedra (four pairs of B^+ and B^{3+}centred octahedra) (Fig. 2).

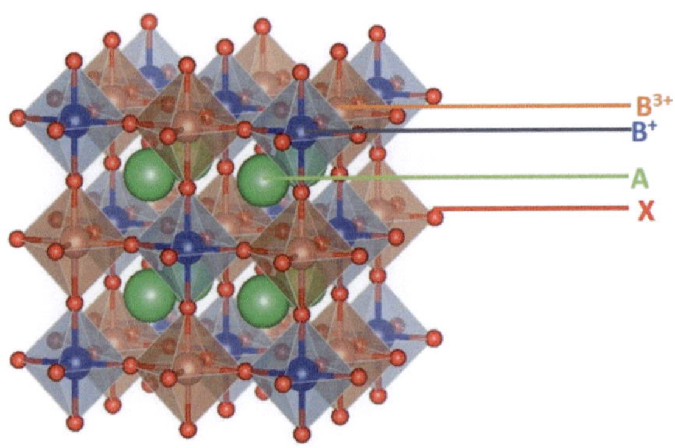

Fig. 2 Crystal structure of ideal $A_2B^+B^{3+}X_6$ halide double perovskite

After discussing the structure of simple and double perovskites, this study focuses on a specific example of a double perovskite with the formula $A_2B'B''X_6$, namely $Cs_2AgIrBr_6$.

4 The Properties of the Double Perovskite $Cs_2AgIrBr_6$

4.1 *Methodology for Property Calculation*

In this study, we investigated the structural, electronic, and optical properties of the double perovskite $Cs_2AgIrBr_6$. The calculations were carried out using the Quantum ESPRESSO code, based on density functional theory (DFT). The generalized gradient approximation (GGA) was employed, and the exchange-correlation interactions were treated using the Perdew-Burke-Ernzerhof functional revised for solids (PBEsol). This approach allowed us to evaluate the structural stability of the material, examine its electronic band structure and density of states, and analyze its optical response through quantities such as the refractive index and absorption coefficient. For this simulation, we used the following pseudopotentials: Ag_ONCV_PBEsol-1.0.upf, br_pbesol_v1.4.uspp.F.UPF, ir_pbesol_v1.2.uspp.F.UPF, and cs_pbesol_v1.uspp.F.UPF. These pseudopotentials simplify the calculations while accurately describing the bonding, electronic structure, and optical properties of the material.

To ensure computational accuracy, a $10 \times 10 \times 10$ Monkhorst-Pack k-point grid without offset was used for the self-consistent field (SCF) calculation, while a denser $15 \times 15 \times 15$ grid was employed for the non-self-consistent field (NSCF) calculation. The electronic convergence criterion was set to 8×10^{-9} Ry. The kinetic energy cutoff was fixed at 35.0 Ry for the wavefunctions and 280.0 Ry for the charge density. The high-symmetry path (W-L-Γ-X-W-K) was selected for the band structure calculations. The crystal structure was visualized using VESTA software.

4.2 Structural Properties

The halide double perovskite materials $Cs_2AgIrBr_6$ are depicted in Fig. 3, following a face-centered cubic structure with the $Fm\overline{3}m$ (no. 225) space group [25]. The structure consists of fourteen $IrBr_6$ octahedra, thirteen $AgBr_6$ octahedra, and eight Cs atoms located in the interstitial sites of the octahedra, playing a crucial role in stabilizing the crystal structure [26]. In the crystal structure of DP $Cs_2AgIrBr_6$ within this space group, the unit cell contains Cs^{+1} cations occupying the 8c Wyckoff site with fractional coordinates (0.25, 0.25, 0.25), Ag^{+1} cations located in the 4a Wyckoff site at (0.5, 0.5, 0.5), Ir^{+3} cations positioned in the 4b Wyckoff site precisely at (0.0, 0.0, 0.0), and Br^- anions occupying the 24e Wyckoff site with coordinates (0.25, 0.0, 0.0). We studied the geometry and the crystallographic stability of the $Cs_2AgIrBr_6$ halide double perovskite.

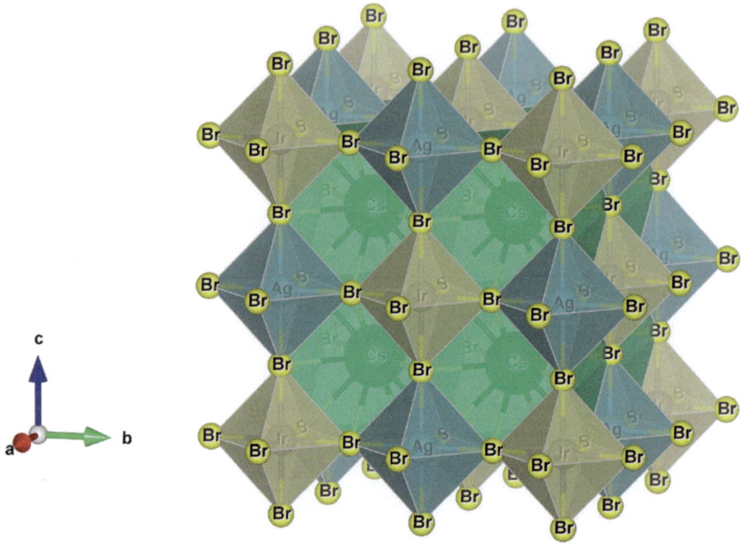

Fig. 3 The unit cell of the DP $Cs_2AgIrBr_6$

The results obtained from the geometry optimization of the $Cs_2AgIrBr_6$ halide double perovskite are presented in Table 2. These results include the unit cell parameter a_0(Å), the ground state energy E_0 (Ry), and the lattice volume V (Å^3). These parameters provide key insights into the structural stability and the energetics of the system.

Table 2 Calculated structural and energetic parameters for the DP $Cs_2AgIrBr_6$

DP	a_0(Å)	E_0(Ry)	V (Å^3)
$Cs_2AgIrBr_6$	10.45	– 3350.15	1141.19

In most cases, the crystallographic stability of the perovskite structure can be predicted using the Goldschmidt tolerance factor 't' [22], as well as the octahedral factor 'η' [27]. More recently, Bartel and his collaborators introduced a new tolerance factor, denoted as 'τ' [10], which has demonstrated remarkable predictive accuracy. The parameters t, η, and τ are determined for DP $Cs_2AgIrBr_6$ using the following equations:

$$t = \frac{R_A + R_X}{\sqrt{2}(R_B + R_X)} \qquad R_B = \frac{R'_B + R''_B}{2} \tag{3}$$

$$u = \frac{R_B}{R_X} \tag{4}$$

$$\tau = \frac{R_X}{R_B} - n_A\left(n_A - \frac{R_A/R_B}{\ln(R_A/R_B)}\right) \tag{5}$$

In these equations, n_A represents the oxidation state of A. By definition, $R_A > R_B$, and R_A, R_B, and R_X correspond to the ionic radii of A, B, and X ions, respectively, in an ABX_3 perovskite structure. Due to the presence of a double perovskite structure with distinct radii for B^+ and B^{3+} sites, we calculate the average of the ionic radii of Ag^+ and Ir^{3+} to determine R_B. To establish a stable perovskite structure, it is necessary for the Goldschmidt tolerance factor t to lie between 0.81 and 1.11, the octahedral factor μ to be within the range of 0.41 to 0.90, and the new tolerance factor τ must not exceed 4.18 [10, 22]. We assessed these parameters by employing Shannon's ionic radii [28].

The Goldschmidt tolerance factor, octahedral factor, and new tolerance factor for the Double Perovskite $Cs_2AgIrBr_6$ have been computed and are presented in Table 3.

Table 3 Geometric parameters and tolerance factors of the DP $Cs_2AgIrBr_6$

DP	Ionic radius of cations (Å)		Ionic radius of anion (Å)	Tolerance factor (t)	Octahedral factor (u)	New tolerance factor (τ)
$Cs_2AgIrBr_6$	r_{Cs}	$\frac{r_{Ag}+r_{Ir}}{2}$	r_{Br}			
	1.88	0.92	1.96	0.94	0.47	4

The structural analysis of $Cs_2AgIrBr_6$, based on the data from Tables 2 and 3, confirms both its thermodynamic and crystallographic stability. The optimized structural parameters including a reasonable lattice constant, large unit cell volume, and highly negative total energy—demonstrate that the compound is thermodynamically stable. Moreover, the geometric factors such as the Goldschmidt tolerance factor ($t = 0.94$), the octahedral factor ($\mu = 0.47$), and the new tolerance factor ($\tau = 4$) fall within the accepted stability ranges for perovskite materials. These combined results validate the formation of a stable double perovskite structure and emphasize the potential of $Cs_2AgIrBr_6$ as a promising candidate for future research and applications in the field of optoelectronic materials.

4.3 *Electronic Properties*

The band structure of the double perovskite $Cs_2AgIrBr_6$ is presented in Fig. 4. The calculations were performed along the high-symmetry points of the Brillouin zone, namely W (0.5000, 0.2500, 0.7500), L (0.5000, 0.5000, 0.5000), Γ (0.0, 0.0, 0.0), X (0.5000, 0.0000, 0.5000), W (0.5000, 0.2500, 0.7500), and K (0.3750, 0.3750, 0.7500). The analysis shows that the valence band maximum (VBM) is located at the L point, while the conduction band minimum (CBM) appears at the Γ point, indicating an indirect band gap (L→ Γ) of 0.56 eV. Despite this indirect nature, the relatively small band gap enables efficient absorption in the near-infrared region, making the material attractive for specific photovoltaic applications such as IR photodetectors or tandem solar cells in combination with wide band gap materials. Furthermore, $Cs_2AgIrBr_6$, as a lead-free double perovskite, provides an environmentally friendly alternative to conventional lead-based perovskites. Its structure incorporates iridium, a heavy element that introduces strong spin–orbit coupling effects, which can positively influence its electronic and optical properties. In summary, despite having an indirect band gap, $Cs_2AgIrBr_6$ is confirmed to be a semiconductor and remains a promising candidate for advanced optoelectronic devices, particularly in infrared sensing technologies.

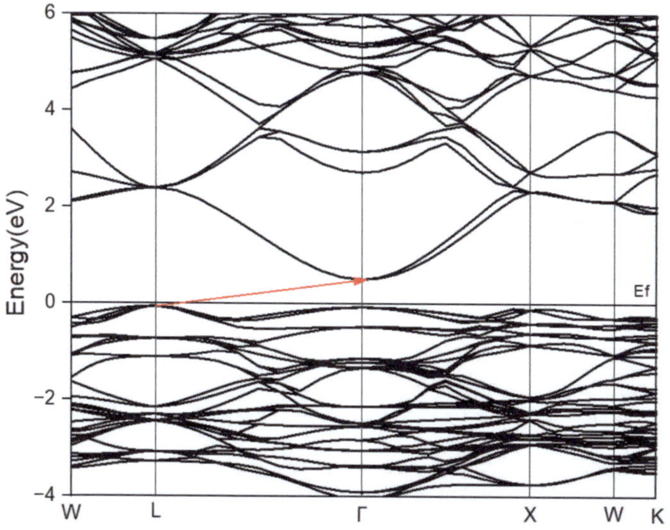

Fig. 4 Band structure of $Cs_2AgIrBr_6$

The total Density of States (TDOS) graph of $Cs_2AgIrBr_6$ illustrates how the electronic states are distributed across energy levels. The DOS provides a global understanding of the electronic structure of the material, showing where electrons are likely to be present. The peaks represent energies with a high number of available states, while the gaps indicate forbidden regions. The Fermi energy has been shifted to 0 eV and serves as the reference for distinguishing the valence and conduction bands. Around this Fermi level, a clear band gap is observed, confirming the semiconducting nature of $Cs_2AgIrBr_6$. The total DOS is essen tial for identifying the general electronic behavior of the material and provides a foundation for the Projected Density of States (PDOS), which gives deeper insight by showing the contributions of individual atoms and orbitals to the electronic structure (Fig. 5).

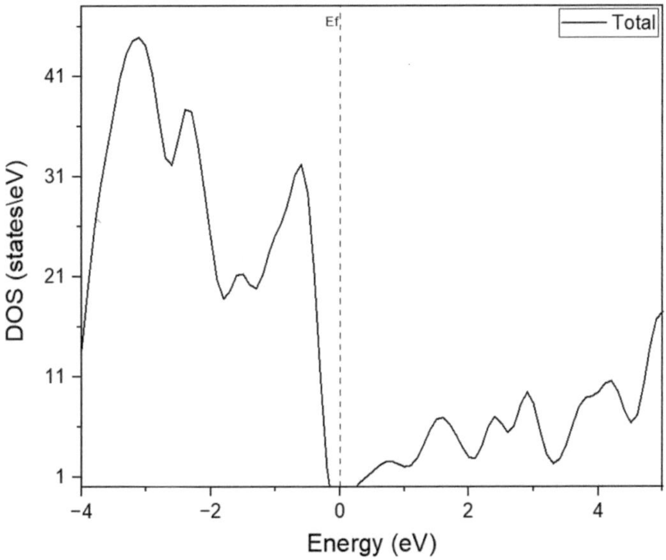

Fig. 5 Density of states of $Cs_2AgIrBr_6$

The PDOS graph of $Cs_2AgIrBr_6$ shows the role of the atomic s, p, and d orbitals of Cs, Ag, Ir, and Br to the material's density of states. The Fermi energy has been shifted to 0 eV (indicated by the dashed vertical line) near the Fermi energy the valence band is mainly composed of Br-p (from about − 1 to 0 eV) and Ir-d orbital (from around − 2 to 0 eV). Small peaks of Cs-d at − 0.5 eV and Ag-d orbitals at − 0.2 eV are also observed in this region, but their impact remains marginal. The conduction band is dominated by Ir-d orbitals, with a slight contribution from Cs-p states at higher energies. The principal interband transitions, which lead to optical excitations, occur from the Ag-d and Br-p states in the valence band to the Ir-d states in the conduction band, enhancing the optoelectronic potential of the material. This configuration supports efficient separation of photogenerated charges, making $Cs_2AgIrBr_6$ a promising candidate for high-efficiency solar cells (Fig. 6).

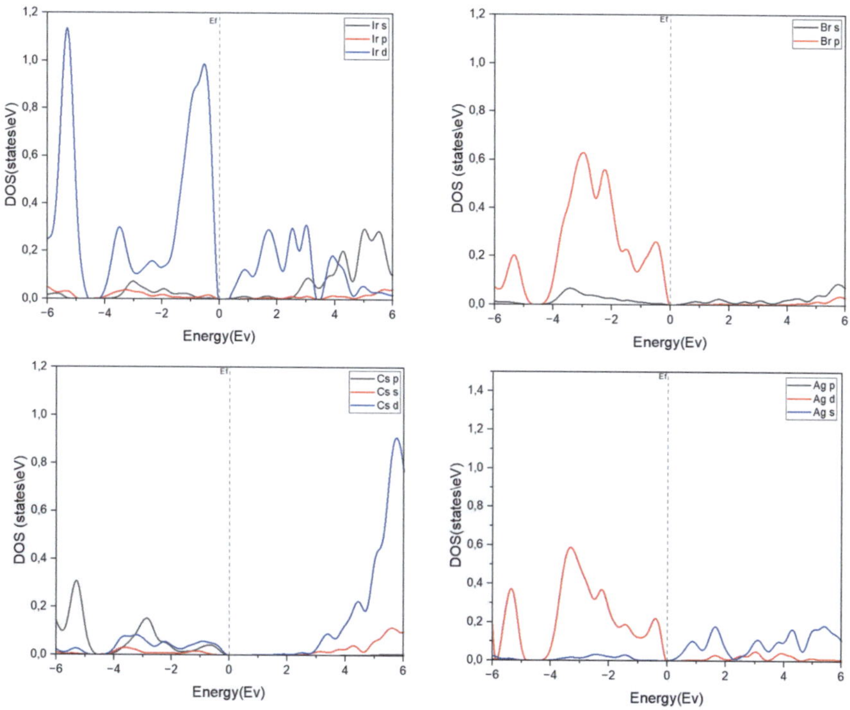

Fig. 6 Projected density of states of $Cs_2AgIrBr_6$

4.4 *Optical Properties*

To evaluate the optoelectronic potential of the $Cs_2AgIrBr_6$ material, a detailed study of its optical properties was conducted through the analysis of several physical quantities that describe the material's response to photon interaction. The resulting graphs include the absorption coefficient, refractive index, extinction coefficient, reflectivity, and energy loss function. These parameters provide essential information on how the material behaves across different photon energy ranges, allowing us to assess its capabilities in terms of light absorption, transmission, reflection, and energy dissipation. Interpreting these plots is crucial to determining the suitability of this material for optoelectronic applications such as solar cells, photodetectors, and photonic devices.

The optical constants refractive index n (ω), extinction coefficient K(ω), absorption coefficient $\alpha(\omega)$, reflectivity R(ω), and energy loss function L(ω) are fundamental parameters that characterize the interaction of a material with electromagnetic radiation. These quantities are systematically derived from the complex dielectric function of the material expressed by the equation $\varepsilon(\omega) = \varepsilon_1(\omega) + i\varepsilon_2(\omega)$ where $\epsilon_1(\omega)$ is the real part of dielectric function follows the Kramer-Kronig relationship and $\epsilon_2(\omega)$ is The imaginary part of the dielectric function is derived from the appropriate momentum matrix elements between the occupied and the unoccupied wave functions within the selection rules over the Brillouin zone. These optical constants are represented by the following equation [29, 30]:

$$n(\omega) = \left[\frac{\varepsilon_1(\omega) + \sqrt{\varepsilon_1^2(\omega) + \varepsilon_2^2(\omega)}}{2} \right]^{1/2} \tag{6}$$

$$K(\omega) = \left[\frac{-\varepsilon_1(\omega) + \sqrt{\varepsilon_1^2(\omega) + \varepsilon_2^2(\omega)}}{2} \right]^{1/2} \tag{7}$$

$$\alpha(\omega) = \frac{2\omega}{c} k(\omega) \tag{8}$$

$$R(\omega) = \frac{(n-1)^2 + k^2}{(n+1)^2 + k^2} \tag{9}$$

$$L(\omega) = \frac{\varepsilon_2(\omega)}{\varepsilon_1^2(\omega) + \varepsilon_2^2(\omega)} \tag{10}$$

The complex refractive index (n + i k) is a vital property of solid materials, providing insights into the speed of light propagation and potential applications in optoelectronics [31].

The real part of the complex refractive index, n(ω), is presented in Figure (a) for the material $Cs_2AgIrBr_6$. The graph shows that the static index is around 3.03, which falls within the ideal range for optoelectronic applications (generally between 2 and 4), showing that the material can effectively interact with light even at low energies. As the energy increases, the refractive index reaches a maximum of 4.6 at an energy of 2.4 eV, corresponding to the visible region, reflecting a very strong photon-matter interaction at this energy. This optical response, combining an appropriate static index and a high peak in the active spectral zone, confirms that $Cs_2AgIrBr_6$ is a promising material for devices such as solar cells, photodetectors, and other optoelectronic systems requiring strong light confinement and effective light management.The extinction coefficient K (ω) represents the imaginary part of the complex refractive index the graph presented in figure (b) shows significant absorption in the visible region, with a maximum peak around 2.6 eV, indicating strong interaction with light in this energy range. This high absorption continues with several additional peaks up to approximately 6 eV, corresponding to the near-ultraviolet region, followed by another marked absorption region between 7 and 9 eV in the deeper ultraviolet. This distribution of absorption peaks suggests that $Cs_2AgIrBr_6$ has good capability to capture light across a wide spectral range, which is promising for its application in optoelectronic devices, particularly photovoltaic cells.

The absorption coefficient graph is a fundamental tool for assessing the optical behavior of a material, particularly its interaction with light across various energy ranges. In semiconductors, the absorption coefficient, $\alpha(\omega)$, plays a pivotal role in describing how photons are absorbed and how electron-hole pairs are generated—an essential mechanism for the operation of photovoltaic devices such as solar cells. The efficiency of optoelectronic and photovoltaic devices is closely linked to this property, as the generation and transport of charge carriers rely directly on light absorption [32] the graphe For the material $Cs_2AgIrBr_6$ presentid in figure (c). The first significant peak appears at 2.4 eV, located in the visible region. The absorption spans a broad range—from 1.2 eV to approximately 10 eV—covering both the visible and ultraviolet domains. Notably, strong absorption peaks are observed between 7 and 9 eV, likely associated with intense electronic transitions stemming from a high density of states. Moreover, the material exhibits remarkably high absorption coefficients, reaching up to $1.8 \times 10^6\ cm^{-1}$, which confirms its excellent light-harvesting capability. An optical band gap of 1.97 eV was obtained for the studied double halide perovskites, indicating that the material begins to absorb efficiently in the visible region of the electromagnetic spectrum. This value was determined using the Tauc plot method [33], which is based on the analysis of the absorption coefficient spectrum and is widely used to evaluate the electronic properties of semiconductors. The method relies on the Tauc relation, expressed as:

$$(\alpha h\nu)^{1/\eta} = A(h\nu - E_g)$$

where α is the absorption coefficient, h is Planck's constant, ν is the photon frequency, A is a material-dependent constant, and E_g is the optical band gap. The exponent η reflects the nature of the optical transition: $\frac{1}{2}$ for direct allowed transitions and 2 for indirect allowed transitions. Although widely used, the Tauc method must be applied with care, as incorrect use may result in inaccurate estimations of the band gap. The graphe of the Tauc plot presented in Figure (d) These features make $Cs_2AgIrBr_6$ a highly promising candidate for advanced optoelectronic applications, including photovoltaic cells, photodetectors, and other light-responsive technologies.

The reflectivity graph of the $Cs_2AgIrBr_6$ material as a function of photon energy presented in figure (e) shows that the static reflectivity at low frequency (or zero energy) can be estimated by extrapolation to approximately 0.3, suggesting that about 30% of very low-energy light is reflected. This gives an indication of the material's optical response in the far-infrared region. Then, a relatively high reflectivity is observed in the visible range, with a maximum of around 0.48 between 2.5 and 3 eV, indicating that the material reflects a significant portion of the incident light in this range. A marked decrease in reflectivity occurs between 3.5 and 6 eV, reaching a minimum close to 0.1–0.15, which reflects better light absorption—a favorable behavior for photovoltaic applications. At higher energies, particularly between 7 and 9 eV, reflectivity increases again with peaks around 0.45, corresponding to notable interaction with UV photons.

The energy loss spectrum of the $Cs_2AgIrBr_6$ material presented in Figure (f) shows a main peak around 4.2 eV, indicating a significant plasmon excitation associated with collective electron oscillations. Additional, less intense peaks appear near 5.0 eV and beyond 8.5 eV, suggesting further electronic transitions between the valence and conduction bands. The moderate variation in the 6–8 eV region reflects low energy dissipation, which may correspond to partial transparency of the material. This optical behavior highlights the ability of $Cs_2AgIrBr_6$ to interact effectively with electromagnetic radiation. These results confirm that $Cs_2AgIrBr_6$exhibits a complex energy response, with absorption phenomena related to plasmons and electronic transitions, which is crucial for optoelectronic applications, especially in the fields of photodetection and solar cells (Fig. 7).

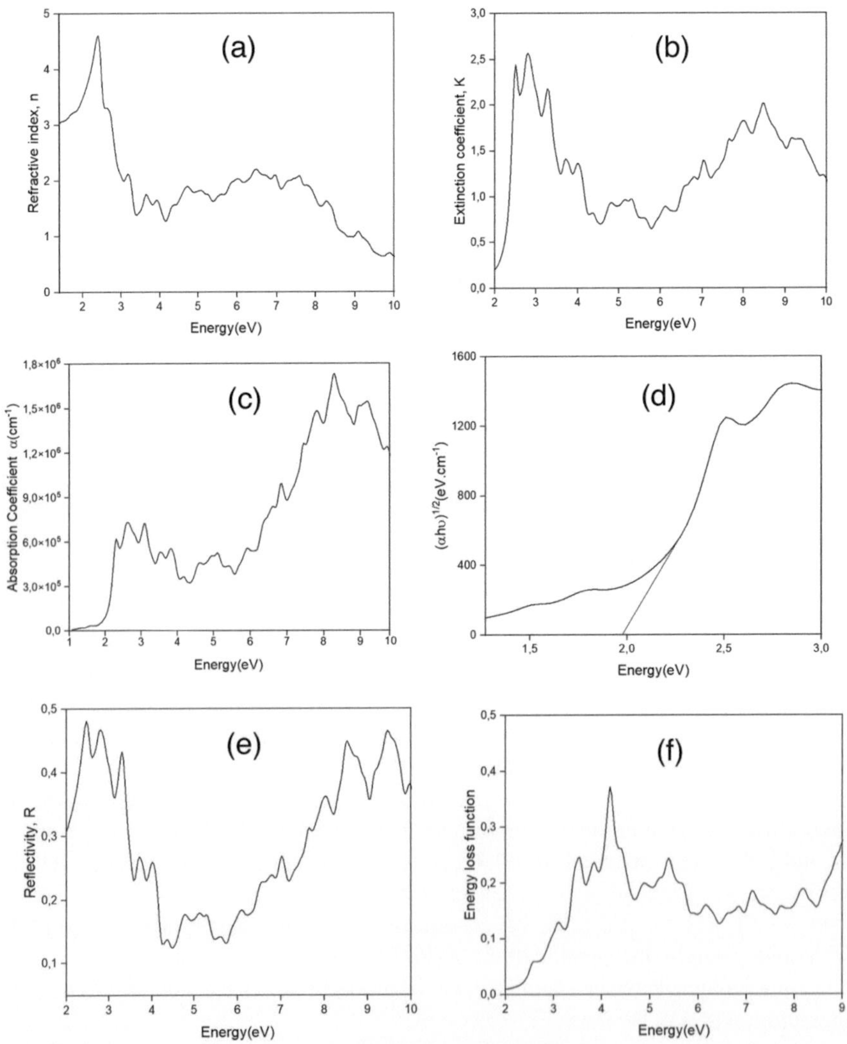

Fig. 7 Optical properties of the double perovskite $Cs_2AgIrBr_6$

5 Conclusion

The results obtained in this numerical study confirm the promising potential of the lead-free double perovskite $Cs_2AgIrBr_6$. Structurally, its cubic configuration of type $Fm\bar{3}m$ is confirmed to be stable based on satisfactory geometric tolerance factors. From an electronic perspective, the material exhibits an indirect band gap of 0.56 eV, highlighting its semiconducting nature. The density of states analysis reveals a strong contribution from Ir-d and Br-p orbitals in the formation of the valence and conduction bands.

Optically, the compound shows broad absorption across the visible and ultraviolet regions, with a very high absorption coefficient and an optical band gap of 1.97 eV. These characteristics demonstrate an excellent ability to harvest light, making $Cs_2AgIrBr_6$ particularly suitable for advanced photovoltaic technologies and photonic devices. Thus, this material positions itself as an environmentally friendly and efficient alternative to conventional lead-based perovskites.

References

1. NREL: Best Research-Cell Efficiency Chart (2022)
2. Kojima, A., Teshima, K., Shirai, Y., Miyasaka, T.: Organometal halide perovskites as visible light sensitizers for photovoltaic cells. J. Am. Chem. Soc. **131**(17), 6050–6051 (2009)
3. Filip, M.R., Hillman, S., Haghighirad, A.A., Snaith, H.J., Giustino, F.: Band gaps of the lead free halide double perovskites $Cs_2BiAgCl_6$ and $Cs_2BiAgBr_6$ from theory and experiment. J. Phys. Chem. Lett. **7**, 2579–2585 (2016)
4. McClure, E.T., Ball, M.R., Windl, W., Woodward, P.M.: Cs_2AgBiX_6 (X = Br, Cl): new visible light absorbing, lead-free halide perovskite semiconductors. Chem. Mater. **28**, 1348–1354 (2016)
5. Li, Z., Kavanagh, S., Napari, M., Palgrave, R.G., Abdi-Jalebi, M., Andaji-Garmaroudi, Z., Hoye, R.L.Z.: Bandgap lowering in mixed alloys of $Cs_2Ag(Sb_xBi_{1-x})Br_6$ double perovskite thin films. arXiv preprint arXiv:2007.00388 (2020)
6. Rose, G.: De novis quibusdam fossilibus quae in montibus Uraliis inveni-untur. typis AG Schadii (1839)
7. Wells, H.L., Johnston, W.R.: Über die Ammoniumbleihalogenide. Zeitschrift für anorganische Chemie **4**(1–2), 117–127 (1893)
8. Goldschmidt, V.M.: Die gesetze der krystallochemie. Naturwis-senschaften **14**(21), 477–485 (1926)
9. Akkerman, Q.A., Manna, L.: What defines a halide perovskite? ACS Energy Lett. **5**(2), 604–610 (2020)
10. Bartel, C.J., et al.: New tolerance factor to predict the stability of perovskite oxides and halides. Sci. Adv. **5**(2), eaav0693 (2019)
11. Kay, H.F., Bailey, P.C.: Structure and properties of $CaTiO_3$. Acta Crystallogr. **10**(3), 219–226 (1957)
12. Megaw, H.D.: Origin of ferroelectricity in barium titanate and other perovskite-type crystals. Acta Crystallogr. **5**(6), 739–749 (1952)
13. Shannon, R.D., Bierstedt, P.E.: Single-crystal growth and electrical proper-ties of BaPbO3. J. Am. Ceram. Soc. **53**(11), 635–636 (1970)
14. Mazet, L., et al.: A review of molecular beam epitaxy of ferroelectric Ba-TiO3 films on Si, Ge and GaAs substrates and their applications. Sci. Technol. Adv. Mater. **16**(3), 036005 (2015)

15. Møller, C.: Crystal structure and photoconductivity of caesium plumbo-halides. Nature **182**(4647), 1436–1436 (1958)
16. Weber, D.: CH3NH3PbX3, ein Pb (II)-system mit kubischer perowskit-struktur/CH3NH3PbX3, a Pb (II)-system with cubic perovskite structure. Zeitschrift für Naturforschung B **33**(12), 1443–1445 (1978)
17. Wasylishen, R.E., Knop, O., Macdonald, J.B.: Cation rotation in methylammonium lead halides. Solid State Commun. **56**(7), 581–582 (1985)
18. Poglitsch, A., Weber, D.: Dynamic disorder in methylammoni-umtrihalogenoplumbates (II) observed by millimeter-wave spectroscopy. J. Chem. Phys. **87**(11), 6373–6378 (1987)
19. Mitzi, D.B., Chondroudis, K., Kagan, C.R.: Organic-inorganic electronics. IBM J. Res. Dev. **45**(1), 29–45 (2001)
20. Kojima, A., et al.: Organometal halide perovskites as visible-light sensitizers for photovoltaic cells. J. Am. Chem. Soc. **131**(17), 6050–6051 (2009)
21. Li, T., et al.: Intrinsic defect properties in halide double perovskites for optoelectronic applications. Phys. Rev. Appl. **10**(4), 041001 (2018)
22. Goldschmidt, V.M.: The laws of crystal chemistry. Naturwissenschaften **14**(21), 477–485 (1926)
23. Wei, Y., Cheng, Z., Lin, J.: An overview on enhancing the stability of lead halide perovskite quantum dots and their applications in phosphor-converted LEDs. Chem. Soc. Rev. **48**(1), 310–350 (2019)
24. Luo, J., et al.: Lead-free halide perovskites and perovskite variants as phosphors toward light-emitting applications. ACS Appl. Mater. Interfaces **11**(35), 31575–31584 (2019)
25. Lv, K., et al.: Lead-free silver-antimony halide double perovskite quantum dots with superior blue photoluminescence. Chem. Commun. **55**(98), 14741–14744 (2019)
26. Li, Z., et al.: Warm white-light emitting silica films prepared using lead-free double perovskite QDs. Dalton Trans. **50**(28), 9804–9811 (2021)
27. Li, C., et al.: Formability of ABX_3 (X = F, CL, BR, I) halide perovskites. Struct. Sci. **64**(6), 702–707 (2008)
28. Shannon, R.D.: Revised effective ionic radii and systematic studies of interatomic distances in halides and chalcogenides. Found. Crystallogr. **32**(5), 751–767 (1976)
29. Zhandun, V.S., Zinenko, V.I.: The effect of structural ordering on the magnetic, electronic, and optical properties of the LaPbMnSbO6 double perovskite. J. Alloys Compd. **671**, 184–191 (2016)
30. Jin, Z., et al.: Electronic structure, elastic, optical and thermodynamic properties of cubic perovskite NaBaF3 with pressure effects: First-principles calculations. Results Phys. **22**, 103860 (2021)
31. Schubert, E.F., Kim, J.K., Xi, J.-Q.: Low-refractive-index materials: a new class of optical thin-film materials. Phys. Status Solidi **244**, 3002–3008 (2007)
32. Yang, R., et al.: Photothermal properties and photothermal conversion performance of nano-enhanced paraffin as a phase change thermal energy storage material. Solar Energy Mater. Solar Cells **219**, 110792 (2021)
33. Tauc, J., Menth, A.: States in the gap. J. Non Cryst. Solids **8**, 569–585 (1972)
34. Arbaoui, Z., Boutahir, M., Fakrach, B., et al.: First-principles analysis of the optoelectronic and thermoelectric properties of black phosphorene for energy harvesting applications. Phase Transitions, pp. 1–12 (2025)
35. Diouf, A.A., Lo, B., Mbaye, M., et al.: Current (I)–voltage (V) characteristics of the pre-and post-annealing nitrogen (N)-doped A-plane-oriented zinc oxide (ZnO) and investigation of the electronic properties of the graphene nanomaterial as a suitable candidate for electrodes in the photovoltaics cells. Adv. Mater. Sustain. Energy Eng. **1**, 15–28 (2025)
36. Attar, A.E., Fatimy, A.E., Mrabet, R.E., et al.: Exploring the electronic transport characteristics of single-walled carbon nanotubes with donor-acceptor characteristics, filled with caffeine molecules. Adv. Mater. Sustain. Energy Eng. **1**, 287–298 (2024)
37. Arbaoui, Z., Boutahir, O., El Mrabet, R., et al.: Phosphorene and phosphorene nanoribbons (PNRs). Adv. Mater. Sustain. Energy Eng. **1**, 299 (2025)

38. El Ouardi, O., Boutahir, M., Fakrach, B., et al.: The optoelectronic properties of boron nitride nanotubes of armchair (7, 7) and zigzag (12, 0) types: a theoretical study. In: E3S Web of Conferences. EDP Sciences, p. 00081 (2025)
39. Arbaoui, Z., Boutahir, O., Rahmani, A., et al.: Exploring the electronic properties of armchair and zigzag phosphorene nanotubes. In: AIP Conference Proceedings, p. 020010. AIP Publishing LLC (2024)
40. Boutahir, M., Rahmani, A., Fakrach, B. (eds.): Preface: advanced materials, microscopy and energy. In: AIP Conference Proceedings, p. 010001. AIP Publishing LLC (2024)
41. El Fatimy, A., Boutahir, M., Rahmani, A., et al.: Enhancing organic solar cell efficiency through metalloporphyrin encapsulation: charge transfer and 04shift in single walled carbon nanotubes. In: MesoSchool 2024: 2D Materials, Superconductivity and Superconducting Circuits, Quantum Transport and Hybrid Systems. Theory and Experiments (2024)
42. El Fatimy, A., Boutahir, M., El Attar, A., et al.: The use of 1D carbon nanotube interacting with caffeine molecules for applications in solar cells: structure optimisation, Raman analysis and optoelectronic properties. Discov. Mater. **4**(1), 51 (2024)
43. Elkhattabi, E.M., Boutahir, M., Termentzidis, K., et al. (eds.): Advanced Materials for Sustainable Energy and Engineering: Selected Proceedings of the 2023 International Conference on Advanced Materials for Sustainable Energy and Engineering (ICAMSEE). Springer Nature (2024)
44. El Fatimy, A., Boutahir, O., Rahmani, A., et al.: Density functional theory calculations of optoelectronic properties of individual and encapsulated magnesium porphyrin in carbon nanotubes for organic nanohybrid solar cells. Energy Technol. **12**(5), 2301135 (2024)
45. El Fatimy, A., Boutahir, M., Lopez, J.M., et al.: Excitation energy transfer via electron transfer between a semi-conducting single walled-carbon nanotube and encapsulated zinc porphyrin or photovoltaic application. ChemistrySelect **9**(1), e202302799 (2024)
46. Arbaoui, Z., Boutahir, O., Rahmani, A., et al.: Phosphorene: a promising two-dimensional material for advanced electronic applications (2024)
47. El Fatimy, A., Boutahir, M., Rahmani, A., et al.: van der Waals interactions on semiconducting single-walled carbon nanotubes filled with porphyrin molecules: structure optimisation and Raman analysis. New J. Chem. **48**(18), 8064–8073 (2024)
48. Fatimy, A.E., Boutahir, M., Termentzidis, K., et al.: Electronic transport properties in donor–acceptor single-walled carbon nanotubes filled with caffeine molecules. In: International Conference on Advanced Materials for Sustainable Energy and Engineering, pp. 238–243. Springer Nature Switzerland, Cham (2023)

Ab Initio Study of the Structural and Optoelectronic Properties of $Rb_2AgIrBr_6$ Insights from DFT Calculations

F. Z. El Marraghi and M. Boutahir

Abstract In this work, a novel lead-free double perovskite compound, $Rb_2AgIrBr_6$, is examined through first-principles calculations based on density functional theory (DFT), using the GGA–PBE approximation within the Quantum ESPRESSO framework. Structural optimization confirms the stability of the material in its cubic phase, with key geometric parameters accurately determined. Electronic structure analysis reveals a direct band gap of 0.6 eV, while the optical band gap, estimated via the Tauc plot method, is found to be approximately 0.89 eV indicating strong absorption in the near-infrared region and potential for low-energy optoelectronic applications. A comprehensive set of optical properties, including the absorption coefficient, extinction coefficient, refractive index, reflectivity, and energy loss function, is also evaluated. The results highlight $Rb_2AgIrBr_6$ as a promising material for use in visible and ultraviolet optoelectronic devices, as well as in tandem solar cells requiring narrow-band-gap absorbers.

Keywords Lead-free double perovskite · $Rb_2AgIrBr_6$ · DFT · Electronic and optical band gap · Structural and opto-electronic properties · Photovoltaic applications

1 Introduction

Over the last decade, metal-halide perovskites have reshaped the landscape of photovoltaic research. First identified in 1839 by Gustav Rose [1], these materials Were largely overlooked en français until 2009, when Miyasaka and colleagues demonstrated their first use in solar cells, achieving a power conversion efficiency (PCE)

F. Z. El Marraghi (✉) · M. Boutahir
Equipe des Matériaux Avancés pour la Transition Energetique (MATE), Laboratoire d'Etude des Matériaux Avancés et Applications (LEM2A), École Normale Supérieure de Meknes, Université Moulay Ismail, Meknes, Morocco
e-mail: f.elmarraghi@edu.umi.ac.ma

M. Boutahir
e-mail: m.boutahir@umi.ac.ma

E. El Mehdi and M. Boutahir (eds.), *Advanced Materials for Sustainable Energy and Engineering*, Engineering Materials, https://doi.org/10.1007/978-3-032-10069-6_17

of 3.8% [2]. While the first-generation devices were based on liquid electrolytes and exhibited poor stability, a breakthrough came in 2012 when Snaith's team introduced a solid-state configuration using spiro-OMeTAD, boosting efficiency beyond 10% [3]. Ongoing progress in device engineering approaches rapidly pushed efficiencies over 20% by 2014 [4], and more recently, tandem architectures have achieved record performances up to 29.8% [5]. Yet, despite their remarkable opto-electronic properties—including strong light absorption, long carrier diffusion lengths, and tunable bandgaps—lead-based perovskites face persistent challenges in terms of environmental toxicity and long-term stability [6]. These limitations have motivated a growing shift toward lead-free alternatives, with halide double perovskites (HDPs) emerging as a particularly promising class of materials.

Halide double perovskites (HDPs), are generally formulated as $A_2B^+B^{3+}X_6$, adopt a heterovalent cationic substitution strategy where two Pb^{2+} ions are replaced by a combination of one monovalent (e.g., Ag^+) and one trivalent (e.g., Bi^{3+} or Ir^{3+}) cation [7, 8]. This approach maintains charge neutrality and promotes a more ordered B-site framework, indreasing both structural robustness and electronic tunability [9, 10]. One of the widely explored cases, $Cs_2AgBiBr_6$, has exhibited favorable opto-electronic properties and high chemical stability under ambient conditions [11]. More recently, focus has expanded to other novel HDP compunds, such as $Rb_2AgIrBr_6$, which adopts a rock-salt-ordered face-centered cubic structure (space group Fm-3m) [12–14]. In this architecture, alternating $[AgBr_6]^{5-}$ and $[IrBr_6]^{3-}$ octahedra share corners within a 3D framework, while Rb^+ ions occupy the cuboctahedral voids. This configuration provides promising stability and a possibly adjustable electronic structure characteristics that are highly desirable for next-generation photovoltaic and opto-electronic devices.

Designing such materials involves a detail analysis of structural parameters. The Goldschmidt tolerance factor (t) and octahedral factor (μ) are commonly used to assess the geometric feasibility and thermodynamic stability of perovskite phases. Values of $0.81 \le t \le 1.0$ and $0.44 \le \mu \le 0.9$ are widely accepted as optimal [15, 16]. Notably, a combinatorial analysis by Giustino and Snaith suggested that over 9000 possible $A_2B^+B^{3+}X_6$ compositions exist, with at least 600 likely to be stable yet still unexplored [17].

Despite their potential, HDPs also exhibit certain intrinsic constraints. Most of them exhibit relatively wide and indirect bandgaps, which can reduce photovoltaic efficiency [18]. Moreover, challenges related to defect formation, film quality, and charge carrier dynamics continue to limit device performance [19–21]. Several strategies have been proposed to overcome these challenges, including film processing optimizations (e.g., thermal annealing, anti-solvent engineering) [22, 23], bandgap modulation via structural transitions or doping [24, 25], and interface passivation techniques aimed at improving charge extraction and reducing recombination [26]. Encouragingly, experimental studies have shown that doping HDPs with conductive additives such as MXenes ($Ti_3C_2T_x$) or performing hydrogen plasma treatments can enhance light absorption and charge transport properties, leading to PCEs above 6% in optimized $Cs^2AgBiBr^6$ based devices [27, 28]. Beyond solar energy conversion,

HDPs are being explored for their potential in a wide range of optoelectronic applications, including UV photodetectors [29, 30], X-ray detection [31], light-emitting diodes (LEDs) [32, 33], and photocatalytic CO_2 reduction [34]. These materials offer a rare combination of non-toxic nature and functional versatility. Furthermore, unlike their lead-containing analogues, certain HDPs show impressive resistance to environmental stressors such as heat, light, and humidity. For instance, Bi^- and Rb-based double perovskites preserve 95% of their initial performance after prolonged exposure to moisture [35], with some prototypes already showing operational lifespans exceeding 20 years [36].

Against this backdrop, the present study assesses the structural and opto-electronic properties of $Rb_2AgIrBr_6$, using density functional theory (DFT) to model its band structure and stability. This work aims to assess its viability as a next generation, lead-free photovoltaic absorber with high performance and long-term environmental resilience.

2 Computational Methods

First-principles calculations were performed within the framework of density functional theory (DFT) using the *Quantum ESPRESSO* package (version 6.7-MaX). The PBE exchange–correlation functional within the GGA framework was employed to describe electronic interactions. The modeled system corresponds to a cubic double perovskite $Rb_2AgIrBr_6$, comprising 40 atoms per unit cell. Periodic boundary conditions were applied with full lattice flexibility (`ibrav = 0` and `CELL_PARAMETERS` provided explicitly). PAW-type pseudopotentials consistent with the PBE functional were used for Rb, Ag, Ir, and Br atoms. The plane-wave basis was truncated at a kinetic energy cutoff of 55 Ry, with the charge density cutoff set at 440 Ry. A $5 \times 5 \times 5$ Monkhorst–Pack grid was used for Brillouin zone sampling in both SCF and NSCF steps, and a cold smearing of 0.02 Ry was applied to aid convergence during metallicity checks. Geometry optimization was carried out using the BFGS quasi-Newton method, with convergence criteria of 4.0×10^{-4} Ry for total energy and 1.0×10^{-4} Ry/Bohr for atomic forces. The final relaxed structure exhibited a lattice constant of approximately 17.1437 a.u. (9.0721 Å). Symmetry constraints were manually reduced by disabling fractional translations. The system contained 344 valence electrons across 206 Kohn–Sham states. Band structure calculations were performed along selected high-symmetry paths, while total and projected density of states (DOS and PDOS) were computed using a 0.1 eV energy resolution. PDOS was decomposed by element and orbital type (s, p, d) to analyze contributions near the Fermi level.

Table 1 The unit cell parameter a_0, ground state energy E_0 and lattice volume

System	a_0 (Å)	Volume (Å^3)	E_0 (Ry)
$Rb_2AgIrBr_6$	9.07	746.342	– 11525.07

3 Results and Discussion

Structural Parameters:

The main structural parameters of $Rb_2AgSbBr_6$ are summarized in Table 1, including the optimized lattice constant a_0 (Å), unit cell volume V (Å^3), and ground-state energy E_0 (Ry), which reflect the presence of relatively large monovalent and trivalent cations at the B-site positions. The results indicate a thermodynamically favorable configuration).

The crystallographic stability of $Rb_2AgSbBr_6$ in the double perovskite structure can be evaluated using geometric descriptors such as the Goldschmidt tolerance factor, the octahedral factor, and the more recently introduced tolerance factor by Bartel et al. These parameters are widely regarded as reliable indicators for predicting the structural viability of stable perovskite phases [37–39]. Their respective values are determined using Eqs. (1)–(4), as outlined above:

$$t = \frac{R_A + R_X}{\sqrt{2}(R_B + R_X)} \tag{1}$$

$$R_B = \frac{R_{B'} + R_{B''}}{2} \tag{2}$$

$$\mu = \frac{R_B}{R_X} \tag{3}$$

$$\tau = \frac{R_X}{R_B} - n_A \left(n_A - \frac{\frac{R_A}{R_B}}{\ln\left(\frac{R_A}{R_B}\right)} \right) \tag{4}$$

In these expressions, n_A denotes the oxidation state of the A-site cation. The ionic radii used for the calculations are sourced from Shannon's compilation. Specifically, R_A corresponds to the radius of the Rb^+ cation, R_X to that of the Br^- anion, and R_B represents the radius of the B-site cation ($R_{B'} = Ag^+$ and $R_{B''} = Ir^{3+}$), as defined in Eq. (2). In this study, is calculated as the arithmetic mean of the ionic radii of Ag^+ and Ir^{3+} (Fig. 1).

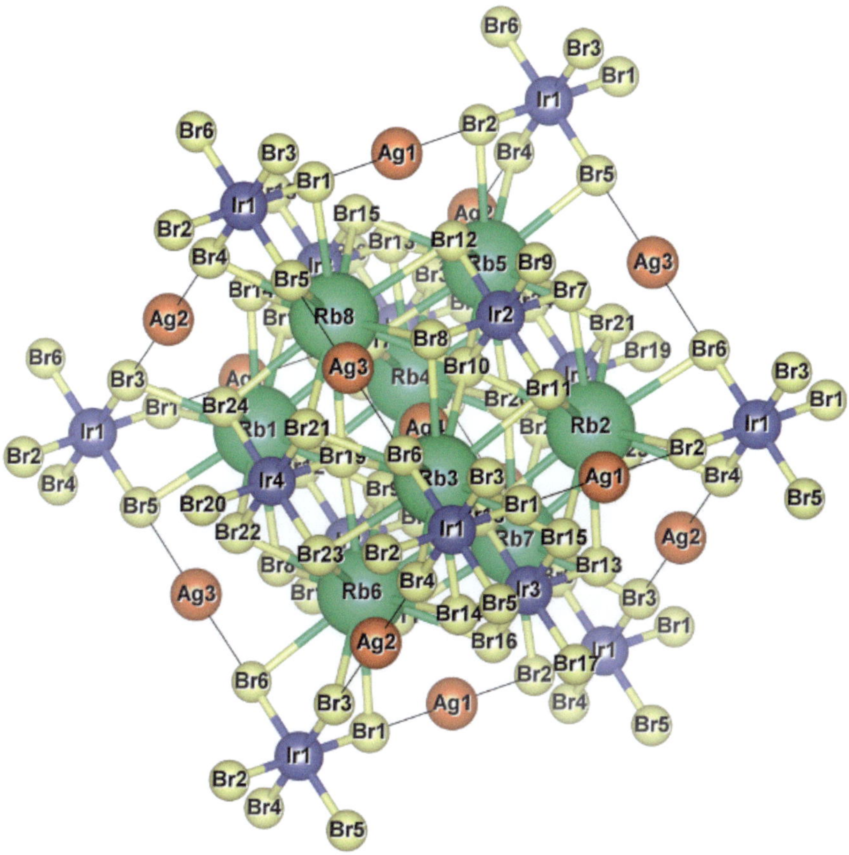

Fig. 1 Crystal structure of $Rb_2AgIrBr_6$, using VESTA software

The structural stability of $Rb_2AgIrBr_6$ was evaluated using Shannon's ionic radii [40]. The obtained values are $t = 0.904$, $\mu = 0.469$, and $\tau = 4.12$, fall within the accepted stability ranges ($0.81 \leq t \leq 1.11$, $0.41 \leq \mu \leq 0.90$, and $\tau \leq 4.18$) [37–39], confirming the geometric feasibility of the perovskite structure (Table 2).

Table 2 Shannon's ionic radii (r), Goldschmidt tolerance factor (t), octahedral factor (μ), and new tolerance factor (τ) for $Rb_2AgIrBr_6$

DP	Ionic radius of cations (Å)	Ionic radius of Br^- (Å)	Goldschmidt factor (t)	Octahedral factor (μ)	New factor (τ)
$Rb_2AgIrBr_6$	$R_A = R_{R_b} =$ 1.72 R_B = (rAg + rIr)/2 = 0.92	$R_X = R_{B_r} =$ 1.96	0.90	0.47	4.12

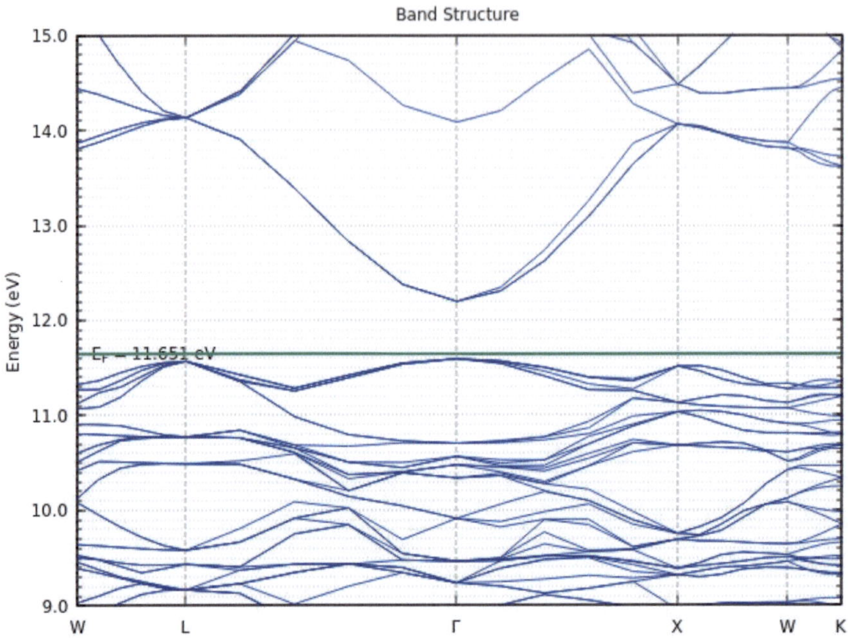

Fig. 2 Band structure of $Rb_2AgIrBr_6$

Electronic Properties:

To gain a deeper understanding of the intrinsic electronic properties of the $Rb_2AgIrBr_6$ double perovskite, a comprehensive analysis of its band structure, total density of states (TDOS), and orbital-resolved projected density of states (PDOS) is conducted. This investigation elucidates the nature of the band gap, the distribution of electronic states in the vicinity of the Fermi level, and the specific contributions of atomic orbitals to both the valence and conduction bands. These electronic features are critical for assessing the material's suitability for charge transport and optoelectronic applications.

Bande Structure:

To investigate the electronic structure of $Rb_2AgIrBr_6$, its band structure was calculated and is presented in Fig. 2. The results indicate a direct band gap of approximately 0.6 eV at the Γ point, suggesting strong potential for efficient photon absorption and emission within the mid-infrared spectral range. This band gap highlights the relevance of this lead-free double perovskite for optoelectronic and infrared sensing technologies. Additionally, the noticeable curvature of both the valence and conduction bands implies low effective masses and, consequently, high charge carrier mobility—an essential characteristic for enhancing the performance of infrared photodetectors and thermophotovoltaic devices (Table 3).

Table 3 The calculated values of the Fermi energy, the Highest Occupied Molecular Orbital (HOMO), the Lowest Unoccupied Molecular Orbital (LUMO), the band gap, the exchangecorrelation functional, and the band gap nature for $Rb_2AgIrBr_6$

Compound	E_{Fermi} (eV)	HOMO (eV)	LUMO (eV)	Band gap (eV)	Functional	Band gap nature
$Rb_2AgIrBr_6$	11.651	11.6	12.2	0.6	GGA PBE	Direct

To obtain a more comprehensive understanding of the electronic states and orbital contributions near the Fermi level, a detailed analysis of the total and projected density of states (DOS/PDOS) was carried out for $Rb_2AgIrBr_6$.

Total Density of States (DOS):

Figure 3 illustrating the total electronic density of states (Total DOS) reveals essential aspects of the material's electronic nature. A distinct discontinuity is observed around the Fermi level, characterized by a region where the DOS is exactly zero. This discontinuity corresponds to an energy band gap of approximately 0.6 eV, centered at [value] eV. This moderate band gap indicates that the material behaves as a semiconductor, unlike metallic materials where the DOS remains non-zero at the

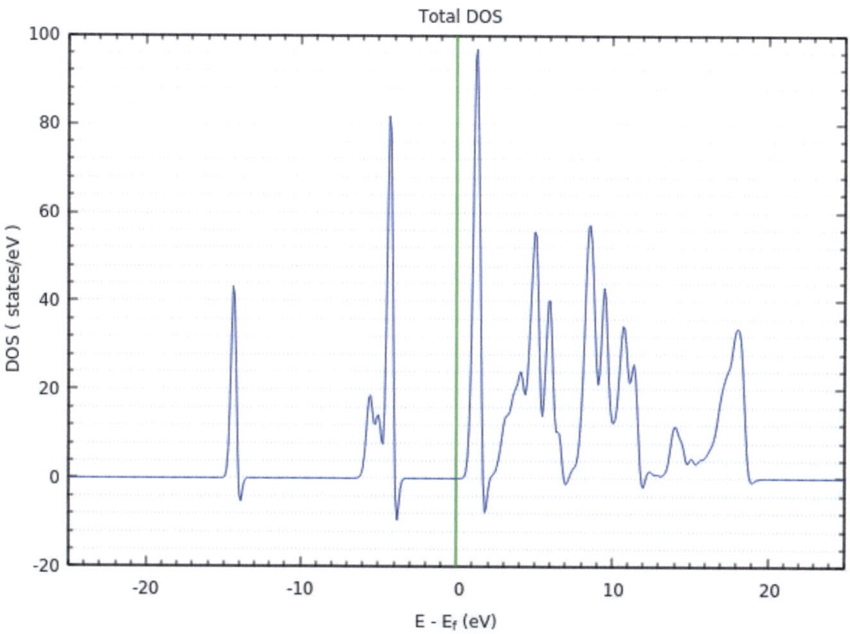

Fig. 3 Total density of states (DOS) of $Rb_2AgIrBr_6$

Fermi level. The absence of states around fermi level, confirms that no electronic states are available at this energy, thereby excluding metallic conductivity.

In addition, several sharp peaks are noticeable in both the valence band (negative energies) and the conduction band (positive energies), reflecting a complex band structure with strong contributions from localized electronic states. Such a profile is typical of materials that may exhibit optical activity in the infrared or near-infrared regions.

Projected Density of States (PDOS):

The projected density of states (PDOS) of $Rb_2AgIrBr_6$, as depicted in Fig. 4, provides a detailed orbital-resolved view of the electronic structure, highlighting the specific contributions from each atomic species. The Fermi energy is set at 11.651 eV, marked by the vertical dashed line.

In the valence band, the dominant features arise primarily from:

- Ir–p orbitals, with a strong peak around –38 eV,
- Rb–s states located near –14 eV,
- Moderate contributions from Ag–p and Br–s orbitals, both situated between –6 eV and –4 eV.

Conversely, the conduction band is characterized by:

- A notable Rb–p contribution around 1 eV,
- Prominent Ir–d states extending from approximately 3 eV to 16 eV,
- Ag–d orbitals contributing significantly in the range 2 eV to 12 eV,
- Minor involvement of Br–p orbitals between 2 eV and 11 eV.

In summary, the valence band is mainly formed by Ir–p and Rb–s orbitals, along with smaller contributions from Ag–p and Br–s. The conduction band, on the other hand, is primarily composed of Ir–d, Ag–d, and Rb–p states, with a slight contribution from Br–p orbitals. This distribution underlines the dominant contribution of Ir orbitals to both the valence and conduction bands, indicating that Ir-related states play a key role in charge transport and are likely responsible for the band-edge electronic transitions in $Rb_2AgIrBr_6$.

The estimated band gap of 0.6 eV corresponds to light absorption in the upper near-infrared (NIR) region, bordering the mid-infrared (MIR), which translates to a wavelength of approximately 2065 nm.

These characteristics make $Rb_2AgIrBr_6$ a promising candidate for infrared optoelectronic applications, such as infrared sub-cells in multi-junction photovoltaic systems, thermophotovoltaic (TPV) converters for energy recovery, and IR sensors for gas detection or thermal imaging technologies.

Optical Properties:

The double perovskite $Rb_2AgIrBr_6$, characterized by a direct band gap of approximately 0.6 eV, exhibits promising optical properties that make it a strong candidate for low-energy optoelectronic and infrared technologies. To gain deeper insight

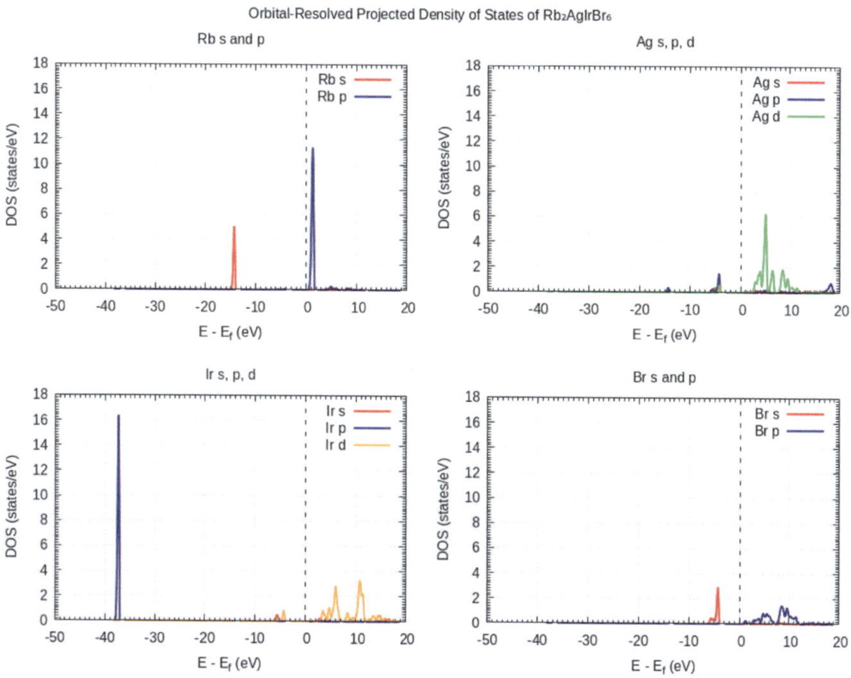

Fig. 4 Orbital-resolved projected density of states of $Rb_2AgIrBr_6$

into its interaction with electromagnetic radiation, we analyze essential optical parameters derived from the complex dielectric function, which is defined as follows [41]:

$$\varepsilon(\omega) = \varepsilon_1(\omega) + i\varepsilon_2(\omega) \quad (5)$$

where:

$\varepsilon_1(\omega)$ is the real part, representing the dispersive response of the material (i.e., how it stores electric energy).

$\varepsilon_2(\omega)$ is the imaginary part, describing energy losses due to photon absorption and electronic transitions.

Based on this formalism, several key optical quantities are extracted, including the absorption coefficient, extinction coefficient, refractive index, reflectivity, and energy-loss function. These parameters offer valuable insights into the fundamental mechanisms of light matter interaction, encompassing photon absorption, electromagnetic wave propagation, surface reflectance, and electronic energy dissipation. In addition, a dedicated analysis is performed to estimate the optical band gap using the Tauc method, derived from the absorption coefficient spectrum. This evaluation is essential for assessing the potential of $Rb_2AgIrBr_6$ as a promising material for photovoltaic and optoelectronic applications.

Absorption coefficient:

The absorption coefficient $\alpha(\omega)$, and the optical band gap energy E_g , are two fundamental parameters for assessing the optoelectronic performance of materials like $Rb_2AgIrBr_6$. Upon photon absorption, electrons can be excited from occupied states in the valence band to unoccupied states in the conduction band, initiating electronic transitions that govern the material's optical response.

The relationship between E_g and $\alpha(\omega)$, as expressed in Eq. (6) [42, 43], enables quantification of the material's light absorption capability and provides a deeper understanding of its optical characteristics across a range of photon energies. It is important to acknowledge that this relationship is influenced by extrinsic factors such as crystal quality and temperature. In this context, our investigation of $Rb_2AgIrBr_6$ focuses on elucidating the interplay between its absorption behavior and electronic structure, with the goal of optimizing its suitability for photovoltaic and optoelectronic device applications.

$$\alpha(\omega) = \frac{4\pi k(\omega)}{\lambda} \tag{6}$$

where:

$k(\omega)$: Extinction coefficient (describes how strongly the material absorbs light at a given frequency).

λ: Wavelength of the incident light (typically in meters or nanometers).

ω: Angular frequency of the incident light .

Figure 5 illustrates the variation of the absorption coefficient $\alpha(\omega)$ as a function of photon energy for the double perovskite $Rb_2AgIrBr_6$, calculated from its complex dielectric function. A clear onset of absorption is observed around 2.0 eV, corresponding to the visible range (approximately 620 nm), suggesting the occurrence of direct electronic transitions near the band edge. With increasing photon energy, the absorption coefficient rises steadily and reaches a strong maximum near 8.5 eV, situated in the deep-ultraviolet (UV) region, around 120 nm. This broad spectral response from the visible to the deep-UV demonstrates the material's pronounced interaction with high-energy photons.

Such optical behavior highlights the potential of $Rb_2AgIrBr_6$ for various UV and deep-UV optoelectronic applications, including solar-blind photodetectors, short-wavelength LEDs, and UV imaging sensors. Additionally, the sharp increase in $\alpha(\omega)$ at the absorption edge indicates a low density of sub-bandgap defect states, which is advantageous for enhancing light absorption and the efficient generation of charge carriers.

The efficiency of optoelectronic and photovoltaic devices is largely governed by their ability to absorb incident photons, as this directly impacts both carrier generation and transport [44].

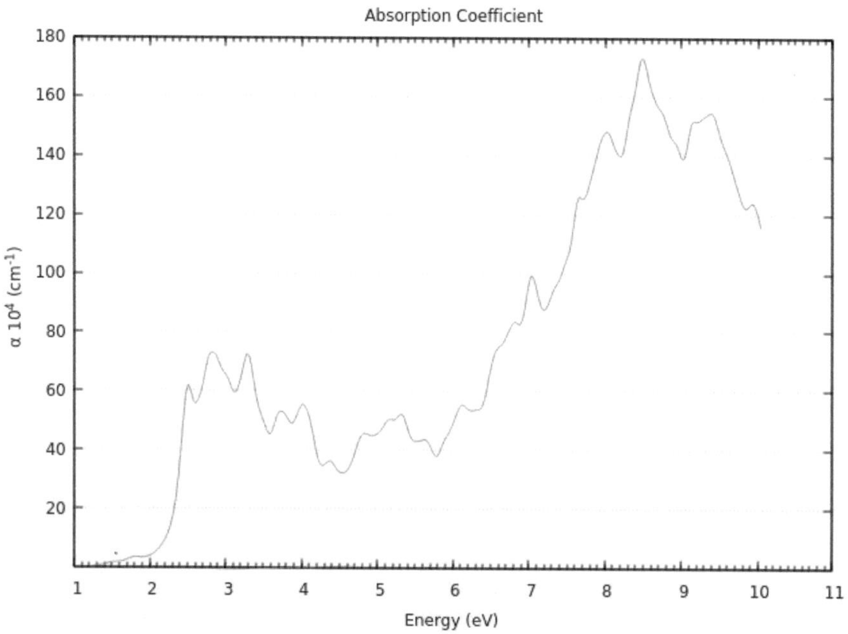

Fig. 5 Optical absorption coefficient spectrum as a function of photon energy of $Rb_2AgIrBr_6$

Estimation of the Optical Band Gap:

To determine the optical band gap of $Rb_2AgIrBr_6$, I employed the Tauc method, which is widely used to analyze the absorption edge of semiconductors [45]. This method is based on the Tauc relation [46, 47]:

$$(\alpha h\nu)^{1/n} = (h\nu - E_g) \tag{7}$$

where α is the absorption coefficient, hν is the photon energy, A is a constant, E_g is the optical band gap, and n depends on the nature of the electronic transition. For a direct allowed transition, n=1/2 [48, 49]. In this study, we plotted $(\alpha h\nu)^2$ as a function of hν, and the linear portion of the curve was extrapolated to the energy axis to estimate the band gap.

As depicted in Fig. 6, the optical band gap of approximately 0.89 eV was extracted by extrapolating the linear region of the Tauc plot to its intersection with the photon energy axis. This value lies within the near-infrared (NIR) region, corresponding to a wavelength of approximately 1390 nm. This result is in good agreement with the electronic band gap obtained from DFT-based calculations, indicating a consistent description of the material's electronic and optical properties. The small difference between the optical and electronic band gaps can be attributed to many-body interactions and to the intrinsic limitations of the exchange–correlation functional employed in the DFT method [50].

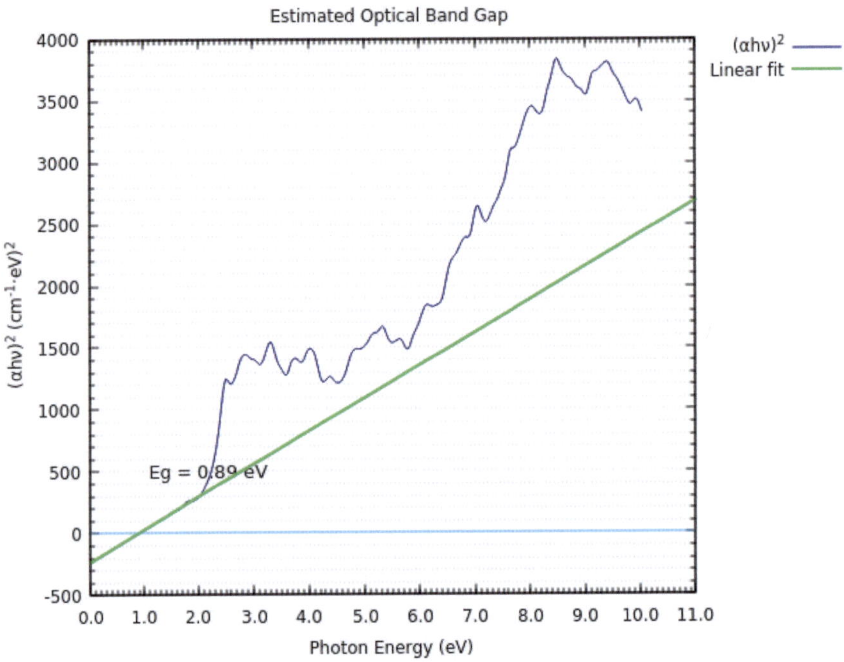

Fig. 6 Tauc plot for optical band gap estimation of $Rb_2AgIrBr_6$

Extinction Coefficient:

Figure 7 presents the variation of the extinction coefficient $k(\omega)$ as a function of photon energy. This parameter corresponds to the imaginary part of the complex refractive index and serves as an indicator of the material's photon absorption capability. The extinction coefficient is defined as follows [42, 43]:

$$k(\omega) = \left[-\frac{\varepsilon_1(\omega)}{2} + \sqrt{\left(\frac{\varepsilon_1(\omega)}{2}\right)^2 + \left(\frac{\varepsilon_2(\omega)}{2}\right)^2} \right]^{1/2} \tag{8}$$

where c is the speed of light, ω is the angular frequency, and $\mathrm{Im}[\tilde{n}(\omega)]$ denotes the imaginary part of the complex refractive index.

The behavior of $k(\omega)$ across different energy ranges provides insight into the strength of absorption processes within $Rb_2AgIrBr_6$, further elucidating its suitability for optoelectronic applications.

The extinction coefficient spectrum exhibits several distinct features across different energy ranges. Notably, pronounced peaks are observed between 2.5 and 4.0 eV, corresponding to the visible spectral range (approximately 496 nm to 310 nm), as well as within the 7.0 to 9.4 eV interval, which lies in the far ultraviolet (UV-C)

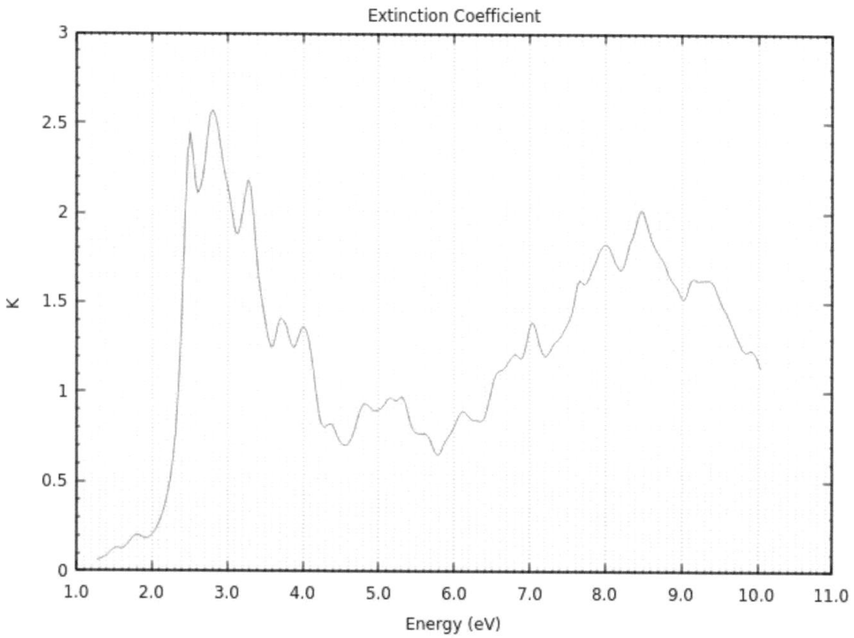

Fig. 7 Energy-resolved analysis of the extinction coefficient

region, with wavelengths spanning 177 nm to 132 nm. A local maximum is particularly evident at 2.8 eV, situated near the boundary between the visible and near-UV regions, indicating significant interband electronic transitions in this zone. In contrast, a marked dip appears around 4.3 eV (near-UV, 288 nm), likely reflecting a reduction in absorption associated with a lower density of available electronic states at that energy. Beyond 6.0 eV (deep-UV, 207 nm), the extinction coefficient increases again, showing multiple oscillations and culminating in a pronounced peak at 8.45 eV, highlighting the onset of high-energy transitions within the electronic band structure.

These spectral features underscore the complex interaction between incident photons and the electronic states of the material. The intensity and energetic positions of the absorption peaks are particularly important for evaluating the material's potential in optoelectronic applications, especially in the UV and near-UV ranges.

In summary, the material demonstrates strong absorption in two main spectral regions:

- Between 2.0 and 4.0 eV, encompassing the tail end of the visible spectrum and extending into the near and mid-UV (UV-A and UV-B),
- Between 7.0 and 9.5 eV, corresponding to the far-UV (UV-C) region.

These optical characteristics suggest that the material is well-suited for applications involving visible and ultraviolet radiation, such as photodetectors, UV filters, or optical coatings. Moreover, its moderate absorption within the visible range particularly between 2.0 and 3.0 eV may also support its use in photovoltaic or light-harvesting technologies that require efficient absorption of visible light.

Refractive Index:

The refractive index is a fundamental parameter that characterizes the optical response of a material, particularly its ability to bend and transmit light. It is intrinsically linked to the complex dielectric function and reflects how electromagnetic waves propagate through the medium. In the present study, the real part of the refractive index was computed using density functional theory (DFT) within the framework of the generalized gradient approximation (GGA), employing the Perdew–Burke–Ernzerhof (PBE) exchange–correlation functional. The refractive index is determined according to the following relation [51]:

$$n(\omega) = \left[\frac{\varepsilon_1(\omega)}{2} + \sqrt{\left(\frac{\varepsilon_1(\omega)}{2}\right)^2 + \left(\frac{\varepsilon_2(\omega)}{2}\right)^2} \right]^{1/2} \tag{9}$$

As shown in Fig. 8, the refractive index exhibits a pronounced peak around 2.4 eV, which corresponds to the blue edge of the visible spectrum (approximately 517 nm in wavelength), reaching a maximum value of approximately 4.6. This peak indicates a strong optical response associated with interband electronic transitions occurring in this energy range.

At lower photon energies, starting from 1.3 eV—near the near-infrared (NIR) region, around 954 nm, and representing the lower limit of the plotted spectrum—the refractive index begins at a value close to 3 and increases gradually toward the observed maximum. However, since the plot does not extend to zero photon energy, the static refractive index (i.e., the value at zero energy) is not directly visible and would need to be estimated either by extrapolation or through separate calculation.

Beyond the peak, the refractive index decreases progressively with increasing photon energy. This behavior is consistent with the typical dispersion characteristics observed in semiconductors and insulators, reflecting the reduced contribution of interband transitions as the photon energy moves away from regions of strong electronic absorption.

Reflectivity:

The comprehensive knowledge of both the real and imaginary components of the complex dielectric function enables the evaluation of additional optical parameters, notably the reflectivity. This quantity provides insight into the fraction of incident light that is reflected at the material's surface and is calculated using the following relation [42, 43]:

$$R(\omega) = \frac{(n(\omega) - 1)^2 + k(\omega)^2}{(n(\omega) + 1)^2 + k(\omega)^2} \tag{10}$$

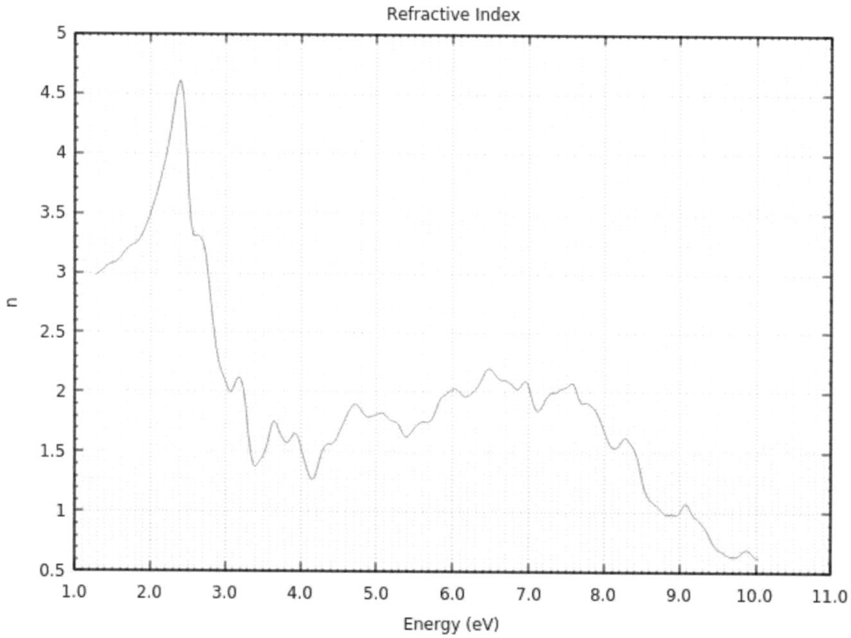

Fig. 8 Optical refractive index dispersion across the photon energy range

As illustrated in Fig. 9, the reflectivity spectrum reveals a pronounced energy-dependent behavior, characterized by multiple peaks and troughs across the studied range. At low photon energies, specifically below 1.5 eV (corresponding to the infrared region, with wavelengths exceeding approximately 827 nm), the static reflectivity is around 0.25, indicating moderate reflection in this region. The reflectivity reaches a maximum of approximately 0.48 near 2.5 eV, which lies within the visible spectrum (wavelength around 496 nm), suggesting a strong reflection response in this part of the spectrum. A significant drop occurs near 4.35 eV, within the near-ultraviolet region (UV-A/B, approximately 285 nm), where the reflectivity falls below 0.14. This dip implies increased transparency or higher absorption at this energy. Beyond 6.1 eV, in the deep ultraviolet (deep-UV) region around 203 nm, the reflectivity begins to rise again, exhibiting several oscillations and culminating in a pronounced peak at approximately 9.45 eV (far-UV, 131 nm), likely due to high-energy electronic transitions.

These spectral features are closely linked to the electronic structure of the material and play a key role in assessing its potential for optoelectronic applications. Nonetheless, the relatively moderate reflectivity observed throughout the visible and ultraviolet regions could present a limitation for photovoltaic applications, where enhanced absorption and reduced reflectivity are typically desired to improve light-harvesting efficiency.

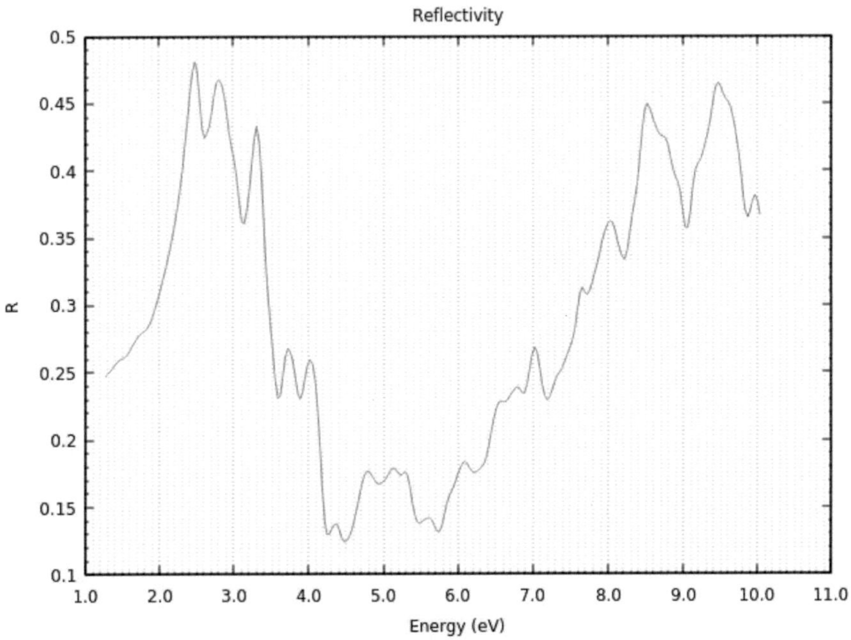

Fig. 9 Reflectivity spectrum over the investigated energy range

Energy Loss Function:

Figure 10 depicts the variation of the energy loss function $L(\omega)$ as a function of photon energy (in eV). The energy loss function, which provides insight into the energy dissipated by fast electrons traversing the material, is mathematically defined as the imaginary part of the inverse of the complex dielectric function [42, 43]:

$$L(\omega) = \frac{\varepsilon_2(\omega)}{\varepsilon_1^2(\omega) + \varepsilon_2^2(\omega)} \tag{11}$$

This function quantifies the energy dissipated by high-energy electrons as they interact with the material, providing valuable insight into collective excitations such as plasmons and interband electronic transitions.

As shown in Fig. 10, the energy loss function spectrum exhibits several prominent features across a wide photon energy range. Notably, strong peaks are observed between 2.5 and 5.8 eV, with a local maximum centered around 4.2 eV, suggesting pronounced electronic excitations in this region. These features are commonly associated with plasmon resonances or intense interband transitions, where the material undergoes significant energy dissipation. A noticeable dip near 6.5 eV may correspond to a temporary reduction in energy loss, potentially due to a lower density of available electronic states or enhanced optical transparency. Beyond 7.0 eV, the loss function increases steadily, exhibiting multiple oscillations and culminating in

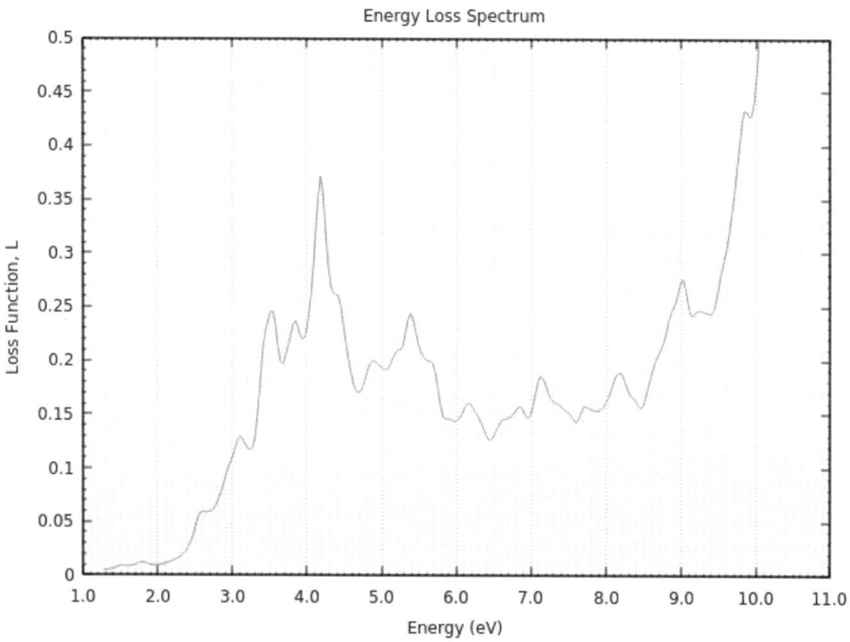

Fig. 10 Energy loss spectrum of $Rb_2AgIrBr_6$

a sharp rise around 10.0 eV. This behavior is indicative of the onset of high-energy transitions, possibly linked to bulk plasmon excitations or strong interband electronic transitions at elevated photon energies.

These spectral characteristics reflect complex interactions between incident high-energy electrons or photons and the electronic structure of the material. The position and intensity of the peaks provide valuable insights into its optoelectronic and photonic properties, particularly in the ultraviolet spectral range.

In summary, the material exhibits significant energy loss in two primary regions:

- Between 2.5 and 5.8 eV, spanning the tail end of the visible spectrum (approximately 496–214 nm) and extending into the near and mid-ultraviolet (UV-A and UV-B),
- Between 8.5 and 10.0 eV, corresponding to the deep ultraviolet region (UV-C).

These features suggest that the material may be suitable for applications such as ultraviolet photodetectors, energy filtering components, or electron energy loss spectroscopy (EELS) systems. However, the relatively low values of the loss function in the visible spectral range (approximately 1.25–2.5 eV) may reduce its effectiveness in conventional photovoltaic or visible light-harvesting devices.

4 Conclusion

Density functional theory (DFT) calculations confirm the structural stability of $Rb_2AgIrBr_6$ in its cubic phase. The material exhibits semiconducting behavior, characterized by a direct electronic band gap of approximately 0.60 eV. The optical band gap, estimated at around 0.89 eV, indicates effective absorption in the near-infrared (NIR) region, which is a key property for optoelectronic applications.

The projected density of states highlights the dominant atomic orbitals contributing to interband electronic transitions, offering insight into the nature of charge carriers. Optical analysis reveals a strong absorption coefficient across the UV and visible spectral ranges, supporting the material's potential for light-harvesting applications. In parallel, the refractive index profile suggests efficient optical confinement, which can enhance light–matter interactions within thin-film architectures. Additional optical parameters such as the extinction coefficient and reflectivity spectra provide further information on photon attenuation and surface reflectance behavior, respectively. The energy loss function indicates minimal excitation losses in the infrared and visible regions, while exhibiting distinct plasmonic features in the ultraviolet range—relevant for UV plasmonic devices and stable high-frequency response.

Altogether, these findings suggest that $Rb_2AgIrBr_6$ is a promising candidate for integration in infrared photodetectors, low-band-gap photovoltaic systems (notably in tandem solar cells), and advanced optoelectronic components. Nevertheless, the material's relatively limited absorption in the infrared and its moderate reflectivity may restrict its performance in conventional single-junction photovoltaic devices, reinforcing its relevance in more specialized or multi-junction architectures.

5 Critical Discussion and Outlook

This study, based on first-principles calculations within the framework of density functional theory (DFT), employed the Quantum ESPRESSO package using the GGA–PBE exchange–correlation functional to explore the structural, electronic, and optical properties of the lead-free double perovskite compound $Rb_2AgIrBr_6$. Structural optimization confirmed the dynamic stability of the material in its cubic phase, supporting its theoretical feasibility. Electronic band structure calculations revealed a direct band gap of approximately 0.60 eV, while the optical band gap, estimated via the Tauc method, was found to be around 0.89 eV. These values place the compound within the near-infrared (NIR) spectral range, highlighting its relevance for low-energy optoelectronic and infrared applications. The projected density of states provided insight into the orbital contributions near the Fermi level, indicating significant hybridization. The optical response functions—including the absorption coefficient, extinction coefficient, refractive index, reflectivity, and energy loss function—demonstrated strong interaction with both visible and ultraviolet radiation, effective light confinement in thin films, and moderate reflectivity within the visible range. Together, these properties point to promising behavior for applications such as tandem solar cells, UV photodetectors, and light-harvesting optoelectronic systems.

Despite these encouraging results, several methodological limitations remain. The use of the GGA–PBE functional, while effective for structural predictions, is known to underestimate band gap values due to its approximate treatment of exchange and correlation effects. A more accurate description of the electronic structure would require the implementation of hybrid functionals (such as HSE06) or many-body perturbation approaches, including the GW approximation. Moreover, the present study did not account for spin–orbit coupling (SOC), which is particularly important for heavy elements such as iridium. SOC can significantly influence the band structure and lead to additional band splitting near the valence and conduction band edges. Its inclusion in future investigations is crucial to obtain a more complete and accurate understanding of the electronic and optical properties.

In addition, vibrational and thermal stability were not addressed in this work. Phonon dispersion analysis and *Ab Initio* molecular dynamics simulations would be essential for assessing lattice dynamical behavior and resistance to thermal degradation—key criteria for practical applications. Furthermore, the effects of intrinsic defects such as vacancies and antisite substitutions should be considered, as they may influence charge carrier recombination and transport mechanisms.

From the perspective of device integration, determining the absolute energy levels of the valence and conduction bands relative to the vacuum level or relevant redox potentials is essential for evaluating the compatibility of $Rb_2AgIrBr_6$ with typical charge transport materials. Modeling heterojunction interfaces with compounds such as TiO_2, ZnO, or Spiro-OMeTAD, along with full device simulations, would help bridge the gap between atomistic predictions and practical optoelectronic performance. Ultimately, experimental efforts involving synthesis techniques such as solution processing, hydrothermal methods, or vapor-phase deposition will be necessary to validate the predicted material properties and to assess real-world considerations such as phase purity, crystallinity, and morphological control.

In conclusion, $Rb_2AgIrBr_6$ emerges as a promising lead-free double perovskite with a narrow band gap, strong absorption in the NIR and UV regions, and favorable optoelectronic characteristics. With further theoretical refinement and experimental verification, this compound holds significant promise for next-generation infrared detectors, tandem photovoltaic systems, and UV-selective optoelectronic devices, contributing to the development of sustainable and high-performance materials for future energy and photonic technologies.

References

1. Rose, G.: Ueber die Mineralien des Urals. Ann. Phys. **124**(1), 85–100 (1839)
2. Miyasaka, T., Kojima, Y., Teshima, K.: Organometal halide perovskites as sensitizers for solar cells. J. Am. Chem. Soc. **131**(17), 6050–6051 (2009)

3. Snaith, H., et al.: Perovskite solar cells: stabilization and performance optimization. Nat. Photonics **6**(8), 511–517 (2012)
4. Green, M.A.: Solar cell efficiency tables (version 45). Prog. Photovolt.: Res. Appl. **23**(1), 1–9 (2014)
5. NREL: Best Research-Cell Efficiency Chart. National Renewable Energy Laboratory (NREL) (2021) [Online]. Available: https://www.nrel.gov/pv/cell-efficiency.html (2025)
6. Ning, Z., et al.: Long-term stability and efficiency enhancement in lead-free double perovskites. Adv. Mater. **31**(48), 1905021 (2019)
7. Moller, C.K.: Crystal structure and optical properties of complex perovskite compounds. Nature **182**, 1436–1437 (1958)
8. Tan, Z., et al.: Recent advances in lead-free double perovskites. Adv. Mater. **32**, 2002443 (2020)
9. Ji, F.:Bandgap Engineering of Lead-Free Halide Double Perovskites. Linköping Studies in Science and Technology, No. 2156, Linköping University, Sweden (2021). ISBN: 978-91-7929-606-3
10. Lozhkina, O.A., et al.: Microstructural analysis and optical properties of the halide double perovskite $Cs_2BiAgBr_6$ single crystals. Chem. Phys. Lett. **694**, 18–22 (2018)
11. Jain, A., Voznyy, O., Sargent, E.H.: High-throughput screening of inorganic perovskites for optoelectronic applications. J. Phys. Chem. C **121**, 7183–7187 (2017)
12. Volonakis, G., et al.: Lead-free halide double perovskites via heterovalent substitution of noble metals. J. Phys. Chem. Lett. **7**, 1254–1259 (2016)
13. Slavney, A.H., et al.: A bismuth-halide double perovskite with long carrier recombination lifetime for photovoltaic applications. J. Am. Chem. Soc. **138**, 2138–2141 (2016)
14. McClure, E.T., et al.: Cs_2AgBiX_6 (X = Br, Cl): new visible light absorbing, lead-free halide perovskite semiconductors. Chem. Mater. **28**, 1348–1354 (2016)
15. Manser, J.S., Christians, A., Kamat, P.V.: Intriguing optoelectronic properties of metal halide perovskites. Chem. Rev. **116**, 12956–13008 (2016)
16. Zhao, X., et al.: Design of lead-free inorganic halide perovskites for solar cells via DFT screening. J. Am. Chem. Soc. **139**, 2630–2638 (2017)
17. Giustino, F., Snaith, H.J.: Toward lead-free perovskite solar cells. ACS Energy Lett. **1**, 1233–1240 (2016)
18. Xiao, Z., et al.: Intrinsic instability of $Cs_2In(I)M(III)X_6$ (M = Bi, Sb; X = halogen) double perovskites: a combined density functional theory and experimental study. J. Am. Chem. Soc. **139**, 6054–6057 (2017)
19. McClure, E.T., et al.: Exploring stability and band gaps of lead-free halide double perovskites. Chem. Mater. **28**, 1348–1354 (2016)
20. Volonakis, G., et al.: $Cs_2InAgCl_6$: a new lead-free halide double perovskite with direct band gap. J. Phys. Chem. Lett. **8**, 772–778 (2017)
21. Yang, X., et al.: Simultaneous power conversion efficiency and stability enhancement of $Cs_2AgBiBr_6$ perovskite solar cells. Adv. Funct. Mater. **30**, 2001557 (2020)
22. Nie, W., et al.: High-efficiency solution-processed perovskite solar cells with millimeter-scale grains. Science **347**, 522–525 (2015)
23. Shadabroo, M.S., et al.: Dimethyl sulfoxide vapor-assisted $Cs_2AgBiBr_6$ homogeneous film deposition for solar cell application. ACS Appl. Energy Mater. **4**, 6797–6805 (2021)
24. Kangsabanik, J., et al.: Double perovskites overtaking the single perovskites in lead-free solar cell design. Phys. Rev. Mater. **2**, 055401 (2018)
25. Caetano, C., et al.: First-principles study of electronic properties of cesium chloride double perovskites. J. Phys. Chem. C **126**, 15065–15071 (2022)
26. Yadav, S.C., et al.: Properties, performance and multidimensional applications of stable lead-free $Cs_2AgBiBr_6$ perovskites. Mater. Today Phys. **26**, 100731 (2022)
27. Lyu, M., et al.: Addressing toxicity of lead: progress and applications of low-toxic metal halide perovskites. Adv. Energy Mater. **7**, 1602512 (2017)
28. Mukhtar, M.W., et al.: Ab-initio study of lead-free double perovskites Cs_2AgZBr_6 (Z = Bi, Sb). J. Solid State Chem. **306**, 122781 (2022)

29. Luo, J., et al.: $Cs_2AgInCl_6$ double perovskite single crystals: parity forbidden transitions and implications. ACS Photonics **5**(2), 398–405 (2017)
30. Wu, C., et al.: Highly efficient and stable self-powered ultraviolet and deep-blue photodetector. Adv. Opt. Mater. (2018). https://doi.org/10.1002/adom.201800811 (2025)
31. Pan, W., et al.: $Cs_2AgBiBr_6$ single-crystal X-ray detectors with a low detection limit. Nat. Photonics **11**(11), 726–732 (2017)
32. Tan, Z., et al.: Highly efficient blue-emitting Bi-doped Cs_2SnCl_6 perovskite variant. Adv. Funct. Mater. **28**(29), 1801131 (2018)
33. Moser, F., Lyu, S.: Luminescence in pure and I-doped AgBr crystals. J. Lumin. **3**(6), 447–458 (1971)
34. Zhou, L., et al.: Synthesis and photocatalytic application of stable lead-free $Cs_2AgBiBr_6$ perovskite nanocrystals. Small **14**(11), 1703762 (2018)
35. Liu, X., Zhang, Y., Li, Z.: Durability and stability of lead-free double perovskites under extreme conditions. Adv. Mater. **34**(12), 2332–2341 (2022)
36. Chen, J., et al.: Comparative study of double perovskite and lead-based perovskite solar cells. J. Mater. Chem. A 5(4), 1020–1030 (2024); Oxford PV: Double Perovskite Solar Cells: A Breakthrough in Stability and Efficiency. Oxford PV (2023)
37. Li, C., et al.: Formability of ABX_3 (X = F, Cl, Br, I) halide perovskites. Acta Crystallogr. B 64, 702–707 (2008). https://doi.org/10.1107/S0108768108032734 (2025)
38. Bartel, C.J., et al.: New tolerance factor to predict the stability of perovskite oxides and halides. Sci. Adv. 5, eaav0693 (2019). https://doi.org/10.1126/sciadv.aav0693 (2025)
39. Goldschmidt, V.M.: Die Gesetze der Krystallochemie. Naturwissenschaften **14**, 477–485 (1926). https://doi.org/10.1007/BF01507527 (2025)
40. Shannon, R.D.: Revised effective ionic radii and systematic studies of interatomic distances in halides and chalcogenides. Acta Crystallogr. A **32**, 751–767 (1976). https://doi.org/10.1107/S0567739476001551 (2025)
41. Gautam, L.K., Haneef, H., Junda, M.M., Saint John, D.B., Podraza, N.J.: Approach for extracting complex dielectric function spectra in weakly-absorbing regions. Thin Solid Films **571**, 548–553 (2014)
42. Zhandun, V.S., Zinenko, V.I.: The effect of structural ordering on the magnetic, electronic, and optical properties of the $LaPbMnSbO_6$ double perovskite. J. Alloys Compd. **671**, 184–191 (2016)
43. Jin, Z., et al.: Electronic structure, elastic, optical and thermodynamic properties of cubic perovskite $NaBaF_3$ with pressure effects: first-principles calculations. Results Phys. **22**, 103860 (2021)
44. Yang, R., et al.: Photothermal properties and photothermal conversion performance of nano-enhanced paraffin as a phase change thermal energy storage material. Sol. Energy Mater. Sol. Cells **219**, 110792 (2021)
45. Tauc, J., Grigorovici, R, Vancu, A.: Optical properties and electronic structure of amorphous germanium. Phys. Status Solidi B **15**(2), 627–637 (1966). https://doi.org/10.1002/pssb.19660150224 (2025)
46. Brahmi, B.N.: Étude ab-initio des propriétés structurales et électroniques des alliages ternaires des semi-conducteurs II-VI à base de zinc. Université Abou-Bekr Belkaid Tlemcen, Algérie, Thèse de doctorat (2013)
47. Flores, M., Recio, J.M., Francisco, E., Blanco, M.A., Martin Pendas, A.: First-principles calculations of .... Phys. Rev. B **66**, 144112 (2002)
48. Fox, M.: Optical Properties of Solids, 2nd edn., Oxford University Press (2010)
49. Adachi, S.: Optical Constants of Crystalline and Amorphous Semiconductors: Numerical Data and Graphical Information. Springer (1999)
50. Yu P., Cardona, M.: Fundamentals of Semiconductors: Physics and Materials Properties, 4th edn. Springer (2010)
51. Pankove, J.I.: Optical Processes in Semiconductors. Dover, New York (1971)

Thermal Conductivity of Single-Walled Carbon Nanotubes (SWCNTs)

Chichaoui Bader, Ourada Haddou, and Mourad Boutahir

Abstract This study explores the thermal conductivity of single-walled carbon nanotubes (SWCNTs), renowned for their exceptional properties and wide range of applications. Following an overview of their discovery, synthesis methods (arc discharge, laser ablation, chemical vapor deposition), and atomic structure and chirality, the focus shifts to how these characteristics influence their thermal behavior. Thermal transport, which is primarily phonon-mediated, is examined through various theoretical models (Boltzmann transport equation, ballistic transport, etc.) and computational approaches (molecular dynamics simulations, density functional theory), with particular attention to scattering mechanisms and key factors such as temperature and geometry. Non-equilibrium molecular dynamics (NEMD) simulations performed using the LAMMPS software further investigate the effects of length, diameter, and temperature on thermal conductivity. The results align with theoretical predictions and underscore the critical role of chirality and thermal saturation effects. This work highlights the significant potential of SWCNTs for advanced thermal management in nanoelectronic devices.

Keywords Thermal conductivity · Molecular dynamics · LAMMPS

1 Introduction

The emergence of nanotechnology has revolutionized the development of advanced materials with unprecedented physical and chemical properties. Among these, Single-Walled Carbon Nanotubes (SWCNTs) have attracted significant attention due to their unique one-dimensional structure and exceptional mechanical, electrical, and thermal characteristics. Composed of a single rolled graphene sheet, these cylindrical nanostructures offer remarkable potential for a wide range of applications, including nanoelectronics, energy storage, composite materials, and biomedicine.

C. Bader (✉) · O. Haddou · M. Boutahir
Laboratoire d'Etude Des Materiaux Avancés Et Applications (LEM2A), Université Moulay Ismail, FSM-FPE-ESTM, Zitoune, Meknès, Morocco
e-mail: b.chichaoui@edu.umi.ac.ma

E. El Mehdi and M. Boutahir (eds.), *Advanced Materials for Sustainable Energy and Engineering*, Engineering Materials, https://doi.org/10.1007/978-3-032-10069-6_18

Since their discovery by Sumio Iijima in 1991, SWCNTs have been the focus of extensive research aiming to understand their atomic structure, improve synthesis techniques, and explore their functional properties. One of their most notable features is that their electronic behavior depends on chirality, allowing them to behave either as metals or as semiconductors. This characteristic makes them highly valuable for technological applications.

In the field of sustainable energy and advanced engineering, understanding the thermal transport behavior of SWCNTs is essential for their integration into nanoscale heat management systems. Numerical simulation plays a crucial role in exploring such complex physical phenomena by enabling researchers to model nanoscale systems and analyze their behavior under controlled conditions. Among various computational techniques, molecular dynamics (MD) is widely recognized as a powerful method for investigating material properties at the atomic and molecular levels.

Specifically, non-equilibrium molecular dynamics (NEMD) has become a standard technique for studying heat transport in nanostructures. When applied to SWCNTs, NEMD simulations offer valuable insights into their intrinsic thermal conductivity and contribute to a deeper understanding of heat transfer mechanisms at the nanoscale.

In this study, we use the LAMMPS software (Large-scale Atomic/Molecular Massively Parallel Simulator), an open-source and highly flexible MD simulation package, to investigate the thermal conductivity of SWCNTs. By applying a temperature gradient and analyzing the resulting heat flux, we rely on Fourier's law to characterize the thermal transport properties of these nanostructures. This computational approach supports the design of next-generation nanomaterials and demonstrates the key role of numerical simulation in advancing sustainable and high-performance technologies.

2 Carbon Nanotube Structure

2.1 Chirality and Structural Classification of (SWCNTs)

Single-Walled Carbon Nanotubes (SWCNTs) are formed by rolling a graphene sheet (composed of carbon atoms arranged in a two-dimensional hexagonal lattice) into a cylindrical structure. The atomic structure of graphene, characterized by strong covalent bonds and a delocalized π-electron system, imparts exceptional electronic and mechanical properties to SWCNTs [1] (Fig. 1).

When graphene is rolled into a tube, the resulting SWCNT maintains this hexagonal arrangement, with its diameter determined by the rolling vector and its length theoretically unlimited [2] (Fig. 2).

A key parameter influencing the properties of SWCNTs is **chirality**, defined by the chiral vector (Ch) and the chiral angle (θ), which dictate how the graphene sheet

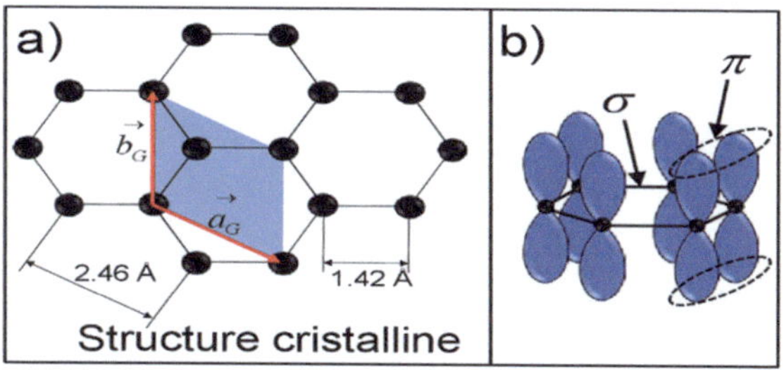

Fig. 1 Crystalline structure of graphene. **a** crystalline structure of graphene. **b** schematic diagram of σ-bonds in the graphene plane and π-orbitals perpendicular to this plane

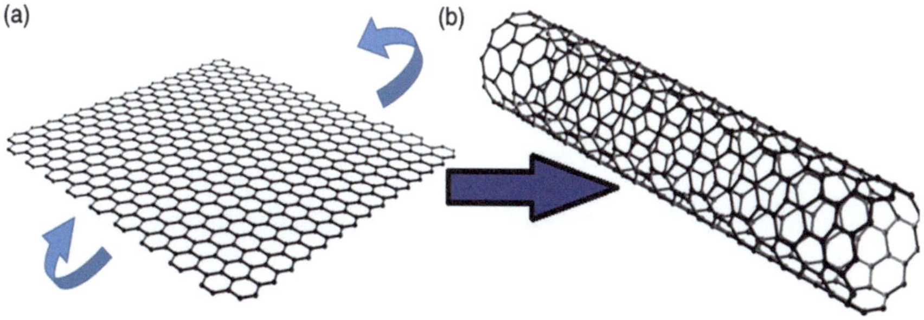

Fig. 2 From graphene sheet to nanotube, **a** diagram of a graphene sheet. **b** a single-walled carbon nanotube (SWCNT)

is wrapped [3]. Based on their chiral indices (n, m), SWCNTs are classified into three types:

- **Zigzag (m = 0, θ = 0°)**: Can be metallic or semiconducting.
- **Armchair (n = m, θ = 30°)**: Always metallic and highly conductive.
- **Chiral (n ≠ m ≠ 0, 0° < θ < 30°)**: Display variable electronic properties, metallic if (n − m) is a multiple of three, otherwise semiconducting [4] (Fig. 3).

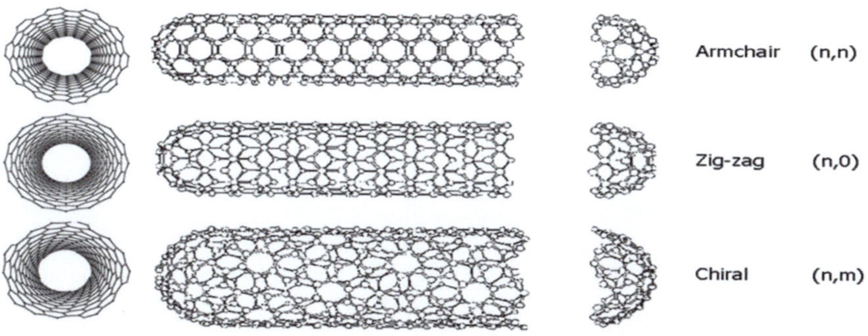

Fig. 3 The three chiralities of carbon nanotubes

2.2 Physical Properties

The length of SWCNTs depends on the synthesis method, ranging from nanometers to several centimeters. Chemical Vapor Deposition (CVD) techniques can produce ultra-long tubes, whereas arc discharge and laser ablation typically yield shorter ones [5, 6]. The tube diameter is determined by the chiral vector and follows the relation:

$$\mathbf{d_t = \frac{a}{\pi}\sqrt{n^2 + nm + m^2}}$$

where a is the graphene lattice constant [3].

2.3 Thermal Conductivity

SWCNTs exhibit **exceptionally high thermal conductivity**, primarily governed by phonon transport rather than electronic conduction [7]. Their **axial thermal conductivity** $k_{\parallel}$, expressed as:

$$\boldsymbol{k_{\parallel} = C_v . v_{g\parallel} . l_{\parallel}}$$

can reach values of **3000–6600 W.m^{-1}.K^{-1}** at room temperature [8]. At lower temperatures (10–100 K), ballistic transport dominates.

In contrast, **radial thermal conductivity** $\boldsymbol{k_{\perp}}$ is much lower, typically between **1–10 W.m^{-1}.K^{-1}** [7], due to weak interlayer van der Waals interactions. This pronounced **thermal anisotropy** is analogous to that of graphene and carbon-based composites.

3 Thermal Conductivity of SWCNTs: Models and Approaches

3.1 Phononic Thermal Conduction in (SWCNTs)

Thermal energy can be transferred through conduction, convection, or radiation, with conduction being the dominant mode in solids such as Single-Walled Carbon Nanotubes (SWCNTs) (Balandin 2011). This section focuses on understanding the mechanisms of thermal conduction within SWCNTs, beginning with a brief review of heat transport in crystalline lattices [8].

- **Phonon-Mediated Heat Transport**:

In solids, thermal conduction occurs via either free electrons or lattice vibrations known as phonons. Electronic thermal conduction is typical in metals, where free electrons carry heat, following the Wiedemann–Franz law. In contrast, phonon-mediated conduction dominates in insulators and semiconductors, where heat is transported by vibrational energy within the crystal lattice. This mechanism is highly dependent on phonon interactions and the crystalline structure, often achieving very high thermal conductivities in materials such as graphene.

- **Phononic Conduction in SWCNTs**:

  Thermal transport in SWCNTs is primarily phononic due to several key factors:

- **Low Free Electron Density**: SWCNTs are often semiconducting, resulting in minimal contribution from free electrons to heat conduction (Balandin, 2011).
- **Highly Ordered Structure**: The quasi-perfect one-dimensional crystalline arrangement of SWCNTs facilitates efficient phonon propagation with minimal scattering.
- **Strong Covalent Bonding**: The sp2 hybridized carbon atoms form exceptionally rigid bonds, allowing for rapid transmission of lattice vibrations [8].
- **Long Phonon Mean Free Path**: Phonons in SWCNTs exhibit minimal diffusion, enabling exceptionally high thermal conductivities, wiche surpass those of copper and diamond **Invalid source specified** (Balandin, 2011).

3.2 Phonon Transport Models in (SWCNTs)

Thermal conduction in solids occurs primarily through two mechanisms: atomic or molecular vibrations and the movement of free electrons. The dominant mechanism depends on the material properties. In Single-Walled Carbon Nanotubes (SWCNTs), heat conduction is mainly driven by lattice vibrations, or phonons [8]. Phonons, which can be viewed as either waves or quasi-particles, are studied using various theoretical and computational models that predict their interactions and influence on thermal conductivity. These models range from classical analytical approaches to advanced quantum-based numerical simulations.

- **Boltzmann Transport Model**:

The phonon Boltzmann Transport Equation (BTE) (Mizuno, 1964) describes the time evolution of the phonon distribution function, considering group velocity and collision effects such as phonon–phonon, phonon-defect, and phonon-boundary scattering.

- **Diffusive Thermal Transport and Fourier's Law**:

At the macroscopic scale, Fourier's law, assuming a diffusive regime where phonons undergo multiple scattering events, often describes phonon transport. However, this approximation fails at the nanoscale, where ballistic effects become significant (Kaiser 2017).

- **Ballistic Phonon Transport**:

When material dimensions approach or fall below the phonon, mean free path, heat conduction transitions to a ballistic regime (Majumdar 1993). In this case, phonons propagate across the material with minimal scattering, and thermal conductivity becomes size-dependent, invalidating Fourier's law.

- **Einstein and Debye Models**:

– **Einstein Model**: Assumes that all atoms oscillate at the same characteristic frequency, providing a simple way to estimate specific heat at low temperatures (Einstein 1907).
– **Debye Model**: Improves upon Einstein's model by introducing a continuous phonon frequency spectrum up to a cutoff frequency (Debye 1912), predicting the T^3 dependence of specific heat at low temperatures and aligning with the Dulong-Petit law at high temperatures.

- **Advanced Numerical Methods**:
  Modern computational power enables sophisticated simulations of phonon transport:

– **Molecular Dynamics (MD)**: Simulates atomic motion by solving Newton's equations with realistic interatomic potentials, capturing non-linear effects and enabling direct calculation of thermal conductivity (Hyzorek 2019).
– **Monte Carlo for BTE**: A stochastic method for numerically solving the BTE by simulating phonon trajectories and interactions (Peraud 2011).
– **Density Functional Theory (DFT)**: Quantum mechanical method to calculate phonon properties and interactions from first principles, providing key parameters for thermal transport modeling (Ponce 2016).

3.3 Phonons as Waves and Quasi-Particles: Implications for Thermal Transport in SWCNTs

Phonons, the quantum representation of lattice vibrations, can be described either as elastic waves or as quasi-particles, depending on the analytical context. Both representations are crucial for understanding phonon transport and its role in the thermal conductivity of Single-Walled Carbon Nanotubes (SWCNTs).

- **Phonons as Elastic Waves**:

In a crystalline solid, phonons can be modeled as waves characterized by a wave vector $\vec{k}$ and angular frequency $\boldsymbol{\omega}$. In a one-dimensional monoatomic lattice, atomic displacements follow harmonic oscillations around equilibrium positions, governed by a dispersion relation $\omega(k) = 2\sqrt{\frac{C_1}{m}}\left|\sin\left(\frac{ka}{2}\right)\right|$, where C_1 is the spring constant and m the atomic mass. The group velocity $v_g = \frac{d\omega}{dk}$ determines how energy is transmitted through the lattice and vanishes at the edges of the Brillouin zone. In multi-atomic unit cells, phonon spectra split into acoustic and optical branches, the former dominating low-frequency transport and the latter appearing at higher frequencies.

- **Phonons as Quasi-Particles**:

In the quasi-particle picture, phonons represent discrete energy packets $E = \left(n + \frac{1}{2}\right)\hbar\omega$. Within the harmonic approximation, phonons propagate without interaction, implying infinite thermal conductivity, which is and an unrealistic scenario. Real materials exhibit anharmonicity in atomic vibrations, introducing phonon–phonon interactions that lead to scattering and finite thermal conductivity.

– Anharmonicity and Phonon–Phonon Interactions:

Anharmonic terms in the interatomic potential, captured by cubic and quartic force constants, enable phonon fusion ($\omega_1 + \omega_2 \rightarrow \omega_3$) and fission ($\omega_1 \rightarrow \omega_2 + \omega_3$) processes. These interactions obey conservation of energy and crystal momentum, distinguishing between Normal (N) and Umklapp (U) processes. Umklapp processes introduce thermal resistance and limit phonon mean free paths, thus reducing thermal conductivity.

– Impact on SWCNTs:

In SWCNTs, where thermal transport is dominated by acoustic phonons, anharmonicity plays a critical role. At low temperatures, phonon transport remains nearly ballistic due to weak scattering. However, as temperature increases, anharmonic interactions become significant, reducing phonon lifetimes and leading to a diffusive regime with saturated thermal conductivity.

3.4 Phonon Vibrational Modes in Carbon Nanotubes

The vibrational modes of phonons in carbon nanotubes (CNTs) are fundamental to their electronic, thermal, and optical properties. These modes originate from the quantized lattice vibrations and are influenced by the cylindrical geometry and boundary conditions of the nanotube (Saito 2013). The phonon modes in CNTs are derived from those of graphene but exhibit modifications due to quantization around the tube circumference.

Dispersion relations, correlating vibrational frequency (ω) with the axial wave vector, describe phonon vibrations in CNTs ($\vec{k}$). In graphene, there are seven phonon branches: three acoustic (longitudinal LA, transverse TA, and flexural ZA) and four optical (longitudinal LO, transverse TO, flexural ZO, and the radial breathing mode RBM).

In CNTs, phonon modes are classified into **acoustic** and **optical** categories:

- **Acoustic Modes**:
  - *Longitudinal Acoustic (LA)*: Atomic vibrations parallel to the nanotube axis.
  - *Transverse Acoustic (TA)*: Atomic oscillations perpendicular to the axis.
  - *Flexural Acoustic (ZA)*: Out-of-plane bending vibrations, significant for thermal transport.
- **Optical Modes**:
  - *Radial Breathing Mode (RBM)*: Collective radial vibrations sensitive to nanotube diameter; prominent in Raman spectroscopy.
  - *Longitudinal Optical (LO) and Transverse Optical (TO) Modes*: Vibrations in and around the axis, strongly coupled to electronic properties.
  - *Flexural Optical (ZO) Mode*: Out-of-plane vibrations with phase opposition, contributing to electron–phonon interactions.

The **phonon dispersion relations** and their **spectroscopic signatures** are critical for characterizing CNTs. Raman and infrared (IR) spectroscopy enable the identification of specific phonon modes, particularly the RBM and the G-band (LO/TO modes), which are key indicators of nanotube structure and diameter (Dresselhaus 2001).

3.5 *Thermal Conductivity Mechanisms in SWCNTs*

The thermal transport in Single-Walled Carbon Nanotubes (SWCNTs) is governed by phonon propagation and scattering processes. The phonon mean free path (MFP) (the average distance a phonon travels before scattering) plays a critical role in determining the material's thermal conductivity. Several mechanisms influence phonon transport:

- **Phonon–Phonon Scattering**:

Phonon–phonon interactions, particularly Umklapp processes, are the primary source of thermal resistance at high temperatures by disrupting the net momentum of heat flow. Normal processes conserve momentum and thus have a limited impact on thermal conductivity (Fig. 4).

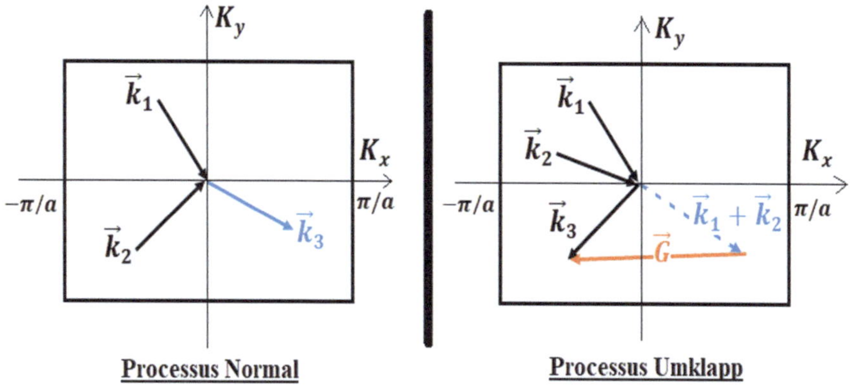

Fig. 4 Three-phonon interaction mechanisms in a 2D lattice

- **Phonon-Defect Scattering**:

Crystalline defects (including impurities, dislocations, vacancies, interstitials, and grain boundaries) scatter phonons, reducing their MFP and consequently lowering thermal conductivity. Interfaces introduce additional thermal resistance, characterized by the Kapitza resistance.

- **Phonon-Boundary Scattering**:

In nanoscale systems like SWCNTs, phonon-boundary scattering becomes significant when the system size approaches the phonon MFP. The Casimir model predicts a size-dependent reduction of thermal conductivity, while the Ziman model incorporates surface roughness effects via a specular reflection parameter.

- **Thermal Transport Regimes**:

  Depending on the relation between phonon MFP and nanotube length, SWCNTs exhibit different thermal conduction regimes:

  - **Ballistic regime (MFP $\gg$ L)**: Phonons propagate without significant scattering, enabling high thermal conductivity, typically at low temperatures.
  - **Diffusive regime (MFP $\ll$ L)**: Enhanced phonon–phonon scattering at high temperatures leads to Fourier-like thermal transport with reduced conductivity (Fig. 5).

A continuous transition from ballistic to diffusive transport occurs with increasing temperature or system length, influenced by structural quality and boundary conditions. In short nanotubes, ballistic conduction dominates, while longer or defect-rich nanotubes exhibit diffusive transport.

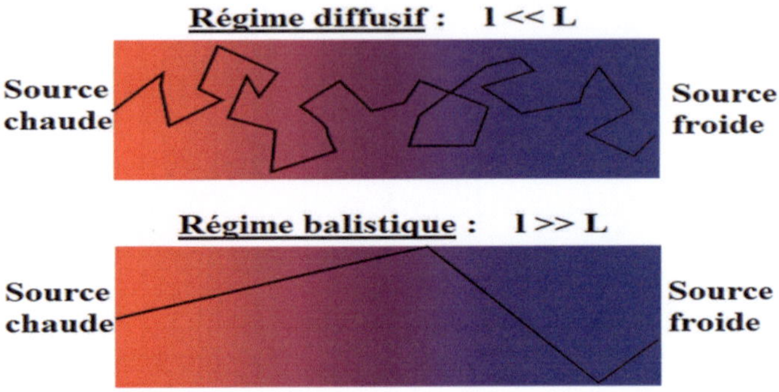

Fig. 5 Thermal transport regimes

3.6 *Factors Influencing Thermal Conductivity in SWCNTs*

The thermal conductivity of Single-Walled Carbon Nanotubes (SWCNTs) is governed by several intrinsic and extrinsic factors, including geometry, temperature, and structural quality.

- **Diameter and Length Effects**:

The diameter of SWCNTs modulates phonon dispersion and mean free path (MFP). Larger diameters support more phonon modes and reduce phonon–phonon scattering, enhancing thermal conductivity. In contrast, smaller diameters increase scattering and lower conductivity. Similarly, tube length determines transport regimes: short tubes exhibit ballistic transport (length-independent conductivity), while longer tubes show diffusive transport where conductivity increases with length until saturation due to phonon–phonon scattering [8].

- **Temperature Dependence**:

Thermal conductivity in SWCNTs is highly temperature-sensitive. At low temperatures ($T \ll T_D$), conduction is ballistic with long phonon MFP and conductivity increases rapidly ($k \propto T^3$). Near the Debye temperature ($T \approx T_D$), most phonon modes are activated, and thermal conductivity peaks. At high temperatures ($T \gg T_D$), Umklapp scattering dominates, causing conductivity to decrease ($k \propto T^{-1}$) [8].

- **Structural Defects and Impurities**:

Defects such as vacancies, dislocations, and junctions disrupt phonon propagation; reduce phonon MFP, and lower thermal conductivity. This effect is particularly significant at high temperatures. Impurities further introduce phonon-impurity scattering and alter phonon modes, diminishing transport efficiency. The higher the impurity concentration, the more pronounced the reduction in conductivity.

3.7 Experimental Methods for Measuring Thermal Conductivity of Carbon Nanotubes

The measurement of thermal conductivity in carbon nanotubes (CNTs) is challenging due to their nanoscale size and manipulation difficulties. Key experimental methods include:

- **Micro-Thermometer Technique**
  - **Principle**: This technique measures the thermal conductivity of individual carbon nanotubes (SWCNTs) by detecting temperature variations along the nanotube. The nanotube is suspended between two micro-thermometers, one acting as a heat source and the other as a temperature sensor. The heat propagation along the nanotube allows determination of thermal conductivity based on the temperature difference (Fig. 6).
  - **Advantages**: Measures thermal conductivity at the nanotube scale, offering precise data for small dimensions and effects of length, diameter, and defects on conductivity.
  - **Limitations**: Requires careful manipulation, precise positioning, and may experience parasitic effects (e.g., contact resistance).

- **Laser Flash Method**
  - **Principle**: A laser pulse heats the surface of a sample, and thermal diffusivity is measured via the temperature rise on the opposite face. Thermal conductivity is then calculated using diffusivity, specific heat capacity, and density (Fig. 7).
  - **Advantages**: Well-suited for nanotube films and composites, providing fast, non-destructive measurements.
  - **Limitations**: Not ideal for individual nanotubes and affected by interfaces or porosity in nanotube assemblies.

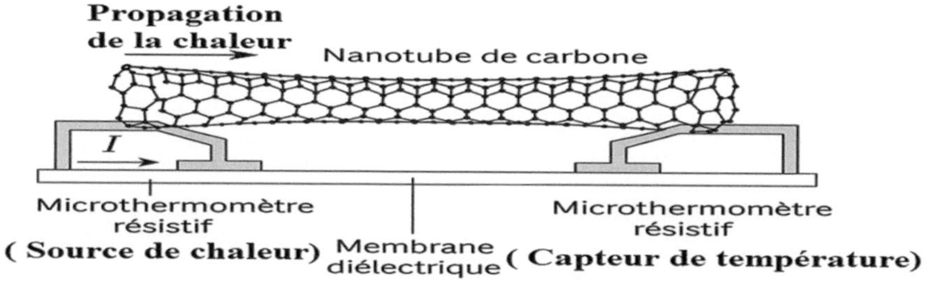

Fig. 6 Diagram of thermal conductivity measurement of a carbon nanotube using the micro-thermometer technique

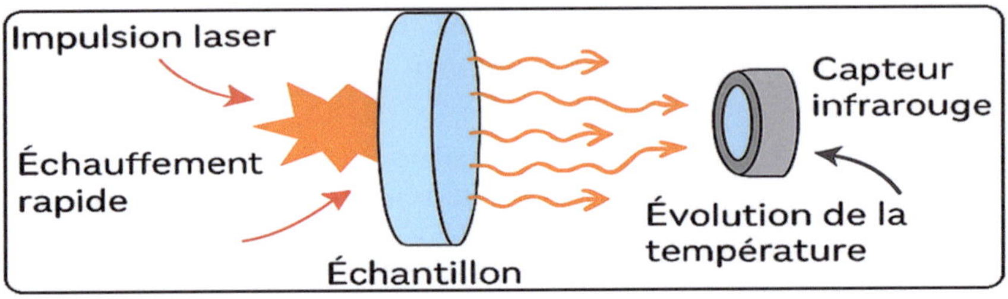

Fig. 7 Schéma de la méthode du laser flash pour mesurer la diffusivité thermique

- **Molecular Dynamics (MD) Simulations**
  - **Principle**: MD simulations model atomic behavior by solving interatomic forces, focusing on phonon transport within CNTs. This method helps analyze the effects of defects, length, and diameter on thermal conductivity.
  - **Advantages**: Provides detailed analysis of phonon transport mechanisms and can be coupled with other techniques (e.g., Density Functional Theory).
  - **Limitations**: Highly dependent on empirical potentials and limited to nanoscale systems due to computational costs.
- **Density Functional Theory (DFT) Simulations**
  - **Principle**: DFT, an ab initio method, calculates electronic and structural properties without empirical parameters. It helps analyze phonon transport and interactions in CNTs.
  - **Advantages**: Precise and does not rely on empirical potentials, offering fine-grained analysis of phonon-electron interactions.
  - **Limitations**: High computational cost and challenges in scaling up for large CNT systems or assemblies.

4 Thermal Conductivity Calculation of SWCNTs via Molecular Dynamics Simulation

4.1 Introduction

Thermal conductivity is a fundamental property of materials, playing a critical role across various fields. Understanding and predicting this property at the atomic scale is essential for the design of materials with optimized thermal performance. Among the available computational tools, Molecular Dynamics (MD) simulation is a powerful approach for investigating the thermal behavior of materials at the nanoscale.

MD relies on solving Newton's equations of motion for a system of atoms interacting through a given interatomic potential. This method enables tracking the temporal evolution of the system and extracting thermophysical quantities (such as thermal conductivity) using statistical mechanics formulations, notably the Green–Kubo relation or the Non-Equilibrium Molecular Dynamics (NEMD) method.

In this work, we present the methodology for calculating thermal conductivity using NEMD to measure the thermal conductivity of single-walled carbon nanotubes (SWCNTs), employing the Large-scale Atomic/Molecular Massively Parallel Simulator (LAMMPS) software. We also detail the obtained results and analyze the influence of factors such as nanotube length, diameter, and temperature on thermal conductivity.

4.2 *LAMMPS Software*

LAMMPS is an open-source atomistic simulation software developed by Sandia National Laboratories. It is designed for large-scale Molecular Dynamics simulations, leveraging parallel computing architectures efficiently. LAMMPS enables modeling a wide range of materials (including solids, liquids, and gases) and can simulate systems containing millions of atoms. It supports various empirical and semi-empirical interatomic potentials and provides tools to compute thermodynamic, mechanical, and transport properties. Owing to its flexibility, compatibility with standard formats (e.g., VMD, OVITO, Atomsk), and extensive module library, LAMMPS has become a reference tool in materials science, chemistry, nanoscience, and biophysics research.

4.3 *Molecular Dynamics (MD)*

- **Overview**

Molecular Dynamics (MD) is a numerical simulation method used to study the behavior of a particle system (atoms or molecules) by solving Newton's equations of motion. Each particle evolves under the influence of forces derived from interatomic potentials that define the interactions between particles. This approach offers a microscopic view of the system's time evolution, providing detailed insights into its structural, thermodynamic, and transport properties. MD is particularly useful for exploring nanoscale phenomena such as diffusion, thermal conduction, phase transitions, and is an essential tool in materials science, chemistry, biophysics, and nanotechnology.

- **Types of MD Simulations**

  Various types of MD simulations are classified based on boundary conditions, calculation methods, or objectives. The main types include:

- **Classical MD**: uses empirical potentials to model atomic interactions.
- **Ab initio MD**: incorporates quantum mechanical calculations (often via Density Functional Theory, DFT) for a more accurate description of interactions but at a high computational cost.
- **Temperature or pressure-controlled MD**: employs thermostats (e.g., Nosé-Hoover, Berendsen) and barostats to simulate specific thermodynamic conditions (NVT, NPT ensembles, etc.).
- **Non-Equilibrium MD (NEMD)**: models non-equilibrium processes such as thermal conduction, shear, or diffusion by imposing a gradient or flux. The choice of simulation type depends on the phenomenon under study and the desired level of precision.

- **Non-Equilibrium Molecular Dynamics (NEMD)**

NEMD is a simulation method used to study the transport properties of materials, such as thermal conductivity, viscosity, or diffusion, under non-equilibrium conditions. Unlike Equilibrium MD (EMD), where properties are derived from natural fluctuations in an equilibrium system, NEMD deliberately imposes a flux (of heat, mass, or momentum) or a gradient (of temperature, concentration, or velocity).

To compute thermal conductivity, a temperature gradient is typically imposed between two regions of the system, and the resulting heat flux is measured. The relationship between these two quantities enables the application of Fourier's law to determine thermal conductivity (Fig. 8):

$$\boldsymbol{k} = -\frac{\boldsymbol{q}}{\nabla \mathbf{T}}$$

where:

k = thermal conductivity $W.m^{-1}.K^{-1}$.
q = heat flux $W.m^{-2}$.
∇T = temperature gradient $K.m^{-1}$.

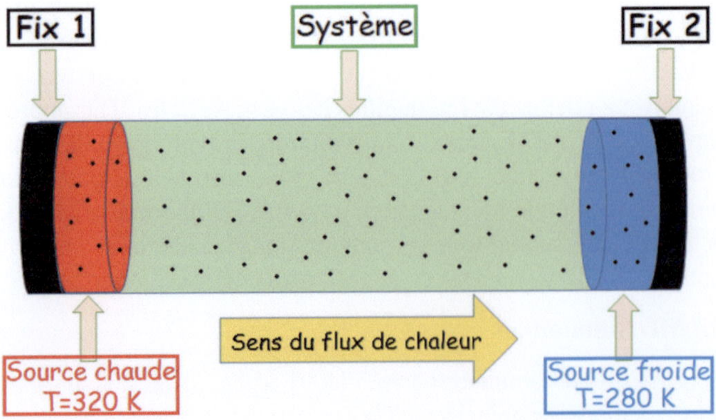

Fig. 8 Schematic of the NEMD protocol for thermal conductivity measurement

This method offers an intuitive and direct approach for studying transport phenomena, particularly in nanostructured materials such as nanotubes, nanowires, and thin films.

5 Results and Discussion

5.1 *Effects of Carbon Nanotube Length*

Figure 9 illustrates the thermal conductivity k (in $W.m^{-1}K^{-1}$) as a function of carbon nanotube (CNT) length (in nm) for three distinct chiralities ((10,10) *armchair*, (15,3) *chiral*, and (17,0) *zigzag*) all with approximately the same diameter ($d \simeq 1, 33$ nm). Simulations were performed at a constant temperature ($T = 200$ K), enabling a rigorous comparison of the effect of chirality on thermal transport.

- Influence of Length: Ballistic-to-Diffusive Transition

For all chiralities, thermal conductivity increases significantly with CNT length, particularly for lengths below 15 nm.

This behavior reflects the well-documented transition from a **ballistic regime** (where phonons traverse the nanotube with minimal scattering) to a **ballistic-diffusive regime**. The growth rate of k decreases as length increases, indicating the onset of saturation due to enhanced phonon scattering.

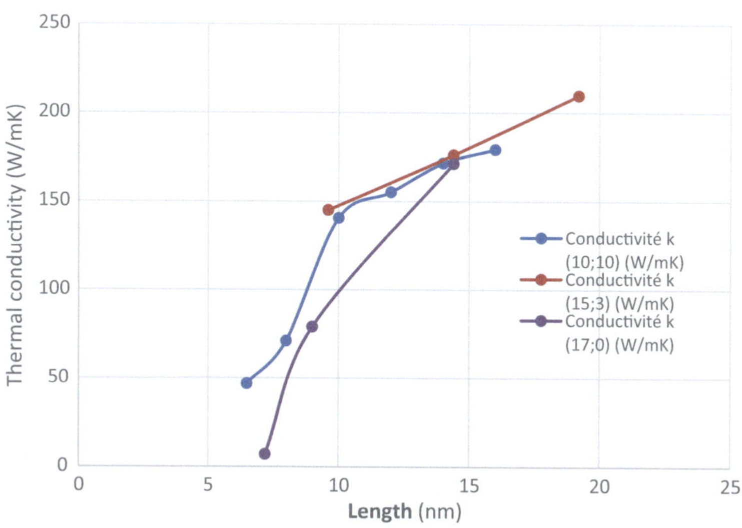

Fig. 9 Thermal conductivity k of a (10,10), (15,3) and (17,0) carbon nanotube as a function of sample length

- Comparison between Chiralities
  - The (15,3) CNT consistently exhibits the highest thermal conductivity, exceeding 210 $W.m^{-1}K^{-1}$ at 20 nm.
  - The (10,10) CNT follows a slightly lower trend, reaching approximately 180 $W.m^{-1}K^{-1}$ at 16 nm.
  - The (17,0) CNT shows the lowest thermal conductivity, around 171 $W.m^{-1}K^{-1}$ at 14.5 nm.
- Interpretation

  These results demonstrate that chirality significantly influences phonon dispersion and thermal transport mechanisms:

  - The (15,3) CNT likely benefits from enhanced coupling of longitudinal acoustic phonon modes, leading to superior thermal transport.
  - The (10,10) CNT, despite its high conductivity, may experience phonon mode limitations due to its perfect symmetry.
  - The (17,0) CNT performs less effectively, likely due to its zigzag structure and possible semiconducting nature.
- Consistency with the Literature

  These findings align well with previous Non-Equilibrium Molecular Dynamics (NEMD) studies:

  - Markussen [9] and [10] also reported a strong dependence of SWCNT thermal conductivity on length at small dimensions.
  - Chirality-related differences at constant diameter are further supported by [11], who observed that chiral SWCNTs can exhibit higher thermal conductivities than their zigzag or armchair counterparts (Fig. 10).

5.2 *Effect de Diameter*

The graph presents the thermal conductivity k (in $W.m^{-1}K^{-1}$) as a function of diameter D (in nm) for three types of single-walled carbon nanotubes (SWCNTs) (Armchair, Zigzag, and Chiral) all with identical length ($L \approx 7.6$ nm) and at constant temperature ($T = 200$ K). This setup enables a rigorous comparison of the impact of chirality on thermal transport.

- Influence of Diameter

For all three types of CNTs, thermal conductivity increases with diameter.

The results show that the thermal conductivity k of SWCNTs increases with diameter D, regardless of chirality. This trend is primarily due to reduced phonon scattering in larger-diameter nanotubes, which enhances heat transport.

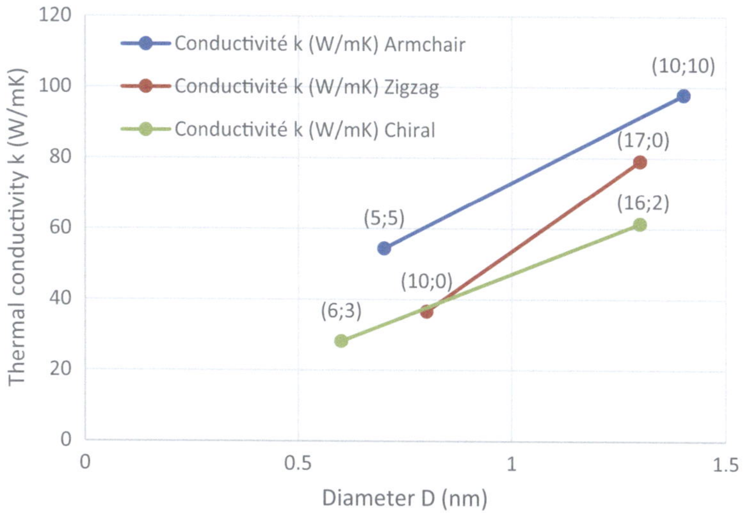

Fig. 10 Thermal conductivity as a function of diameter for different types of carbon nanotubes

- Comparison between Chiralities
  - **Armchair (n = m)**: Exhibits the highest thermal conductivity at comparable diameters; e.g., (10,10) reaches approximately 100 $W.m^{-1}K^{-1}$.
  - **Zigzag (m = 0)**: Displays intermediate thermal conductivity; e.g., (17,0) around $80W.m^{-1}K^{-1}$.
  - **Chiral**: Shows the lowest thermal conductivity at a given diameter; e.g., (16,2) reaches approximately $62W.m^{-1}K^{-1}$.

This behavior is expected: symmetric structures such as Armchair promote more efficient phonon (heat carrier) transport, whereas chiral structures introduce additional phonon scattering and disorder, thereby reducing thermal conductivity.

- Impact of Chirality

The results clearly demonstrate that chirality significantly influences the thermal properties of CNTs. Armchair nanotubes exhibit the highest thermal conductivity, followed by Zigzag, with Chiral CNTs showing the lowest values at equivalent diameters.

- Comparison with the Scientific Literature
  - **Diameter Dependence**: Studies such as [7, 8] have reported that CNT thermal conductivity increases with diameter, partly due to reduced phonon scattering, which is consistent with the trends observed here.

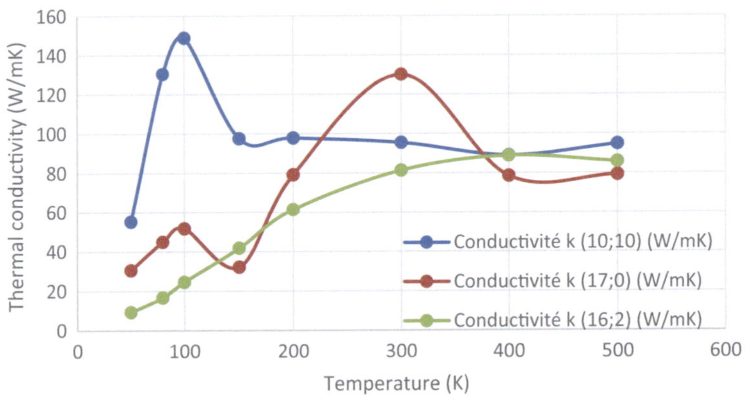

Fig. 11 Thermal conductivity (in W/m.K) as a function of temperature (in K) of carbon nanotube (10,10), (15,3) and (17,0)

- **Effect of Chirality**: CNT chirality directly affects phonon dispersion, which strongly influences thermal conductivity. Armchair CNTs, with their high symmetry and metallic nature, generally support more efficient thermal transport. In contrast, Zigzag and especially Chiral CNTs (often semiconducting) possess more dispersive or localized phonon modes, which hinder heat propagation and thus lower thermal conductivity.
- **Numerical Values**: The reported k values range from approximately 30 to 100 $W.m^{-1}K^{-1}$, which is realistic for short SWCNTs at room temperature. At the macroscopic scale, or for long, well-aligned SWCNTs, much higher values (> 3000 $W.m^{-1}K^{-1}$) have been reported [8]. However, at the atomic scale and under realistic conditions (including defects, curvature, and environmental effects such as substrate or air), significantly lower values are typically observed.

- Conclusion

  The graph highlights:

  - A clear dependence of thermal conductivity on diameter.
  - A hierarchy of CNT types: **Armchair > Zigzag > Chiral** (Fig. 11).

5.3 *Effects of Temperature*

The graph illustrates the evolution of thermal conductivity k (in $W.m^{-1}K^{-1}$) as a function of temperature T (in K) for three types of single-walled carbon nanotubes (SWCNTs) (Armchair (10,10), Zigzag (17,0), and Chiral (16,2)) all with identical diameter ($d \simeq 1,33$ nm) and length ($L \simeq 7,6$ nm). This setup highlights the combined effects of temperature and chirality on thermal transport.

- Graph Analysis
  - **Armchair (10,10) Behavior**: Thermal conductivity increases sharply, reaching a peak of approximately 150 $W.m^{-1}K^{-1}$ around 100 K. It then drops rapidly and stabilizes around 95 $W.m^{-1}K^{-1}$ between 200 and 500 K. This pattern suggests a rapid saturation of thermal transport, characteristic of highly ordered structures.
  - **Zigzag (17,0) Behavior**: Thermal conductivity exhibits an irregular growth marked by several secondary peaks, with a notable maximum of about 130 $W.m^{-1}K^{-1}$ around 300 K. Beyond this peak, conductivity decreases and stabilizes around $W.m^{-1}K^{-1}$, indicating more complex phonon scattering dynamics and less ordered thermal transport.
  - **Chiral (16,2) Behavior**: Thermal conductivity increases steadily with temperature, reaching a plateau near 90 $W.m^{-1}K^{-1}$ above 400 K. This profile is characterized by the absence of pronounced peaks or irregularities, though its overall conductivity remains lower than that of the other two types.

- Comparison with the Scientific Literature
- **Temperature Dependence**: According to studies by [7, 8], CNT thermal conductivity increases with temperature at low temperatures due to enhanced phonon excitation. However, at higher temperatures, Umklapp phonon–phonon scattering becomes dominant, leading to saturation or even a decrease in thermal conductivity. This mechanism accounts for the observed peaks in the (10,10) and (17,0) CNTs, followed by a plateau or decline in k.
- **Effect of Chirality**: Armchair CNTs exhibit the highest thermal conductivity, attributed to their highly symmetric and metallic structure, which promotes more efficient phonon transport. In contrast, Zigzag and especially Chiral configurations introduce a greater number of diffusive or localized phonon modes, which hinder optimal heat conduction. These observations align well with the literature, particularly the findings of [3, 12], which emphasize the critical influence of structural symmetry on CNT thermal conductivity.
- **Conclusion**

  This graph confirms that:

- The thermal conductivity of carbon nanotubes strongly depends on both temperature and chirality.
- At low temperatures, k increases sharply, reaching a maximum before saturating or decreasing.
- Armchair (10,10) CNTs maintain superior thermal performance, followed by Zigzag (17,0), and then Chiral (16,2), consistent with both theoretical and experimental results reported in the literature.

6 Conclusion

This study has provided deeper insight into the thermal behavior of single-walled carbon nanotubes (SWCNTs), which rank among the most promising nanomaterials due to their exceptional mechanical, electrical, and thermal properties.

In the initial phase, we examined the atomic structure, chirality, synthesis methods, and fundamental properties of SWCNTs. These structural characteristics directly influence the nanotubes' heat transport capabilities, particularly as a function of diameter, length, and chirality type (armchair, zigzag, chiral).

We then conducted a detailed analysis of the physical mechanisms governing thermal conduction, which is primarily phonon-mediated. Various theoretical and computational models (such as Boltzmann transport, Debye theory, molecular dynamics (MD), and density functional theory (DFT)) were explored to simulate and quantify heat transport. The distinction between ballistic and diffusive transport regimes was clearly established, along with the critical roles of phonon–phonon (Umklapp), phonon-defect, and phonon-boundary scattering processes in limiting thermal conductivity.

Finally, through molecular dynamics simulations using the LAMMPS software, we demonstrated the influence of nanotube length, diameter, temperature, and chirality on thermal conductivity. The results confirm that armchair nanotubes exhibit the highest conductivity, and that thermal saturation at elevated temperatures arises from the intensification of Umklapp scattering processes.

This study highlights that optimizing the thermal properties of SWCNTs requires precise control over their structural parameters and operating conditions. The findings pave the way for advanced applications in nanoscale thermal management, including nanocomposites, high-density electronic devices, and smart thermal sensors.

References

1. Geim, A., et al.: The rise of graphene. Nat. Mater. 183–191 (2007). https://doi.org/10.1038/nmat1849
2. Iijima, S.: Helical microtubules of graphitic Carbon. Nature 56–58 (1991). https://doi.org/10.1038/354056a0
3. Saito, R., Dresselhaus, G.: Physical Properties of Carbon Nanotubes.(1998)
4. Reich, S., et al.: Carbon nanotubes: basic concepts and physical properties (2004)
5. Li, W., et al.: Large-scale synthesis of aligned carbon nanotubes. Science 276–278 (2004)
6. Zhang, X., et al.: Ultralong single-wall carbon nanotubes. Nature 703–703 (2006)
7. Berber, S., et al.: Unusually high thermal conductivity of carbon nanotubes. Phys. Rev. Lett. 84(20), 4613 (2000)
8. Pop, E., et al.: Thermal conductance of an individual single-wall carbon nanotube above room temperature. Nano Letters 96–100 (2006)
9. Markussen, T.J.-P.: Length dependence of thermal conductance in atomic chains. Phys. Rev. B **79**(3), 035415 (2009)
10. Aksamija, Z.: Thermal conductivity of silicon nanostructures: effects of geometry and disorder. Appl. Phys. Lett. **98**(14), 141919 (2011)

11. Shiomi, J.M.: Non-equilibrium molecular dynamics of heat conduction in single-walled carbon nanotubes. Physica E **29**(1–2), 104–109 (2006)
12. Chang, C.W.: Solid-state thermal rectifier. Science **314**(5802), 1121–1124 (2008)
13. Baughman, R., et al.: Carbon nanotubes—the route toward applications. Science 787–792 (2002)
14. Guo, T.: Chem. Phys. Lett. **243**(1–2), 49 (1995)
15. Kumar, R., et al.: Temperature ranges for carbon nanotube growth via CVD. Mater. Sci. Eng. Rep. 22, 58–65 (2020)
16. Vaccarini, L.: Synthetic metals. Nature (1999)
17. Maser, W.: Carbon **40**(10), 1685 (2002)
18. Smith, J., et al:. Synthesis of carbon nanotubes by chemical vapor deposition (CVD) using metal catalysts. J. Nanotechnol. Res. 15(4), 234–245 (2018)
19. Treacy, M., et al.: Exceptionally high young's modulus observed for individual nanotubes. Nature 678–680 (1996)

Theoretical Analysis of the Optical Properties of Vanadium Dioxide Using Drude and Lorentz Models

Mohammed Zouini, Abderrahim Ben Chaib, El Mehdi El Khattabi, Mourad Boutahir, and Kalpana Sahoo

Abstract This chapter offers an in-depth theoretical analysis of the optical properties of vanadium dioxide (VO_2), a material known for its metal–insulator transition. Relying exclusively on classical models—the Drude model for free electrons, the Lorentz model for bound electrons, and their combination in the Drude-Lorentz model—we describe the analytic expressions of the complex dielectric function as well as the resulting optical quantities, such as the refractive index, the absorption coefficient, etc. reflectivity and transmittivity. This study provides a better understanding of the fundamental optical behavior of VO_2 based on established physical principles, without the need for numerical or experimental calculations.

1 Introduction

Materials are now at the heart of technological development because the knowledge and control of the macroscopic and microscopic properties of these materials allows them to be used more efficiently and to develop new materials for better adaptation to specific nanotechnology applications; which leads us to a significant evolution towards the manipulation of cells and materials at the nanoscale [1–10].

Vanadium dioxide (VO_2) is a remarkable material due to its ability to undergo a reversible transition from a metallic state to an insulating state. This transition leads

M. Zouini (✉) · A. B. Chaib
Sidi Mohamed Ben Abdellah University ENS, Fez, Morocco
e-mail: mohammedzouini@gmail.com

M. Zouini · E. M. El Khattabi
Systems and Sustainable Environment Laboratory, Private University of Fez UPF, Fez, Morocco

Department of Science, École Normale Supérieure, University Moulay Ismail Meknes, Meknes, Morocco

M. Boutahir
École Normale Supérieure, University Moulay Ismail Meknes, Meknes, Morocco

K. Sahoo
School of Civil Engineering, KIIT Deemed to Be University, Bhubaneswar, Odisha, India

E. El Mehdi and M. Boutahir (eds.), *Advanced Materials for Sustainable Energy and Engineering*, Engineering Materials, https://doi.org/10.1007/978-3-032-10069-6_19

to dramatic changes in its electrical and optical properties, making it particularly interesting for applications in the fields of optoelectronics, sensors and optically switched devices [11–20].

In this chapter, we are exclusively interested in the theoretical analysis of the optical properties of VO_2. We adopt an approach based on the classical models of condensed matter physics, namely the Drude model for free electrons, the Lorentz model for bound oscillators, and their synthesis in the Drude-Lorentz model. These models make it possible to interpret the phenomena of polarization, absorption, refraction and reflection of light in VO_2 as a function of its electronic structure.

The objective of this study is to lay the analytical foundations necessary to understand the optical behavior of VO_2 without resorting to experimental results or numerical simulations. We thus develop the fundamental relations between the complex dielectric function and the different optical quantities, integrating the Kramers–Kronig relations to link the real and imaginary parts of the response functions.

2 Modeling of Optical Properties by the Drude-Lorentz Model

In radiation, the electric field is probably coupled to the dipole moments associated with the movements of charged entities such as phonons, polarons or plasmon and thus allows them to be revealed. This response is called permittivity or dielectric function related to the optical conductivity of the material.

The electric field and the electric displacement are related by the relation; $\vec{D} = \varepsilon\varepsilon_0 \cdot \vec{E}$ where ε is the dielectric function and ε_0 is the dielectric constant of the vacuum. In addition, the dielectric function can be expressed as a function of susceptibility χ such as:

$$\varepsilon(\omega) = 1 + \chi(\omega) \tag{1}$$

3 Drude Model

Drude's (1900) model [21, 22] is based on the kinetic theory of electrons in a metal which assumes that the material has immobile positive ions and non-interactive electron gas. This simple model uses the classical mechanical theory of the free electron. It was constructed in order to explain the transport properties of conduction electrons in metals [23, 24]. Similarly, he describes the optical properties of metals well.

Indeed, the equation of the motion of free electrons is defined by the equation resulting from the fundamental relation of the following dynamics:

$$m\frac{d^2\vec{R}}{dt^2} = -m\Gamma_d\frac{d\vec{R}}{dt} - e\vec{E} \tag{2}$$

Where e: effective charge and Γ_d: collision frequency.

The solution of this equation is written as a plane wave:

$$\vec{R} = \vec{R}_0\exp\left[i(\omega t - \vec{q}\cdot\vec{r})\right] \tag{3}$$

Substituting in the previous equation of motion, we have the solution:

$$\vec{R} = \frac{-e\cdot\vec{E}}{m\left(-\omega^2 + i\Gamma_d\cdot\omega\right)}$$

$$\vec{R} = \frac{1}{m}\cdot\frac{-e\cdot\vec{E}}{\left(-\omega^2 + i\Gamma_d\cdot\omega\right)} \tag{4}$$

The induced dipole moment $\vec{d}$ is related to $\vec{R}$ by this relationship:

$$\vec{d} = -e\cdot\vec{R}$$

$$\vec{d} = \frac{e^2}{m}\cdot\frac{\vec{E}}{\left(-\omega^2 + i\Gamma_d\cdot\omega\right)} \tag{5}$$

The polarizability $\alpha(\omega)$ is given by where: $d(\omega) = \alpha(\omega)\cdot E(\omega)$

$$\alpha(\omega) = \frac{e^2}{m}\cdot\frac{1}{\left(-\omega^2 + i\Gamma_d\cdot\omega\right)} \tag{6}$$

The polarization per unit volume is given by:

$$\vec{P} = N\cdot\alpha(\omega)\cdot E(\omega) = \varepsilon_0\cdot\chi(\omega)\cdot E(\omega) \tag{7}$$

where N is the oscillator density per unit volume

Susceptibility is deduced from the previous equation: $\chi(\omega)$

$$\chi(\omega) = \left(\frac{N\cdot e^2}{\varepsilon_0 m}\right)\cdot\frac{1}{\left(-\omega^2 + i\Gamma_d\cdot\omega\right)}$$

$$\chi(\omega) = \frac{\omega_p^2}{\left(-\omega^2 + i\Gamma_d\cdot\omega\right)} \tag{8}$$

plasma frequency $\omega_p = \sqrt{\frac{Ne^2}{\varepsilon_0 \mu}}$.

The dielectric function is then given by this relation:

$$\varepsilon(\omega) = 1 + \chi(\omega) = 1 + \frac{\omega_p^2}{\left(-\omega^2 + i\Gamma_d \cdot \omega\right)} \tag{9}$$

Note that for $\omega \to \infty$, $\varepsilon(\infty) = 1$, then it is possible to rewrite the dielectric function as:

$$\varepsilon(\omega) = \varepsilon_\infty + \frac{\omega_p^2}{-\omega^2 + i\Gamma_d \cdot \omega} \tag{10}$$

$$\varepsilon(\omega) = \varepsilon_1(\omega) + i\varepsilon_2(\omega)$$

The real $\varepsilon_1(\omega)$ and imaginary parts $\varepsilon_2(\omega)$ of the dielectric function are given by:

$$\varepsilon_1(\omega) = \varepsilon_\infty + \frac{\omega_p^2}{-\omega^2 + \Gamma_d^2}$$

$$\varepsilon_2(\omega) = \frac{\omega_p^2 \cdot \Gamma_d}{\omega \cdot \left(\omega^2 + \Gamma_d^2\right)}$$

Behavior of the Drude dielectric function

- ω_p is the frequency of the plasma corresponding to the position of the photon energy where $\varepsilon_\infty(\omega)$ is approximately equal to zero. When ω_p increases, then the amplitude of $\varepsilon_2(\omega)$ and $\varepsilon_2(\omega)$ also increases.
- If $\omega < \omega_p$ then the real part is negative: any electric field cannot penetrate the metal, which is totally reflective. The optical constants of the material are complex.
- If $\omega = \omega_p$ then the real part of the dielectric function is zero. This means that all electrons oscillate in phase over the entire propagation length of the material.
- If, $\omega > \omega_p$ the reflectivity decreases and the metal becomes transparent. The refractive index of the material is almost real.

Since conduction electrons are considered free, the Drude oscillator is an extension of a single Lorentz oscillator to a case where the restoring force and resonant frequency are zero ($\Gamma_0 = 0, \omega_t = 0$).

4 Lorentz Model

The classical theory of Lorentz (1878) [25–27] is based on the classical theory of the interaction between light and matter and it is used to describe the frequency-dependent polarization due to bound charge. The Lorentzian dispersion formula comes from the solution of the equation of an electron bound to a nucleus driven by an oscillating electric field E. The response is equivalent to the conventional mass on a spring that has damping and external driving force. It generates damped harmonic oscillations.

This model can be used to describe the frequency response of many materials and generally shows a strong dispersion around the resonant frequency [28–30]. It is best suited to materials that have bonded electrons, with the possibility of having many oscillators in a given system [31–34]. It is valid for semiconductors and insulators. It is a model of the forced and damped harmonic oscillator (Fig. 1).

If we consider the simplified case of a system of two atoms of reduced mass μ, bound by a constant of force K, in a state of forced vibration with respect to each other because of the coupling of the dipole moment and the electric field $\vec{E}$, the equation of motion is written:

$$\mu \frac{d^2\vec{R}}{dt^2} = -\mu_0 \frac{d\vec{R}}{dt} - K\vec{R} - e\vec{E} \tag{11}$$

where e: effective load and Γ_0: the damping coefficient.

The solution of this equation is written as a plane wave:

$$\vec{R} = \vec{R}_0 \exp\left[i(\omega t - \vec{q} \cdot \vec{r})\right] \tag{12}$$

Substituting in the previous equation of motion, we have the solution:

$$\vec{R} = \frac{-e \cdot \vec{E}}{\mu\left(\frac{K}{\mu} - \omega^2 + i\Gamma_0 \cdot \omega\right)}$$

$$\vec{R} = \frac{1}{\mu} \cdot \frac{-e \cdot \vec{E}}{\left(\omega_t^2 - \omega^2 + i\Gamma_0 \cdot \omega\right)} \tag{13}$$

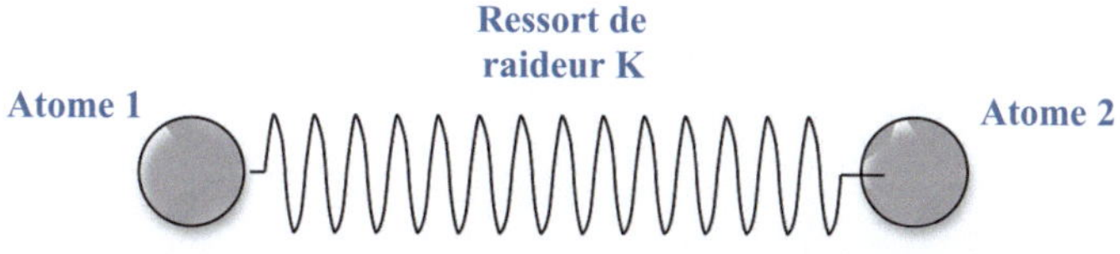

Fig. 1 Simplified representation of the Lorentz oscillator model for two atoms

where $\omega_t = \sqrt{\frac{K}{\mu}}$ is the natural frequency of the oscillator.

The induced dipole moment $\vec{d}$ is related to $\vec{R}$ by this relationship:

$$\vec{d} = -e \cdot \vec{R}$$

$$\vec{d} = \frac{e^2}{\mu} \cdot \frac{\vec{E}}{\left(\omega_t^2 - \omega^2 + i\Gamma_0 \cdot \omega\right)} \tag{14}$$

The polarizability $\alpha(\omega)$ is given by where: $d(\omega) = \alpha(\omega) \cdot E(\omega)$

$$\alpha(\omega) = \frac{e^2}{\mu} \cdot \frac{1}{\left(\omega_t^2 - \omega^2 + i\Gamma_0 \cdot \omega\right)} \tag{15}$$

Taking the sum of the dipole moment of a single atom over all the atoms of a volume, it follows that the polarization per unit volume is given by:

$$\vec{P} = N \cdot \alpha(\omega) \cdot E(\omega) = \varepsilon_0 \cdot \chi(\omega) \cdot E(\omega)$$

where N is the oscillator density per unit volume.

Electrical susceptibility χ is the impulse response of an atomic resonance, obtained from the previous equation:

$$\chi(\omega) = \left(\frac{N \cdot e^2}{\varepsilon_0 \mu}\right) \cdot \frac{1}{\left(\omega_T^2 - \omega^2 + i\Gamma_0 \cdot \omega\right)}$$

$$\chi(\omega) = \frac{\omega_p^2}{\left(\omega_t^2 - \omega^2 + i\Gamma_0 \cdot \omega\right)} \tag{16}$$

plasma frequency $\omega_p = \sqrt{\frac{Ne^2}{\varepsilon_0 \mu}}$.

The dielectric function is then given by this relation:

$$\varepsilon(\omega) = 1 + \chi(\omega) = 1 + \frac{\omega_p^2}{\left(\omega_t^2 - \omega^2 + i\Gamma_0 \cdot \omega\right)} \tag{17}$$

The limits and the dielectric function respectively at low and high frequencies are given by:$\varepsilon_s \varepsilon_\infty$

$$\begin{cases} \varepsilon_s = \varepsilon(\omega \to 0) = 1 + \dfrac{\omega_p^2}{\omega_t^2} \\ \varepsilon_\infty = \varepsilon(\omega \to \infty) = 1 \end{cases}$$

The expression of the dielectric function for a single Lorentz oscillator is given by:

$$\varepsilon(\omega) = \varepsilon_\infty + \frac{(\varepsilon_s - \varepsilon_\infty) \cdot \omega_t^2}{\left(\omega_t^2 - \omega^2 + i\Gamma_0 \cdot \omega\right)} \tag{18}$$

If ε_∞ the dielectric constant is at high frequency, it takes into account the contribution of the high-energy inter-band transition. Generally, $\varepsilon_\infty = 1$ but can be greater than 1 if higher energy oscillators exist and are not taken into account. The constant $\varepsilon_s(\varepsilon_s > \varepsilon_\infty)$ gives the value of the static dielectric function at zero frequency.

The difference $\varepsilon_s - \varepsilon_\infty$ represents the strength of the single Lorentz oscillator. The larger Γ_0 it is, the smaller the width of the peak of the single Lorentz oscillator.

ω_t is the resonant frequency of the oscillator whose energy corresponds to the absorption peak.

If there is more than one oscillator, the dielectric function is assumed to be equal to the sum of the contributions of the different oscillators.

For N Lorentz oscillators:

$$\varepsilon(\omega) = \varepsilon_\infty + \frac{(\varepsilon_s - \varepsilon_\infty) \cdot \omega_t^2}{\left(\omega_t^2 - \omega^2 + i\Gamma_0 \cdot \omega\right)} + \sum_{j=1}^{N} \frac{f_j \cdot \omega_{0j}^2}{\omega_{0j}^2 - \omega^2 + i\gamma_j \cdot \omega} \tag{19}$$

Provided that, $\omega \neq 0$, f_i is the oscillator force defined by:

$$f_j = \frac{e_j^2 \cdot N_j}{\varepsilon_0 \mu_j \cdot \omega_{0j}^2} \tag{20}$$

ω_{0j} (j = 1, 2 … N) is the resonant energy (peak) of an oscillator for a set of several Lorentzian oscillators.

γ_j (j = 1, 2 … N) is the broadening parameter corresponding to the peak energy of each oscillator.

5 Drude-Lorentz Model

The Lorentz oscillator model works well for insulators and semiconductors above the band gap. Drude's model describes the optical properties of metals well but does not take into account the notion of the energy of the optical bandgap E_g.

The combination of the two models is often adequate when the material is somewhat conductive and has a metallic character, hence the Drude-Lorentz model:

$$\varepsilon(\omega) = \varepsilon_\infty + \frac{(\varepsilon_s - \varepsilon_\infty) \cdot \omega_t^2}{\omega_t^2 - \omega^2 + i\Gamma_0 \cdot \omega}$$

$$+\sum_{j=1}^{N}\frac{f_j\cdot\omega_{0j}^2}{\omega_{0j}^2-\omega^2+i\gamma_j\cdot\omega}+\frac{\omega_p^2}{-\omega^2+i\Gamma_d\cdot\omega} \tag{21}$$

6 The Optical Constants of a Semiconductor

The interaction of semiconductors with light is of decisive importance for photonic and optoelectronic devices as well as for the characterization of semiconductor properties. When light hits a semiconductor, reflection, transmission, and absorption are considered, as with any dielectric material.

The semiconductor's response depends largely on the photon energy (or wavelength) of light, and various processes contribute to the dielectric function. The energy and wavelength of a photon are related by: $E = h\upsilon = hc/\lambda$.

In general, in a semiconductor, the propagation of an electromagnetic wave depends on the external pressure and temperature and can be modified by electric or magnetic fields. When a semiconductor has no energy state between the valence band and the conduction band, it is transparent to light characterized by a frequency of ω below the critical frequency ω_0, where $\omega_0 = E_g/\hbar$, E_g the gap energy of the semiconductor and $\hbar$ the Planck constant reduced.

In this section, we will mention the different constants and optical properties of this material.

6.1 *The Refractive Index N*

In optics, the refractive index (also known as the refractive index) of a material is a dimensionless number that describes the speed at which light travels through the material. It is defined as: $n = c/v$.

Let be an electromagnetic plane wave that propagates according to (oz) and polarized according to (ox).

Applying Maxwell's equations, the component E_x of the electric field vector E therefore satisfies the differential equation:

$$\mu\varepsilon\frac{d^2E_x(z)}{dt^2}+4\psi\mu\frac{dE_x(z)}{dt}=c^2\frac{\partial^2E_x(z)}{\partial z^2} \tag{22}$$

where:

c: the speed of light in a vacuum.

μ: permeability ($\mu = 1$).

ε: the dielectric constant of the material.

σ: the electrical conductivity of the semiconductor.

The solution of this equation is of the form:

$$E_x = E_0 \exp\left[i(\omega t - q \cdot z)\right] \tag{23}$$

Furthermore, with respect to the angular frequency of the wave ω, which indicates the angular change per unit of time, and the wavenumber (or angular wavenumber) q, which represents the proportionality between the angular frequency ω and the linear velocity (propagation velocity): $v = \frac{\omega}{q}$.

Equation (23) becomes:

$$E_x = E_0 \exp\left[i\omega\left(t - \frac{z}{v}\right)\right] \tag{24}$$

With v = c/N is the propagation velocity of the wave in this medium, and N = n − ik represents the complex refractive index of the semiconductor.

n: the real part refractive index, it affects the phase of the electromagnetic wave.

k: the extinction coefficient, or the attenuation index, it acts on the amplitude of the wave.

We can write:

$$E_x = E_0 e^{\left(-\omega \frac{kz}{c}\right)} \cdot \exp\left[i\omega\left(t - \frac{nz}{c}\right)\right] \tag{25}$$

We note the exponential decrease in the amplitude of the plane electromagnetic wave.

By substituting this expression in Eq. (22) we have:

$$\varepsilon = i\left(\frac{4\psi}{\omega} - 2nk\right) + n^2 - k^2 \tag{26}$$

Therefore

$$\varepsilon = n^2 - k^2; \quad nk = \frac{2\psi}{\omega} \tag{27}$$

Hence: $n^2 - k^2 = (n + k)(n - k)$

$$\varepsilon^2 = (n + k)^2 (n - k)^2 = \left(n^2 + k^2\right)^2 - \left(\frac{4\psi}{\omega}\right)^2 \tag{28}$$

This led us to find the following equations:

$$n^2 = \frac{\varepsilon}{2} \cdot \left(\sqrt{1 + \left(\frac{4\psi}{\varepsilon\omega} \right)^2} + 1 \right) \tag{29}$$

$$k^2 = \frac{\varepsilon}{2} \cdot \left(\sqrt{1 + \left(\frac{4\psi}{\varepsilon\omega} \right)^2} - 1 \right) \tag{30}$$

6.2 The Absorption Coefficient *α*

In the absorption process, energy is transferred from the electromagnetic field to the semiconductor. In the case of a linear absorption process, when the probability of light absorption is proportional to the incoming intensity, the decrease in intensity in the absorbing medium is exponential (Lambert–Beer law):

$$I = I_0 \exp(-\alpha z) \tag{31}$$

I is the intensity of the electromagnetic wave passing through a thickness z of the semiconductor is proportional to the square of the amplitude, averaged over time over many oscillations of the wave, which is equivalent to:

$$\begin{aligned} I &= |E_x|^2 = \left| E_0 e^{\left(-\omega \frac{kz}{c}\right)} \cdot \exp\left[i\omega\left(t - \frac{nz}{c} \right) \right] \right|^2 \\ I &= |E_x|^2 = E_0 e^{\left(-\omega \frac{kz}{c}\right)} \cdot \exp\left[i\omega\left(t - \frac{nz}{c} \right) \right] \\ &\quad \cdot E_0 e^{\left(-\omega \frac{kz}{c}\right)} \cdot \exp\left[-i\omega\left(t - \frac{nz}{c} \right) \right] \\ I &= E_0^2 \cdot \exp\left(-\frac{2\omega kz}{c} \right) \end{aligned}$$

We ask, $I_0 = E_0^2$

$$I = I_0 \exp\left(-\frac{2\omega kz}{c} \right) \tag{32}$$

And according to Eqs. (31) and (32):

$$\alpha = \frac{2\omega k}{c} \tag{33}$$

The absorption coefficient can be measured by a photomultiplier.

6.3 The Complex Dielectric Constant ε

The dissipation fluctuation theorem links the real part of the dielectric constant $\varepsilon_1(\omega)$ ω and its imaginary part, by the Kramers–Kronig relations $\varepsilon_2(\omega)$ [35], which are relations between the real and imaginary parts of the dielectric function. They are general in nature and are based on the properties of a complex analytic response function $f(\omega) = f_1(\omega) + if_2(\omega)$

$$f(\omega) = \frac{1}{\pi i}\mathrm{P}\left\{\int\limits_{-\infty}^{\infty}\frac{\mathrm{f}(s)}{\mathrm{s}-\omega}\mathrm{ds}\right\} \tag{34}$$

where P{….} is the principal value of Cauchy

We equation the real parts of Eq. (34) to obtain:

$$f_1(\omega) = \frac{1}{\pi}\mathrm{P}\left\{\int\limits_{-\infty}^{\infty}\frac{f_2(s)}{\mathrm{s}-\omega}\mathrm{ds}\right\} = \frac{1}{\pi}\mathrm{P}\left\{\int\limits_{0}^{\infty}\frac{f_2(s)}{\mathrm{s}-\omega}\mathrm{ds} + \int\limits_{-\infty}^{0}\frac{f_2(s)}{\mathrm{s}-\omega}\mathrm{ds}\right\}$$

$$= \frac{1}{\pi}\mathrm{P}\left\{\int\limits_{0}^{\infty}\frac{f_2(s)}{\mathrm{s}-\omega}\mathrm{ds} - \int\limits_{0}^{-\infty}\frac{f_2(s)}{\mathrm{s}-\omega}\mathrm{ds}\right\}$$

$$f_1(\omega) = \frac{1}{\pi}\mathrm{P}\left\{\int\limits_{0}^{\infty}\frac{f_2(s)}{\mathrm{s}-\omega}\mathrm{ds} - \int\limits_{0}^{\infty}\frac{f_2(-s)}{-\mathrm{s}-\omega}\mathrm{d}(-\mathrm{s})\right\}$$

According to the Kramers–Kronig conditions, the function $f_1(\omega)$ is even, and the function $f_2(\omega)$ is odd with respect to the real ω: $f_1(-\mathrm{s}) = f_1(\mathrm{s}); f_2(-\mathrm{s}) = -f_2(\mathrm{s})$; $\frac{1}{\mathrm{s}+\omega} + \frac{1}{\mathrm{s}-\omega} = \frac{2\mathrm{s}}{\mathrm{s}^2-\omega^2}$

$$f_1(\omega) = \frac{1}{\pi}\mathrm{P}\left\{\int\limits_{0}^{\infty}\frac{f_2(s)}{\mathrm{s}-\omega}\mathrm{ds} + \int\limits_{0}^{\infty}\frac{f_2(s)}{\mathrm{s}+\omega}\mathrm{d}(\mathrm{s})\right\}$$

$$f_1(\omega) = \frac{2}{\pi}\mathrm{P}\left\{\int\limits_{0}^{\infty}\frac{\mathrm{s}\cdot f_2(s)}{s^2-\omega^2}\mathrm{ds}\right\} \tag{35}$$

This is one of the Kramers–Kronig relationships. The other relation follows the assimilation of the imaginary parts of Eq. (34):

$$f_2(\omega) = -\frac{1}{\pi}\mathrm{P}\left\{\int\limits_{-\infty}^{\infty}\frac{f_1(s)}{\mathrm{s}-\omega}\mathrm{ds}\right\} = -\frac{1}{\pi}\mathrm{P}\left\{\int\limits_{-\infty}^{0}\frac{f_1(s)}{\mathrm{s}-\omega}\mathrm{ds} + \int\limits_{0}^{\infty}\frac{f_1(s)}{\mathrm{s}-\omega}\mathrm{ds}\right\}$$

$$f_2(\omega) = -\frac{1}{\pi}\mathrm{P}\left\{\int_{-\infty}^{\infty} \frac{f_1(s)}{s-\omega}\mathrm{d}s\right\} = -\frac{1}{\pi}\mathrm{P}\left\{-\int_{0}^{-\infty} \frac{f_1(s)}{s-\omega}\mathrm{d}s + \int_{0}^{\infty} \frac{f_1(s)}{s-\omega}\mathrm{d}s\right\}$$

$$f_2(\omega) = -\frac{1}{\pi}\mathrm{P}\left\{-\int_{0}^{\infty} \frac{f_1(-s)}{-s-\omega}\mathrm{d}(-s) + \int_{0}^{\infty} \frac{f_1(s)}{s-\omega}\mathrm{d}s\right\}$$

$$f_2(\omega) = -\frac{1}{\pi}\mathrm{P}\left\{-\int_{0}^{\infty} \frac{f_1(s)}{s+\omega}\mathrm{d}(s) + \int_{0}^{\infty} \frac{f_1(s)}{s-\omega}\mathrm{d}s\right\}$$

$$f_2(\omega) = -\frac{1}{\pi}\mathrm{P}\left\{\int_{0}^{\infty} \frac{-(s-\omega)+(s+\omega)}{(s+\omega)(s-\omega)} f_1(s)\cdot \mathrm{d}(s)\right\}$$

$$f_2(\omega) = -\frac{1}{\pi}\mathrm{P}\left\{\int_{0}^{\infty} \frac{2\omega}{s^2-\omega^2} f_1(s)\cdot \mathrm{d}(s)\right\}$$

Hence

$$f_2(\omega) = -\frac{2\omega}{\pi}\mathrm{P}\left\{\int_{0}^{\infty} \frac{f_1(s)}{s^2-\omega^2}\mathrm{d}(s)\right\} \tag{36}$$

For the complex dielectric function, it is written as: $\varepsilon(\omega)\varepsilon_1(\omega)+i\varepsilon_2(\omega)$, and according to Kramers–Kronig, the susceptibility can be interpreted as the Fourier transform of the time-dependent polarization in the semiconductor after an infinitely short pulsed electric field, i.e., the impulse response of the polarization: $f(\omega) = \chi(\omega) = \chi_1(\omega) + i\chi_2(\omega)$

$$\chi_1(\omega) = \frac{2}{\pi}\mathrm{P}\left\{\int_{0}^{\infty} \frac{s\cdot\chi_2(s)}{s^2-\omega^2}\mathrm{d}s\right\} \tag{37}$$

$$\chi_2(\omega) = -\frac{2\omega}{\pi}\mathrm{P}\left\{\int_{0}^{\infty} \frac{\chi_1(s)}{s^2-\omega^2}\mathrm{d}(s)\right\} \tag{38}$$

And since $\varepsilon(\omega) = 1 + \chi(\omega)$, we find the Kramers–Kronig relations concern the dielectric constant:

$$\varepsilon_1(\omega) = \frac{2}{\pi}\mathrm{P}\left\{\int_{0}^{\infty} \frac{s\cdot\chi_2(s)}{s^2-\omega^2}\mathrm{d}s\right\} \tag{39}$$

$$\varepsilon_2(\omega) = -\frac{2\omega}{\pi} P\left\{ \int_0^{\infty} \frac{\chi_1(s)}{s^2 - \omega^2} d(s) \right\} \tag{40}$$

With $\begin{cases} \chi_1(s) = \varepsilon_1(s) - 1 \\ \chi_2(s) = \varepsilon_2(s) \end{cases}$

$$\varepsilon_1(\omega) = 1 + \frac{2}{\pi} P\left\{ \int_0^{\infty} \frac{s \cdot \varepsilon_2(s)}{s^2 - \omega^2} ds \right\} \tag{41}$$

$$\varepsilon_2(\omega) = -\frac{2\omega}{\pi} P\left\{ \int_0^{\infty} \frac{\varepsilon_1(s) - 1}{s^2 - \omega^2} d(s) \right\} \tag{42}$$

Equivalent relationships for the refractive index n and the extinction coefficient k will be demonstrated. Remember that the refractive index and susceptibility are linked:

$$n(s) + ik(s) = \sqrt{1 + \chi(s)}$$

For small susceptibilities (i.e. $|\chi(s)| < 1$), the above equation can be simplified using the binomial approximation:

$$n(s) + ik(s) = 1 + \frac{\chi(s)}{2} = 1 + \frac{\chi_1(s)}{2} + i\frac{\chi_2(s)}{2}$$

Putting the real and imaginary parts of the two sides of the equation on an equal footing, we obtain:

$$\begin{cases} n(s) = 1 + \dfrac{\chi_1(s)}{2} \\ k(s) = \dfrac{\chi_2(s)}{2} \end{cases} \Rightarrow \begin{cases} 2n(\omega) - 2 = \chi_1(\omega) \\ 2k(\omega) = \chi_2(\omega) \end{cases}$$

According to Eqs. (37) and (38), the equations of n and k can be concluded in this way:

$$n(\omega) = 1 + \frac{2}{\pi} P\left\{ \int_0^{\infty} \frac{s \cdot k(s)}{s^2 - \omega^2} ds \right\} \tag{43}$$

$$k(\omega) = -\frac{2\omega}{\pi} P\left\{ \int_0^{\infty} \frac{n(s) - 1}{s^2 - \omega^2} d(s) \right\} \tag{44}$$

Since the absorption coefficient is related to the extinction coefficient by it follows: $\alpha(\omega) = \frac{2\omega k(\omega)}{c}$ *and* $k(s) = \frac{c}{2s}\alpha(s)$

$$n(\omega) = 1 + \frac{c}{\pi} P\left\{\int_0^\infty \frac{\alpha(s)}{s^2 - \omega^2} ds\right\} \tag{43}$$

$$\alpha(\omega) = -\frac{4\omega^2}{\pi c} P\left\{\int_0^\infty \frac{n(s) - 1}{s^2 - \omega^2} d(s)\right\} \tag{44}$$

The refractive index n is equal to $n = \sqrt{\varepsilon\mu}$ where ε is the relative permittivity of the material and μ is its relative permeability. Most natural materials are non-magnetic at optical frequencies, i.e. μ is very close to 1, so n is approximately $\sqrt{\varepsilon}$. In this particular case, the complex relative permittivity ε, with the real and imaginary parts ε_1 and ε_2, and the complex refractive index N, with the real and imaginary parts n and k follow the relationship:

$$\varepsilon = \varepsilon_1(\omega) + i\varepsilon_2(\omega) = N^2 \tag{45}$$

$$\begin{aligned} N^2 &= (n - ik)^2 = n^2 - k^2 - 2ink \\ \varepsilon_1(\omega) &= n^2 - k^2; \quad \varepsilon_2(\omega) = 2nk \end{aligned} \tag{46}$$

And consequently we have the relations that link n and k to ε_1 and ε_2:

$$n = \frac{1}{\sqrt{2}}\left\{\varepsilon_1 + \sqrt{\varepsilon_1^2 + \varepsilon_2^2}\right\}^{\frac{1}{2}} \tag{47}$$

$$k = \frac{1}{\sqrt{2}}\left\{-\varepsilon_1 + \sqrt{\varepsilon_1^2 + \varepsilon_2^2}\right\}^{\frac{1}{2}} \tag{48}$$

$$\begin{aligned} \alpha &= \frac{2\omega \cdot k}{c} = \frac{4\pi\nu \cdot k}{c} = \frac{4\pi \cdot k}{\lambda} \\ \alpha &= \frac{4\pi}{\lambda}\frac{1}{\sqrt{2}}\left\{-\varepsilon_1 + \sqrt{\varepsilon_1^2 + \varepsilon_2^2}\right\}^{\frac{1}{2}} \end{aligned} \tag{49}$$

We can thus calculate the variation $n(\omega)$ when the absorption spectrum $\alpha(\omega)$ is known.

6.4 *The Reflectivity R or Reflection Coefficient*

The optical measurements that give the most complete information about the electronic system are the measurements of the reflectivity of light with normal incidence on single crystals. The reflectivity coefficient R is a complex function defined on the surface of the crystal as the ratio between the reflected intensity I_R and the incident intensity I_0:

It is defined as follows:

$$R = \frac{I_R}{I_0} = \frac{\text{l'intensité réflechie}}{l'intensit\acute{e}\ \text{incidente}}$$

The refractive index n and the extinction coefficient K in the crystal are related to the reflectivity at normal incidence by:

$$R = \left| \frac{n + ik - 1}{n + ik + 1} \right|^2$$

where from

$$R = \frac{(n-1)^2 + k^2}{(n+1)^2 + k^2} \tag{50}$$

The reflectivity R undergoes an anomalous dispersion in the infrared and tends to 1 when the frequency ω of the incident wave is practically equal to the plasma frequency of the material, where: $\omega_p = \left(\frac{4\pi n_p e^2}{m^* \varepsilon}\right)^{\frac{1}{2}} n_p$ and: n_P the concentration of charge carriers (electrons, holes, …).

6.5 *Transmittivity T or Transmission Coefficient*

Transmittivity or transmittance coefficient describes the amount of light that passes through a sample without change. In other words, it is light that is not absorbed, scattered, or reflected. In most cases, the scattered and/or reflected light is tiny and insignificant. Transmission is usually measured as a percentage.

This coefficient is defined as the ratio between the intensity of light transmitted I_t through a layer of the z-thickness material and the incident intensity I_0.

$$T = \frac{I_t}{I_0} \tag{51}$$

In general: $T = 1 - R$.

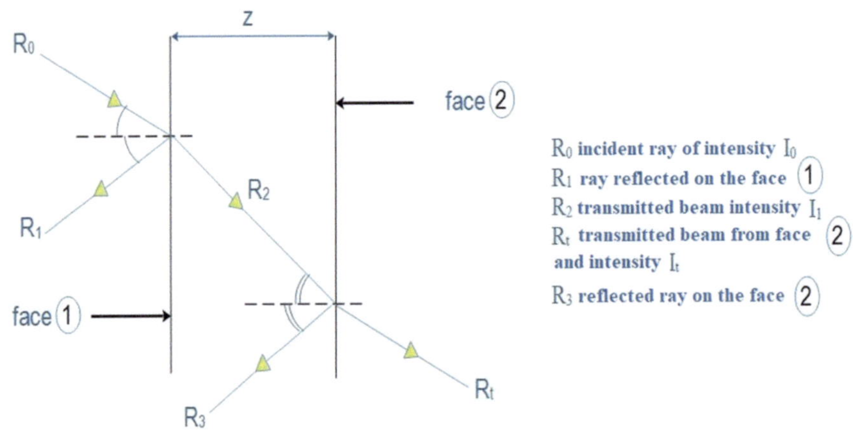

Fig. 2 Vanadium dioxide thin layer

$$T = \frac{(1 - R)^2 \exp(-\alpha z)}{1 - R^2 \exp(-2\alpha z)} \tag{52}$$

Where R is the reflectivity and α is the absorption coefficient of this thin film (Fig. 2).

On the face (1) of this layer we have: therefore $I_0 = I_1 + I_{R_1}$; or $I_{R_1} = RI_0$, $I_1 = (1 - R)I_0$.

On the face (2) of this thin layer, the intensity I which reaches this face is:, and consequently $I = I_1 \exp(-\alpha z)$

$$I = (1 - R)I_0 \exp(-\alpha z)$$

Finally, the transmitted intensity is:

$$I_t = (1 - R)I \cong (1 - R)^2 I_0 \exp(-\alpha z)$$

where from:

$$T = \frac{I_t}{I_0} \cong (1 - R)^2 \exp(-\alpha z) \tag{53}$$

This expression is obtained by neglecting $R^2 \exp(-2\alpha z)$ before 1 in the formula (52).

Therefore, we can calculate the absorption coefficient α and deduce the extinction coefficient k from the spectrum of transmittivity T and reflectivity R of a thin film of known thickness z, using the formula: $k = \frac{\alpha\lambda}{4\pi}$.

7 Conclusion

Based exclusively on theoretical modeling, this chapter demonstrates the ability of the Drude and Lorentz models to accurately account for the optical properties of vanadium dioxide. Drude's model is relevant for describing the behavior of free electrons in the metal phase, while Lorentz's model explains the response of bonded electrons in the semiconductor phase. Their combination makes it possible to reproduce the main characteristics of the complex dielectric function and the associated optical constants. These theoretical results confirm the potential of VO_2 as a material with switchable optical properties, essential for the development of nanotechnological and energy-efficient applications, without the need for numerical calculations or experimental simulations in this phase of the study.

References

1. Li, H., Wang, X., Zhu, X., Duan, X., Pan, A.: Composition modulation in one-dimensional and two-dimensional chalcogenide semiconductor nanostructures. Chem. Soc. Rev. **47**(20), 7504–7521 (2018)
2. Li, H., Wu, X., Liu, H., Zheng, B., Zhang, Q., Zhu, X., et al.: Composition-Modulated Two-Dimensional Semiconductor Lateral Heterostructures via Layer-Selected Atomic Substitution. ACS Nano **11**(1), 9617 (2016)
3. Li, H., Liu, H., Zhou, L., Wu, X., Pan, Y., Ji, W., et al.: Strain-tuning atomic substitution in two-dimensional atomic crystals. ACS Nano **12**(5), 4853–4560 (2018)
4. Morin, F.J.: Oxides which show a metal-to-insulator transition at the neel temperature. Phys. Rev. Lett. **3**, 34–36 (1959). https://doi.org/10.1103/PhysRevLett.3.34
5. Kübler, C., Ehrke, H., Huber, R., Lopez, R., Halabica, A., Haglund, R.F., Leitenstorfer, A.: Coherent structural dynamics and electronic correlations during an ultrafast insulator-to-metal phase transition in VO_2. Phys. Rev. Lett. **99**, 1–4 (2007). https://doi.org/10.1103/PhysRevLett.99.116401
6. Mossanek, R.J.O., Abbate, M.: Optical response of metallic and insulating VO_2 calculated with the LDA approach. J. Phys. Condens. Matter **19**, 346225 (2007). https://doi.org/10.1088/0953-8984/19/34/346225
7. Berezina, O.Y., Velichko, A.A., Lugovskaya, L.A., Pergament, A.L., Stefanovich, G.B.: Metal-semiconductor transition in nonstoichiometric vanadium dioxide films. Inorg. Suppress. **43**, 505–511 (2007). https://doi.org/10.1134/S0020168507050123
8. Cao, J., Gu, Y., Fan, W., Chen, L.Q., Ogletree, D.F., Chen, K., Tamura, N., Kunz, M., Barrett, C., Seidel, J., Wu, J.: Extended mapping and exploration of the vanadium dioxide stress-temperature phase diagram. Nano Lett. **10**, 2667–2673 (2010). https://doi.org/10.1021/NL101457K
9. Zouini, M., Chaib, A.B., Anigrou, Y., Benbrahim, F.Z., El Khattabi, E.M. (2025). A Review of the Dynamics of Metal–Insulator Transitions of Vanadium Dioxide. In: Boutahir, M., El Mehdi, E. (eds) Advanced Materials for Sustainable Energy and Engineering. Engineering Materials. Springer, Cham. https://doi.org/10.1007/978-3-031-75032-8_12
10. El Khattabi, E.M., Zouini, M., Benbrahim, F.Z., Mohammed, O.J. (2025). Energy Performance Optimization in Buildings: A Review of Concepts, Regulations, and Simulation Methods. In: Boutahir, M., El Mehdi, E. (eds) Advanced Materials for Sustainable Energy and Engineering. Engineering Materials. Springer, Cham. https://doi.org/10.1007/978-3-031-75032-8_25

11. Kang, L., Gao, Y., Chen, Z., Du, J., Zhang, Z., Luo, H.: Pt/VO_2 double-layered films combining thermochromic properties with low emissivity. Sol. Energy Mater. Ground. Cells. **94**, 2078–2084 (2010). https://doi.org/10.1016/j.solmat.2010.06.023
12. Kana, J.K., Ndjaka, J.M., Vignaud, G., Gibaud, A., Maaza, M.: Optics Communications **284**(3), 807–812 (2011)
13. Ben Chaib, A., Mdaa, A., Zorkani, I., Jorio, A.: Optical properties of the Vanadium dioxide. J. Adv. Phys. **8**, 2148–2155 (2015)
14. Ben Chaib, A., Mdaa, A., Zorkani, I., Jorio, A.: Optical properties of a thin layer of the Vanadium dioxide at the metal state. J. Adv. Phys. **9**, 1–8 (2015)
15. Ben Chaib, A., Mohammed, Z., Zorkani, I., Mdaa, A., Jorio, A.: Study of the absorptance of a thin layer of the Vanadium dioxide. Ind. J. Sci. Technol. 9 (2016). https://doi.org/10.17485/ijst/2016/v9i48/100111
16. Lappalainen, J., Heinilehto, S., Jantunen, H., Lantto, V.: Electrical and optical properties of metal-insulator-transition VO_2 thin films. J. Electroceram. **22**(1–3), 73–77 (2008)
17. Cavalleri, A., Dekorsy, T., Chong, H.H.W., Kieffer, J.C., Schoenlein, R.W.: Evidence for a structurally-driven insulator- to-metal transition in VO2: A view from the ultrafast timescale. Phys. Rev. B, **70** (2004)
18. Eyert, V.: The metal-insulator transitions of VO_2: a band theoretical approach. Ann. Phys (2002). Vanadium oxide section. In: Gmelin Handbook, vol. 16, p. 482 (1990)
19. Paradis, S., Laou, P., Alain, D., Doped Vanadium dioxide with enhanced infrared modulation. In: Technical Memorandum DRDC-VALCARTIER-TM-2007-002, vol. 4 (2007)
20. Andersson, G., Paju, J., Lang, W., Berndt, W.: Studies on Vanadium oxides. I. phase analysis. Acta Chemica Scandinavica. Danish Chem Soc 8, 1599–606 (1954)
21. Drude, P.: Zur Elektronentheorie der metalle. Annalen der Physik **306**(3), 566 (1900)
22. Drude, P.: Zur Elektronentheorie der Metalle; II. Teil. Galvanomagnetische und thermomagnetische Effecte. Annalen der Physik **308**(11), 369 (1900)
23. Fox, M.: Optical Properties of Solids, 2nd edn., pp. 33–44. Oxford University Press, Oxford (2010)
24. Chuang, S.: Physics of Photonic Devices, 2nd edn., pp. 193–196. Wiley, New York (2009)
25. Lorentz, H.A.: The motion of electrons in metallic bodies. I Proceedings Koninklijke Akademie van Wetenschappen **7**, 438–453 (1905)
26. Lorentz, H.A.: The motion of electrons in metallic bodies. II Proceedings Koninklijke Akademie van Wetenschappen **7**, 585–593 (1905)
27. Lorentz, H.A.: The motion of electrons in metallic bodies. III Proceedings Koninklijke Akademie van Wetenschappen **7**, 684–691 (1905)
28. Rivas, J.G., Snchez-Gil, J.A., Kuttge, M., Bolivar, P.H., Kurz, H.: Optically switchable mirrors for surface plasmon polaritons propagating on semiconductor surfaces. Phys. Rev. B **74**, 245–324 (2006)
29. Huang, Y., Ho, S.T.: Computational model of solid state, molecular, or atomic media for FDTD simulation based on a multilevel multi-electron system governed by Pauli exclusion and Fermi-Dirac thermalization with application to semiconductor photonics. Opt. Express **14**, 3569–3587 (2006)
30. Hryciw, A., Jun, Y.C., Brongersma, M.L.: Plasmonic: electrifying plasmonics on silicon. Nat. Mater. **9**, 3–4 (2010)
31. Lee, K.H., Ahmed, I., Goh, R.S.M., Khoo, E.H., Li, E.P., Hung, T.G.G.: Implementation of the FDTD method based on the Lorentz-Drude model on GPU for plasmonics applications. Prog. Electromagnet. Res. **116**, 441–456 (2011)
32. Janke, C., Rivas, J.G., Bolivar, P.H., Kurz, H.: All optical switching of electromagnetic radiation through subwavelength apertures. Opt. Lett. **30**(18), 2357–2359 (2005)
33. Ahmed, I., Khoo, E., Kurniawan, O., Li, E.: Modeling and simulation of active plasmonics with the FDTD method by using solid state and Lorentz-Drude dispersive model. J. Opt. Soc. Am. B **28**(3), 325–359 (2011)
34. Van Exter, M., Grischkowsky, D.: Optical and electronics properties of doped silicon from 0.1 to 2 THz. Appl. Phys. Lett. **56**(17), 1694–1696 (1990)

35. de Kronig, R.L., Penney, W.G.: Quantum mechanics of electrons in crystal lattices. Proc. Roy. Soc. A: Math. Phys. Eng. Sci. **130**(814), 499–513 (1931)

Comprehensive Review on Dissipated Energy in Vanadium Dioxide and Its Technological Applications

Mohammed Zouini, Abderrahim Ben Chaib, El Mehdi El Khattabi, Mourad Boutahir, and Kalpana Sahoo

Abstract This chapter focuses on the mathematical modeling and exploitation of dissipated energy in Vanadium Dioxide (VO_2), a smart material exhibiting thermochromic phase transition. The study investigates the hysteresis loop observed during heating and cooling in the infrared spectrum, quantifying the dissipated energy in both pure and doped (5% W and 15% Cr) VO_2 thin films. The mathematical models for transmissivity are presented, and the dissipated energy is calculated as a function of temperature and incident photon frequency. The research highlights two main application domains: information technology and energy. In information technology, the dissipated energy is leveraged to increase data storage speed in optical memories and storage devices, with a particular focus on the impact of W-doping on storage time. In the energy sector, VO_2 nanoparticles are explored as nano-additives in fuels to reduce dissipated energy during combustion, thereby improving engine efficiency and reducing CO_2 emissions. The study concludes that n-type doping (5% W) enhances information storage time, while p-type doping (15% Cr) is promising for fuel additives due to its ability to reduce dissipated energy.

Keywords Vanadium dioxide (VO_2) · Hysteresis · Optical memory · Nano-additives · Dissipated energy

M. Zouini (✉) · A. B. Chaib
Sidi Mohamed Ben Abdellah University ENS, Fez, Morocco
e-mail: mohammedzouini@gmail.com

M. Zouini · E. M. El Khattabi
Systems and Sustainable Environment Laboratory, Private University of Fez UPF, Fez, Morocco

Department of Science, École Normale Supérieure, University Moulay Ismail Meknes, Meknes, Morocco

M. Boutahir
École Normale Supérieure, University Moulay Ismail Meknes, Meknes, Morocco

K. Sahoo
School of Civil Engineering, KIIT Deemed to Be University, Bhubaneswar, Odisha, India

E. El Mehdi and M. Boutahir (eds.), *Advanced Materials for Sustainable Energy and Engineering*, Engineering Materials, https://doi.org/10.1007/978-3-032-10069-6_20

1 Introduction

Today, nanotechnology has become a very important part of our daily lives, contributing to the significant improvement of many technological and industrial sectors such as information technology, homeland security, medicine, transportation, energy, food security, environmental science, and many more. Moreover, in the context of scientific and technological developments, nanotechnology has become the main focus of research for all researchers. Many of the benefits of nanotechnology depend on the ability to adapt material structures at scales below 100 nm to obtain nanomaterials composed of nanoparticles, which greatly expands the concepts of materials science. These nanoparticles are included as nano-additives in different fields. Titanium dioxide nanoparticles TiO_2 in sunscreens [1], nanoparticles in food packaging [2]; zinc oxide nanoparticles ZnO in sunscreens, exterior coatings and paints [1, 3]; cerium oxide nanoparticles CeO_2 [4, 5] acting as catalysts for fuels.

This is leading to new discoveries such as smart materials. In particular, there are materials that have the ability to receive information from their environment and interact with this environment by modifying their properties (optical, electrical, mechanical, etc.) following external stimuli. We investigated vanadium dioxide VO_2, a material that is currently being researched for the development of new technologies, and the possibility of integrating nanoparticles of this material into the energy domain, knowing that it is considered a smart material.

In the infrared spectrum, during the heating and cooling of this material, a hysteresis loop appeared. In physics, a hysteresis loop represents an amount of energy dissipated [6, 7] in this material by passing through this loop completely or partially. We will try to make a mathematical modeling of the optical hysteresis loop [8, 9], throughout the VO_2 (semiconductor) $\rightleftarrows$ VO_2 (metal) phase transition, during heating followed by cooling between temperatures $\theta_1 = 30$ °C and $\theta_2 = 100$ °C or between $\theta_1 = 30$ °C and θ_i such that $\theta_1 < \theta_i < \theta_2$.

In this chapter, we discuss in particular the simulation and analysis of the energy dissipated in its pure and doped state with 5% W and 15% Cr of vanadium dioxide, and the exploitation of this dissipated energy in this material which allows it to be a promising material in different fields. In computer science; the use of this dissipated energy as a means of increasing the speed of data storage in optical memories and storage masses.

In the field of energy: the use of vanadium dioxide nanoparticles as nano-additives in fuels to reduce the energy dissipated by combustion.

2 Mathematical Modeling of Energy Dissipated in Vanadium Dioxide

We based ourselves on the bibliographic results referenced by [10, 11], concerning the study of the T transmittivity of a thin film of vanadium dioxide.

According to Fig. 1 and as shown in Scheme 1, we model the mathematical expression of the T transmittivity for the pure and doped state (5% W and 15% Cr) of Vanadium dioxide in different temperature ranges θ.

In the pure state of VO_2, for a temperature of 5 °C <θ< 30 °C, the expression of the transmittivity T, in heating and cooling, is:

$$\begin{cases} T = a\theta + C_1 \\ a \to 0; \quad C_1 = cste \end{cases}$$

For a temperature of 5 °C < θ < 30 °C, the transmittivity is constant and takes the value

$$T = C_1 = 40.3\%$$

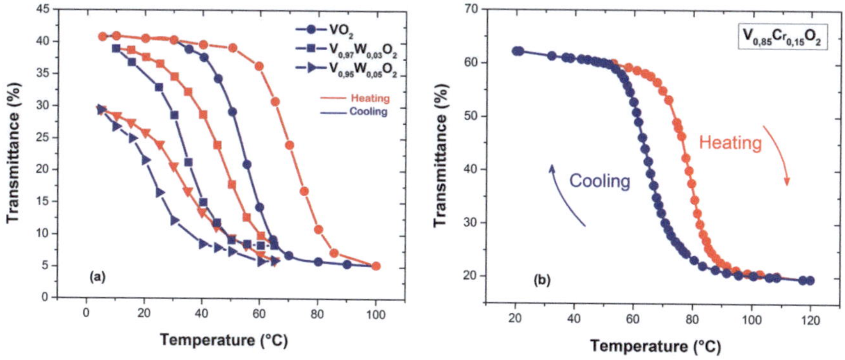

Fig. 1 Transmittivity of a thin film of vanadium dioxide during heating and cooling for λ = 2000 nm. **a** VO_2 W-doped by Batista et al. [10]. **b** VO_2 doped with Cr by Beteille and Livage [11]

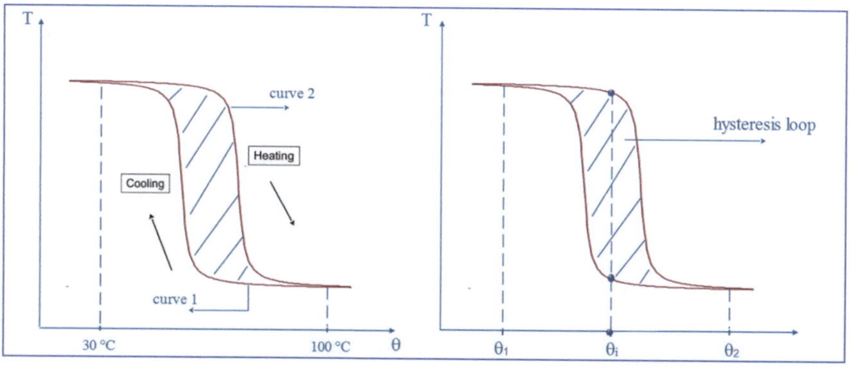

Scheme 1 The variation of the T transmittivity of a thin film of vanadium dioxide with the hysteresis loop

When the limits of the temperature range $\theta_1 < \theta << \theta_2$ take the values $\theta_1 = 30$ °C and $\theta_2 = 100$ °C, the transmittivity curve T becomes exponential, and the mathematical expressions T in heating and cooling are:

$$\text{By heating:} \begin{cases} T_1 = A\exp(-b_1\theta) \\ b_1 = \text{cste} > 0; \quad A = \text{cste} \end{cases}$$

$$\text{Cooling:} \begin{cases} T_2 = B\exp(-b_2\theta) \\ b_2 = \text{cste} > 0; \quad B = \text{cste} \end{cases}$$

where, A, B, b_1 and b_2 are constants that can be determined graphically, from the hysteresis curve:

$$A = 4.100765\%; B = 3.956749\%; b_1 = 0.040547(°C)^{-1}; b_2 = 0.0532(°C)^{-1}.$$

For the VO_2 doped state of 5% W, the temperature limits $\theta_1 < \theta < \theta_2$, takes the values $\theta_1 = 5$ °C and $\theta_2 = 65$ °C, and the mathematical expressions T in heating and cooling are:

$$\text{By heating:} \begin{cases} T_1 = A\exp(-b_1\theta) \\ b_1 = \text{cste} > 0; \quad A = \text{cste} \end{cases}$$

$$\text{Cooling:} \begin{cases} T_2 = B\exp(-b_2\theta) \\ b_2 = \text{cste} > 0; \quad B = \text{cste} \end{cases}$$

where, $A = 0.508193\%$; $B = 0.543532\%$; $b_1 = 0.033509$ $(°C)^{-1}$; $b_2 = 0.04621$ $(°C)^{-1}$

For the state doped VO_2 by 15% Cr, the temperature limits $\theta_1 < \theta < \theta2$, takes the values $\theta_1 = 53$ °C and $\theta_2 = 117$ °C, and the mathematical expressions T in heating and cooling are:

$$\text{By heating:} \begin{cases} T_1 = A\exp(-b_1\theta) \\ b_1 = \text{cste} > 0; \quad A = \text{cste} \end{cases}$$

$$\text{Cooling:} \begin{cases} T_2 = B\exp(-b_2\theta) \\ b_2 = \text{cste} > 0; \quad B = \text{cste} \end{cases}$$

where, $A = 4.045339\%$; $B = 16.135495\%$; $b_1 = 0.036154$ $(°C)^{-1}$; $b_2 = 0.048012$ $(°C)^{-1}$

We calculate the area I between two curves 1 and 2 of the hysteresis loop in the VO_2 material during this up-and-down variation as shown in Scheme. 1, by Eq. (1):

$$|I| = \frac{E_{dissipée}}{h\nu} = \frac{E_{dissipée}}{\hbar\omega} \tag{1}$$

We ask:

$$I_1 = \int_{\theta_1}^{\theta_2} T_1 d\theta; \quad I_2 = \int_{\theta_2}^{\theta_1} T_2 d\theta \tag{2}$$

$$I = I_2 - I_1 \tag{3}$$

Thus:

$$I = -\int_{\theta_1}^{\theta_2} T_1 d\theta + \int_{\theta_2}^{\theta_1} T_2 d\theta \tag{4}$$

$$\begin{aligned}
I &= -\int_{\theta_1}^{\theta_2} A\exp(-b_1\theta)d\theta - \int_{\theta_1}^{\theta_2} B\exp(-b_2\theta)d\theta \\
I &= \frac{A}{b_1}\left[\exp(-b_1\theta)\right]_{\theta_1}^{\theta_2} + \frac{B}{b_2}\left[\exp(-b_2\theta)\right]_{\theta_2}^{\theta_1} \\
I &= \frac{A}{b_1}(\exp(-b_1\theta_2) - \exp(-b_1\theta_1)) + \frac{B}{b_2}(\exp(-b_2\theta_2) - \exp(-b_2\theta_1)) \\
I &= \frac{A}{b_1}\left(e^{-b_1\theta_2} - e^{-b_1\theta_1}\right) + \frac{B}{b_2}\left(e^{-b_2\theta_2} - e^{-b_2\theta_1}\right)
\end{aligned} \tag{5}$$

The energy dissipated in this material during this hysteresis loop is:

$$E_{dissipée} = |I| \cdot h\nu \tag{6}$$

For $\theta_1 < \theta_i < \theta_2$,

$$\begin{cases} 30\,°C < \theta_i < 100\,°C \text{ pour le } VO_2 \text{ pur} \\ 5\,°C < \theta_i < 65\,°C \quad \text{ pour le } VO_2 \text{ W-dopé} \\ 53\,°C < \theta_i < 117\,°C \text{ pour le } VO_2 \text{ Cr-dopé} \end{cases}$$

$$E_{dissipée} = \left|I_2' - I_1'\right| \cdot h\nu,$$

where:

$$I_1' = \int_{\theta_1}^{\theta_i} T_1 d\theta; \quad I_2' = \int_{\theta_2}^{\theta_i} T_2 d\theta \tag{7}$$

$$I' = I'_2 - I'_1 \tag{8}$$

$$I'_1 = \int_{\theta_1}^{\theta_i} A\exp(-b_1\theta)d\theta; \quad I'_2 = \int_{\theta_2}^{\theta_i} B\exp(-b_2\theta)d\theta$$

$$I'_1 = -\frac{A}{b_1}\left[\exp(-b_1\theta)\right]_{\theta_1}^{\theta_i}; \quad I'_2 = -\frac{B}{b_2}\left[\exp(-b_2\theta)\right]_{\theta_2}^{\theta_i}$$

$$I'_1 = -\frac{A}{b_1}(\exp(-b_1\theta_i) - \exp(-b_1\theta_1)); \quad I'_2 = -\frac{B}{b_2}(\exp(-b_2\theta_i) - \exp(-b_2\theta_2))$$

$$I' = \frac{A}{b_1}(\exp(-b_1\theta_i) - \exp(-b_1\theta_1)) - \frac{B}{b_2}(\exp(-b_2\theta_i) - \exp(-b_2\theta_2)) \tag{9}$$

$$E_{dissipée} = h\nu \cdot \left|\frac{A}{b_1}(\exp(-b_1\theta_i) - \exp(-b_1\theta_1)) - \frac{B}{b_2}(\exp(-b_2\theta_i) - \exp(-b_2\theta_2))\right| \tag{10}$$

It can be seen that the dissipated energy $E_{dissipated}$ in vanadium dioxide depends on two parameters: the frequency ν of the incident photons and the temperature θ.

From the expression of the dissipated energy $E_{dissipated}$, we can extract the variation of the transmissibility T according to θ for a fixed frequency ν.

3 Harnessing the Energy Dissipated in Vanadium Dioxide

3.1 Computer Science Sector

3.1.1 Types of Information Storage

Data storage is the process of recording information on a storage medium. There are three main types of media, as shown in Scheme 2:

- Magnetic media
- Optical media
- Electronic media.

Indeed, the storage of information is one of the essential functions of a general-purpose computer.

Usually, a recording can be done with almost any form of energy. For example, electronic data storage requires electrical energy to store and retrieve data.

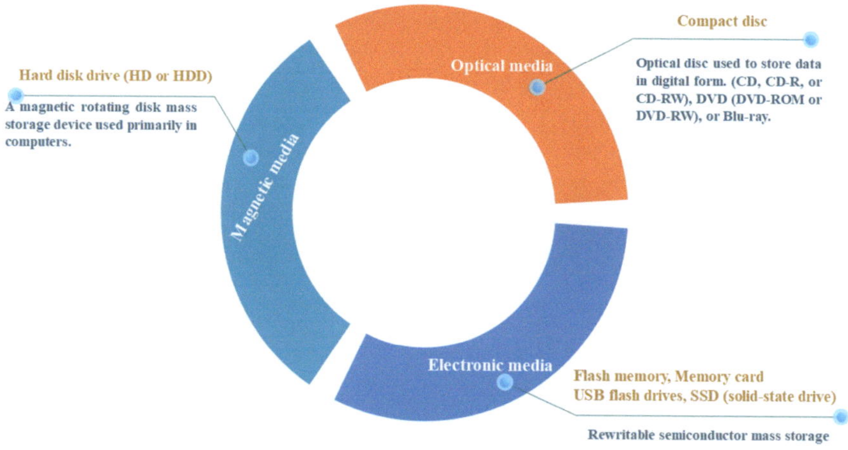

Scheme 2 Types of information storage

3.2 *The Use of Dissipated Energy as a Means of Increasing the Speed of Data Storage in Optical Memories and Storage Masses.*

3.2.1 The Energy Dissipated in Pure VO_2 and VO_2 Boosted by 5% W

According to this mathematical modeling of the functions that limit this optical hysteresis loop concerning the variation of the transmissibility T as a function of temperature θ, we can calculate the dissipated energy E in a thin film of pure vanadium dioxide and doped W or doped Cr [10, 11], in the infrared spectrum for a frequency ν or a given energy ħω, by completely or partially crossing this hysteresis loop as shown in Fig. 1. It is noted that the T transmittivity of this thin film of pure and doped VO_2 varies exponentially between the limits of hysteresis but irreversibly during heating followed by cooling and vice versa; this produced an energy dissipated in this material.

By fixing the frequency ν in the infrared spectrum and varying the temperature θ as shown in Figs. 2, 3, 4 and 5, we deduce that the energy dissipated in intrinsic VO_2 and n-type doped VO_2 by 5% W increases if the temperature θ increases; this result allows us to see the decrease in the amplitude of T transmittivity and to consider the transition phase of this thin film from the semiconductor state to the metallic state and vice versa. This dissipated energy also allows us to consult the deferential electronic transitions in the semiconductor state since it is related to the light energy ħω of the incident photons for the given temperatures, and it causes the excitation of the electrons of the energy bands of pure VO_2 on the one hand, and for the W-doped VO_2, on the other hand, we can control the increase of this dissipated energy as a function of temperature. In the infrared, this dissipated energy E in this hysteresis loop can be converted into other forms of energy that will become useful, such as optical memories or storage ground, infrared detectors, solar water heaters, where it is used for pump trip temperatures that help circulate fluids in thermal networks.

If we perform a frequency of 50 or 1000 times or more of this hysteresis loop regarding the transmissibility T of this thin film of VO_2, we find that the dissipated energy E becomes important in value that we can transform according to our utilitarian or particular needs. It is therefore necessary to know and be able to calculate this energy dissipated in this material during a thermochromic phase transition.

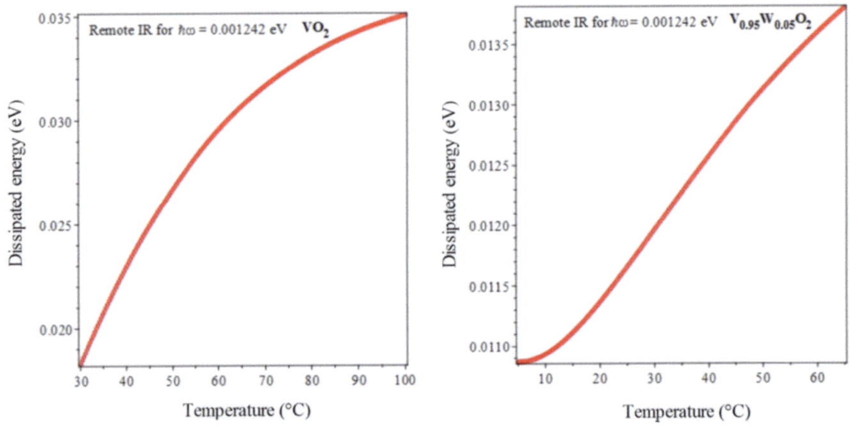

Fig. 2 The energy dissipated in pure VO_2 and VO_2 doped by 5% W as a function of temperature in the far infrared for $\hbar\omega = 0.001242$ eV

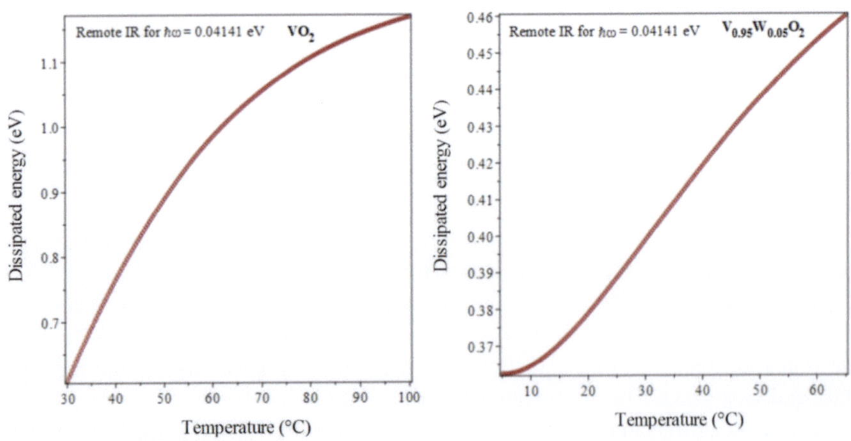

Fig. 3 The energy dissipated in pure VO_2 and VO_2 doped by 5% W as a function of temperature in the far infrared for $\hbar\omega = 0.04141$ eV

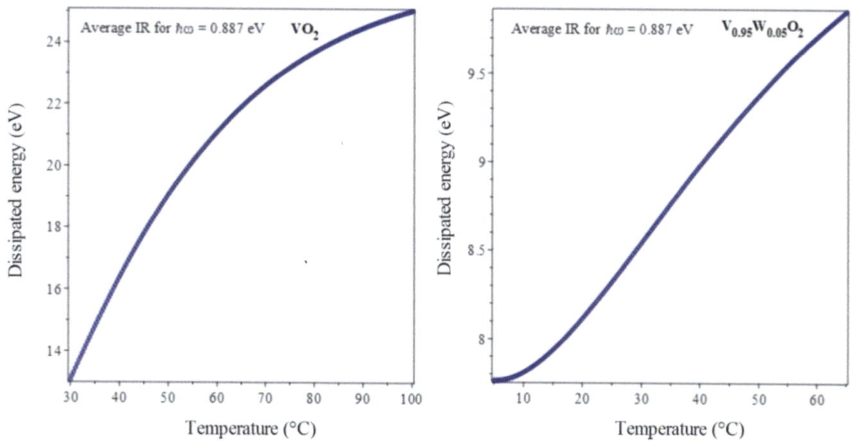

Fig. 4 The energy dissipated in pure VO_2 and VO_2 doped by 5% W as a function of temperature in the mid-infrared for ħω = 0.887 eV

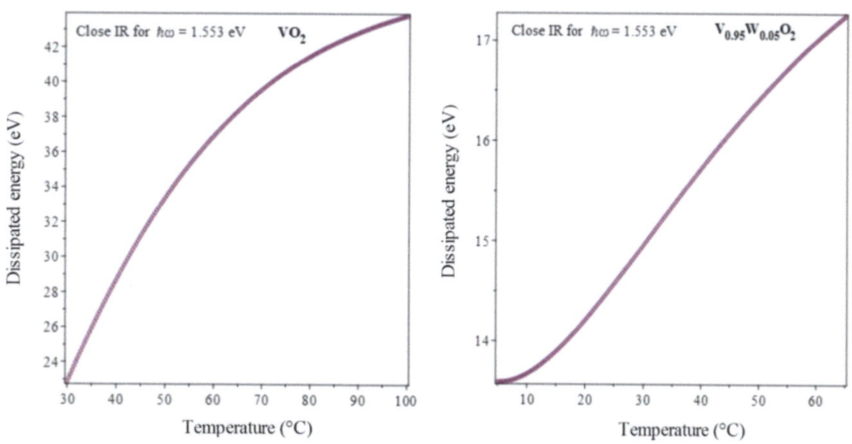

Fig. 5 The energy dissipated in pure VO_2 and VO_2 doped by 5% W as a function of temperature in the near-infrared for ħω = 1.553 eV

3.2.2 The Δt Time of Information Storage in Pure VO_2 and Boosted by 5% to W and 15% to Cr

According to Heisenberg's uncertainty principle [12], $\Delta E \cdot \Delta t \cong \hbar/2$ or $\Delta x \cdot \Delta p \cong \hbar/2$. This energy can be dissipated in optical memories (or storage mass) made by thin films of pure or doped vanadium dioxide in order to be able to calculate the Δt time of information storage. Indeed, if the energy ΔE is low then Δt is large and vice versa. This is achievable in the infrared spectrum, as shown in Tables 1, 2 and 3 which are taken from Figs. 6, 7 and 8 respectively.

Table 1 Some values of the storage time as a function of the energy dissipated in the I. R for pure VO_2

Photon energy in the R.I. ħω (eV)	Dissipated energy E (eV)	Storage time Δt (ns)
0.001242	0.034	2×10^{-5}
0.04141	1.15	5.6×10^{-7}
0.887	25	2.6×10^{-8}
1.553	43	1.5×10^{-8}

Table 2 Some values of the storage time as a function of the energy dissipated in the I. R for W-doped VO_2

Photon energy in the R.I. ħω (eV)	Dissipated energy E (eV)	Storage time Δt (ns)
0.001242	0.013	4.77×10^{-5}
0.04141	0.46	1.43×10^{-6}
0.887	9.85	6.68×10^{-8}
1.553	17.25	3.81×10^{-8}

Table 3 Some values of the storage time as a function of the energy dissipated in the I. R for Cr-doped VO_2

Photon energy in the R.I. ħω (eV)	Dissipated energy E (eV)	Storage time Δt (ns)
0.001242	0.032	2.1×10^{-5}
0.04141	1.041	6.32×10^{-7}
0.887	22.31	3×10^{-8}
1.553	39.07	1.68×10^{-8}

We find that the restitution time Δt increases with type n doping, whereas type p doping does not have a great influence on this time.

It has been observed that doping a thin layer of VO_2 reduces transmittivity. The surface area of the hysteresis loop decreases by the W tungsten-doped VO_2 film, which allows for an increase in the storage time Δt of the information in the storage masses.

For tungsten VO_2 doping, we can see that the dissipated energy increases with the energy of the incident photons in the infrared, and for each zone of the IR, we have:

$$\frac{\Delta t(\text{ W-doped VO}_2)}{\Delta t(\text{pure VO}_2)} \cong 2.5$$

In this study, we noted that the storage time t in a thin film of W-doped vanadium dioxide is the multiple of the storage time in a thin film of pure vanadium dioxide by a factor of about 2.5.

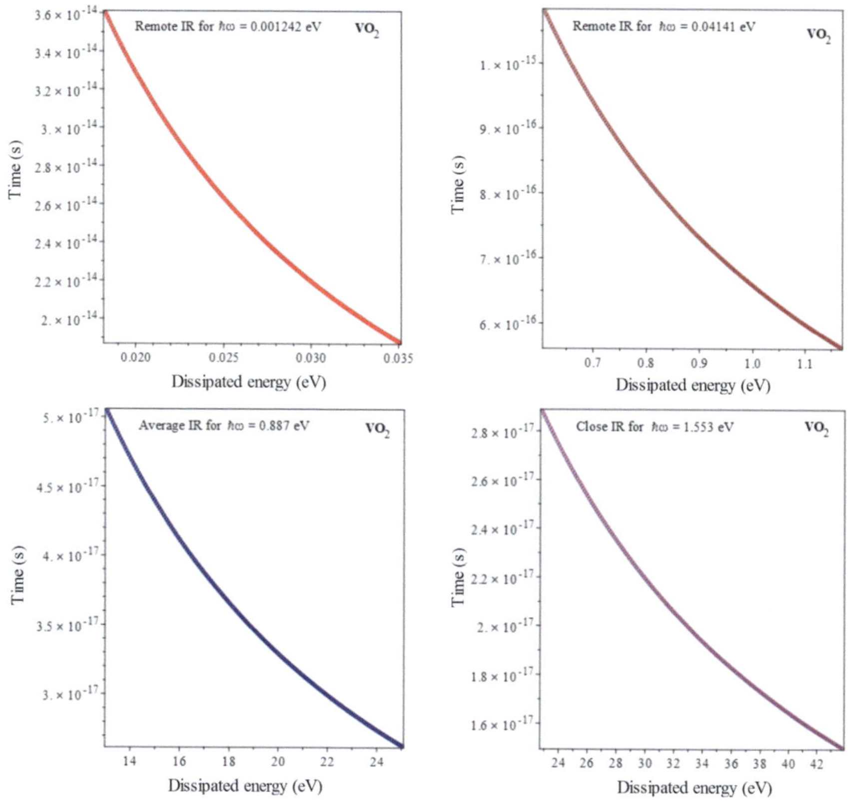

Fig. 6 Variation of the storage time Δt as a function of the energy dissipated in the infrared for pure VO_2

For Cr-doped VO_2, we find that this time Δt is of the same order of magnitude as that of pure VO_2.

For this, we strongly suggest the manufacture of storage masses and optical memories based on the n-type doped with vanadium dioxide (5%W), since we can exploit the increase in this dissipated energy as a function of temperature, as a way to increase the speed of information storage.

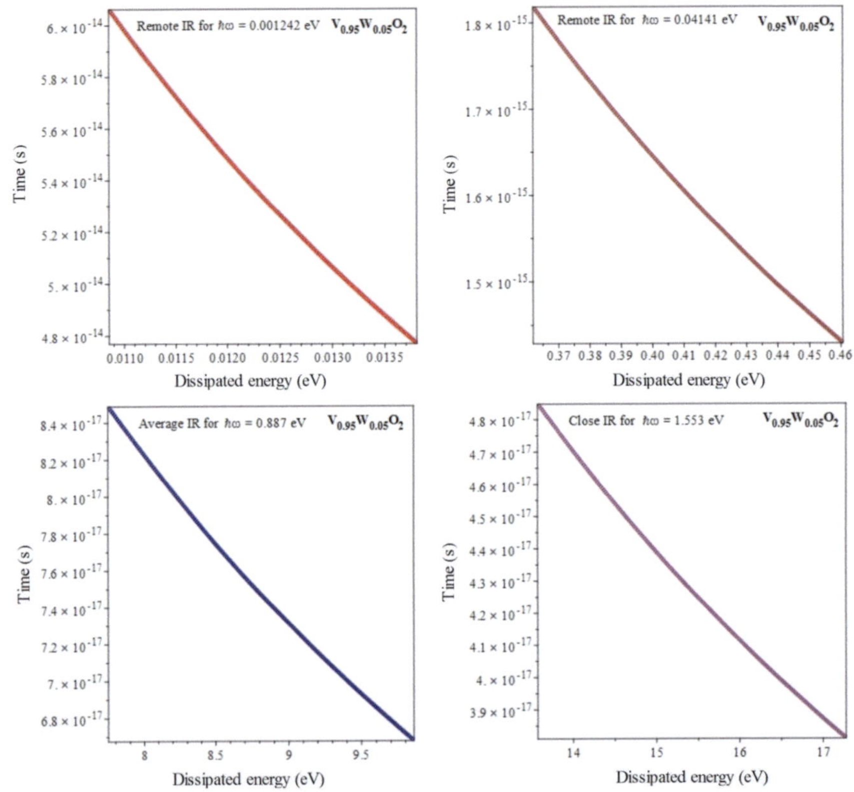

Fig. 7 Variation of the storage time Δt as a function of the energy dissipated in the infrared for W-doped VO_2

4 Energy Sector

4.1 Bibliographical Overviews

Without energy, there is no life, no development. However, today, the high consumption of energy based on fossil resources, which release large quantities of CO_2, is causing a shift towards a large-scale climate imbalance and a significant degradation of our planet. Humanity has therefore taken up the challenge to solve these problems as quickly as possible [13]. A series of international conferences and summits gave rise to a new concept of "Sustainable Development". This concept, as defined by the United Nations General Assembly, is "development that meets the needs of the present without compromising the ability of future generations to meet their own needs". Following the Earth Summit in Rio in 1992, and the Kyoto Protocol in 1997, several objectives to be followed were set at the international level to move towards sustainable development that defines the principles of progress reconciling

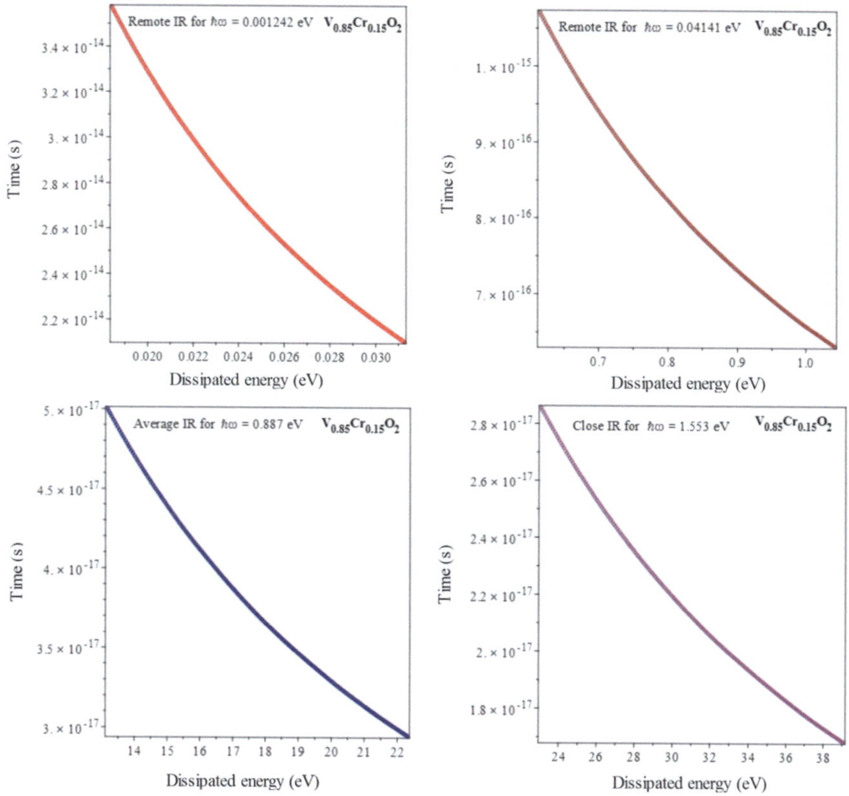

Fig. 8 Variation of storage time Δt as a function of energy dissipated in the infrared for Cr-doped VO_2

the economy, and access to the same resources for all at the social level as well as the preservation of resources for the environment.

The transportation sector is one of the largest sources of greenhouse gas (GHG) emissions globally. This sector accounts for about 24% of global emissions [14, 15]. These emissions come mainly from the combustion of fossil fuels, 90% of which are petroleum-based, which are used by cars, trucks, ships, trains and planes. As a result, it plays an important role in mitigating global climate change.

Currently, this sector has become the research centre for the development of new technologies aimed at reducing these CO_2 emissions. A lot of research has been conducted in the field of fuel production to address growing environmental concerns. Fuel formulations and various additives [16–20] are now designed to improve fuel consumption and reduce harmful emissions. In particular, engine smoke emissions, which result from incomplete combustion, contribute significantly to the greenhouse effect. These emissions can be reduced through the use of additives in the fuel [21] that help promote a fuller, cleaner combustion. As a result, this has positive effects on air quality, especially in urban areas where road traffic is usually intense.

In this context, the possibility of studying the effect of nano-additives in fuels on engine performance has opened up a wide range of potential ways for researchers to reduce fuel consumption and thus reduce CO_2 emissions.

4.2 The Use of Vanadium Dioxide Nanoparticles as Nano-additives in Fuels to Reduce the Energy Dissipated by Combustion

The results obtained showed that doping of n-type or p-type vanadium dioxide reduces the energy dissipated compared to the pure state of this material. When doping is type n, the energy dissipated increases with increasing temperature in the infrared spectrum, due to the electronic mobility in this material. When doping is type p, the energy dissipated decreases compared to the pure state of VO_2 because there are an excess of holes created by this type of doping. It is therefore obvious to consider the use of p-type nanoparticles doped with vanadium dioxide (15% Cr) as a suitable candidate for nano-additives in fuels that can reduce the energy dissipated by combustion and optimize energy efficiency (Figs. 9, 10, 11 and 12).

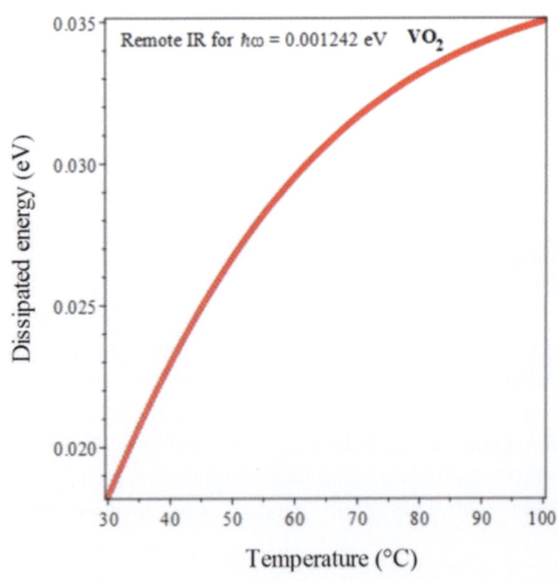

Fig. 9 The energy dissipated in pure VO_2 and VO_2 doped by 15% Cr as a function of temperature in the far infrared for $\hbar\omega = 0.001242$ eV

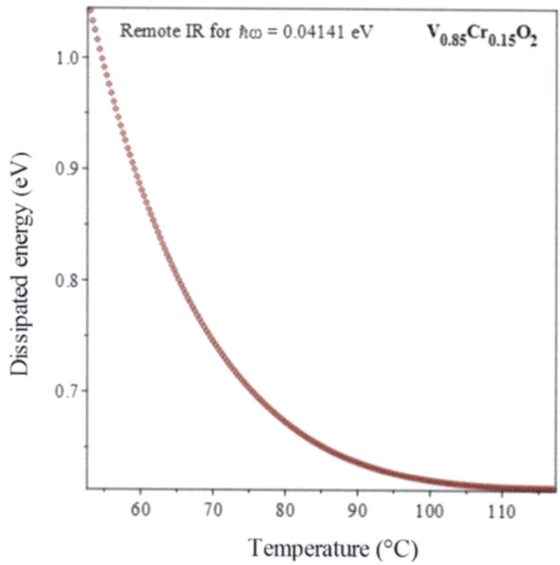

Fig. 10 The energy dissipated in pure VO_2 and VO_2 doped by 15% Cr as a function of temperature in the far infrared for $\hbar\omega$ = 0.04141 eV

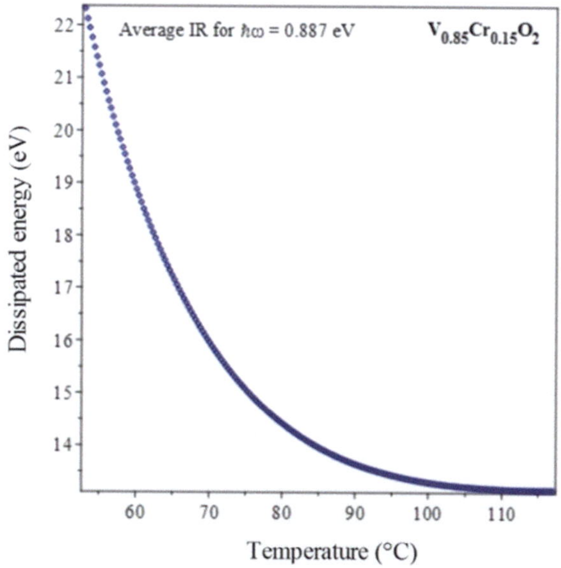

Fig. 11 The energy dissipated in pure VO_2 and VO_2 doped by 15% Cr as a function of temperature in the mid-infrared for $\hbar\omega$ = 0.887 eV

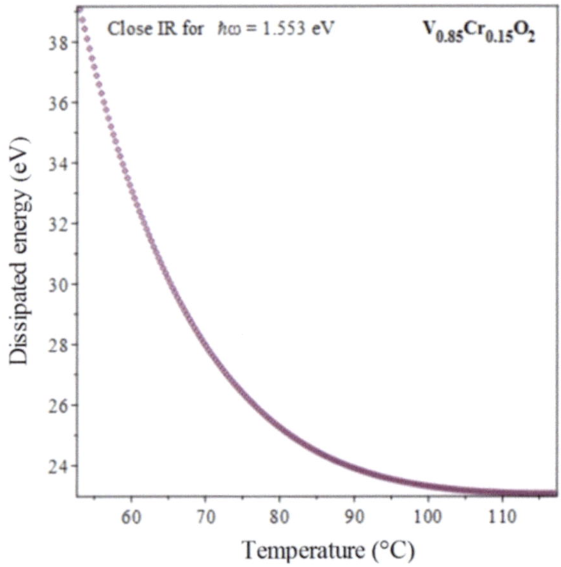

Fig. 12 The energy dissipated in pure VO_2 and VO_2 doped by 15% Cr as a function of temperature in the near-infrared for $\hbar\omega$ = 1.553 eV

5 Conclusion

This journal chapter has explored in depth the mathematical modelling of the energy dissipated within vanadium dioxide (VO_2), both in its pure and doped form (with 5% tungsten and 15% chromium). We have detailed how the cooling operation followed by heating of this material generates a hysteresis loop, the area of which faithfully represents the energy dissipated. This fundamental analysis is crucial to understanding the unique thermochromic properties of VO_2.

The results presented clearly demonstrated that the doping of vanadium dioxide, whether n-type or p-type, has a significant impact on the reduction of the dissipated energy compared to the pure state of the material. It has been shown that n-type doping leads to an increase in dissipated energy with the increase in temperature in the infrared spectrum, a phenomenon attributable to the increased mobility of electrons. Conversely, p-type doping is characterized by a decrease in the energy dissipated compared to pure VO_2, due to the creation of an excess of holes. These observations underline the ability to modulate the energetic properties of VO_2 by controlled doping.

In addition, this review highlighted the considerable potential of VO_2 in two major application areas. First, in the IT sector, the exploitation of the energy dissipated from this material, in the form of thin films, opens up promising prospects for the manufacture of more efficient and faster mass storage devices and optical memories. Secondly, in the energy field, the integration of VO_2 nanoparticles as additives in fuels could play a key role in reducing the energy dissipated due to incomplete combustion, thus contributing to improved engine efficiency. These advancements position vanadium dioxide as a leading smart material for future innovations in data storage and energy optimization.

Disclaimer

References

1. Lu, P.J., Fang, S.W., Cheng, W.L., Huang, S.C., Huang, M.C., Cheng, H.F.: Characterization of titanium dioxide and zinc oxide nanoparticles in sunscreen powder by comparing different measurement methods. J. Food Drug Anal. **26**(3), 1192–1200 (2018)
2. Shivakumar, N., Madhusudan, P., Daniel, S.C.G.K.: Nanomaterials for smart food packaging. In: Handbook of Nanomaterials for Industrial Applications, pp. 260–70 (2018)
3. Tornero, A.C.F., M., Blasco, M.G., Azqueta, M.C., Acevedo, C.F., Castro, C.S., López, S.J.R.: Antimicrobial ecological waterborne paint based on novel hybrid nanoparticles of zinc oxide partially coated with silver. Progr. Org. Coatings **121**, 130–141
4. Dhinesh, B., Isaac JoshuaRamesh Lalvani, J., Parthasarathy, M., Annamalai, K.: An experimental analysis on the influence of fuel borne additives on the single cylinder diesel engine powered by *Cymbopogon flexuosus* biofuel. J. Energy Inst. **90**(4), 634–645
5. Annamalai, M., Dhinesh, B., Nanthagopal, K., SivaramaKrishnan, P., Isaac JoshuaRamesh Lalvani, J., Parthasarathy, M., Annamalai, K.: An assessment on performance, combustion and emission behavior of a diesel engine powered by ceria nanoparticle blended emulsified biofuel. Energy Convers. Manage. **123**, 372–380 (2016)
6. Mossanek, R.J.O., Abbate, M.: Optical response of metallic and insulating VO_2 calculated with the LDA approach. J. Phys. Condens. Matter **19**, 346225 (2007). https://doi.org/10.1088/0953-8984/19/34/346225
7. Kana, J.B.K., Vignaud, G., Gibaud, A., Maaza, M.: Thermally driven sign switch of static dielectric constant of VO_2 thin film. Opt. Mater. (Amst) **54**, 165–169 (2016). https://doi.org/10.1016/j.optmat.2016.02.032
8. Klimov, V.A., Timofeeva, I.O., Khanin, S.D., Shadrin, E.B., Ilinskii, A.V., Silva-Andrade, F.: Hysteresis loop construction for the metal-semiconductor phase transition in vanadium dioxide films. Tech. Phys. **47**(9), 1134–1139 (2002)
9. Gagaoudakis, E., Kortidis, I., Michail, G., Tsagaraki, K., Binas, V., Kiriakidis, G., Aperathitis, E.: Study of low temperature rf-sputtered Mg-doped vanadium dioxide thermochromic films deposited on low-emissivity substrates. Thin Solid Films (2016)
10. C. Batista, Ribeiro, R.M., Teixeira, V.: Synthesis and characterization of VO_2-based thermochromic thin films for energy-efficient windows. Nanoscale Res. Lett. **6** (2011)
11. Beteille, F., Livage, J.: Optical switching in VO_2 thin films. J. Sol-Gel Sci. Technol. (1998)
12. Messiah, A.: Quantum Mechanics—Vol. 1, pp. 114–117, 269–270. Dunod (1995)
13. Almost half of global GDP under actual or intended net zero emission targets. Energy and Climate Intelligence Unit, London, UK, 18th February, 2020. https://eciu.net/news-and-events/press-releases/2020/almost-half-of-global-gdp-under-actual-or-intended-net-zero-emissions-targets
14. Crippa, M., Oreggioni, G., Guizzardi, D., Muntean, M., Schaaf, E., Lo Vullo, E., Solazzo, E., Monforti-Ferrario, F., Olivier, J.G.J., Vignati, E.: Fossil CO_2 and GHG emissions of all

world countries: 2019 report. EUR 29849 EN, Publications Office of the European Union, Luxembourg (2019). https://op.europa.eu/en/publication-detail/-/publication/9d09ccd1-e0dd-11e9-9c4e-01aa75ed71a1/language-en
15. Teter, J., Le Feuvre, P., Gorner, M., Scheffer, S.: Tracking transport. International Energy Agency (IEA), Paris, France (2019). https://www.iea.org/reports/tracking-transport-2019
16. Pandey, A., Nandgaonkar, M., Pandey, U., Suresh, S., et al.: The effect of cerium oxide nanoparticles fuel additive on performance and emission of Karanja biodiesel fueled compression ignition military 585kW heavy duty diesel engine. In: SAE Technical Paper (2018). https://doi.org/10.4271/2018-01-1818
17. Gan, Y., Qiao, L.: Combustion characteristics of fuel droplets with addition of nano and micron-sized aluminium particles. Combust. Flame (2011). https://doi.org/10.1016/j.combustflame.2010.09.005
18. El-Seesy, A., Attia, A.: The effect of Aluminum oxide nanoparticles addition with Jojoba methyl ester-diesel fuel blend on a diesel engine performance, combustion and emission characteristics. Fuel **224**, 147–166 (2018). https://doi.org/10.1016/j.fuel.2018.03.076
19. Ganesh, D., Gowrishankar, G.: Effect of nano-fuel additive on emission reduction in a biodiesel fueled CI engine. In: 2011 International Conference on Electrical and Control Engineering. https://doi.org/10.1109/ICECENG.2011.6058240
20. Snow, S., McGee, J., Miller, D., Bass, V., Schladweiler, M., et al.: Inhaled diesel emissions generated with cerium oxide, nanoparticle fuel additive induce, adverse pulmonary and systemic effects. Toxicol. Sci.: Off. J. Soc. Toxicol. 142. https://doi.org/10.1093/Toxsci/KFU187
21. Chandrasekaran, V., Arthanarisamy, M., Nachiappan, P., Dhanakotti, S., Moorthy, B.: The role of nano additives for biodiesel and diesel blended transportation fuels. Transp. Res. Part D: Transp. Environ. **46**, 145–156 (2016). https://doi.org/10.1016/j.trd.2016.03.015

Study of the Optical Response of Vanadium Dioxide (VO_2) from the Complex Dielectric Constant

Abderrahim Ben Chaib, Mohammed Zouini, El Mehdi El Khattabi, Mourad Boutahir, and Kalpana Sahoo

Abstract The analysis of the optical properties of a material is largely based on the knowledge of its complex dielectric constant, denoted $\varepsilon = \varepsilon_1 - i\varepsilon_2$, where ε_1 represents the real part (related to dispersion) and ε_2 the imaginary part (related to absorption). In the framework of our study, particular attention is paid to the real part ε_1, as it plays a decisive role in the description of the Coulomb interaction within an exciton, especially for ground states such as the 1S exciton. Our work focuses specifically on solid vanadium dioxide (VO_2), a material known for its remarkable electronic and optical properties, particularly around its metal–insulator transition. We focus on the behavior of the 1S exciton, seeking to determine its energy and its influence on the optical properties of the material. To do this, we solve the Schrödinger equation of exciton using a variational approach, based on the effective mass approximation. This method makes it possible to estimate the binding energy of the exciton and to extract a functional expression of ε_1 as a function of the energy E of the exciton. Then, using the Maple software, we proceed to numerical simulations to plot the evolution of the real dielectric constant ε_1 as well as the refractive index n as a function of the energy E of the 1S exciton. These simulations are performed both in the immediate vicinity of the VO_2 bandgap, and well beyond, in order to cover a wide range of photonic energies. From these results, we deduce two fundamental optical quantities: the reflectivity R and the transmittivity T of the massive VO_2, for a normal incidence of photons on the surface of the material. In this context, we consider a weakly absorbent semiconductor state, which simplifies the analytical expressions of

A. B. Chaib · M. Zouini (✉)
Sidi Mohamed Ben Abdellah University ENS, Fez, Morocco
e-mail: mohammedzouini@gmail.com

M. Zouini · E. M. El Khattabi
Systems and Sustainable Environment Laboratory, Private University of Fez UPF, Fez, Morocco

Department of Science, École Normale Supérieure, University Moulay Ismail Meknes, Meknes, Morocco

M. Boutahir
École Normale Supérieure, University Moulay Ismail Meknes, Meknes, Morocco

K. Sahoo
School of Civil Engineering, KIIT Deemed to Be University, Bhubaneswar, Odisha, India

E. El Mehdi and M. Boutahir (eds.), *Advanced Materials for Sustainable Energy and Engineering*, Engineering Materials, https://doi.org/10.1007/978-3-032-10069-6_21

R and T while maintaining an acceptable precision for physical interpretation. This study aims to better understand the optoelectronic behavior of VO_2 in its semiconductor phase, and to demonstrate the importance of excitons in modifying its optical properties, especially in the low-energy domain where these quasiparticles dominate light-matter interactions.

Keywords Massive vanadium dioxide · Dielectric function · Refractive index · Semiconductor · Exciton 1S · Exciton energy · Variational method · Effective mass approximation · Gap energy · Schrödinger's equation · Coulomb interaction · Maple simulation

1 Introduction

Vanadium dioxide (VO_2) is attracting increasing interest in optoelectronics and photonics research due to its reversible, rapid and dramatic phase transition. This material has a well-known transition from the semiconductor to the metallic state at around 68 °C, which takes place in the space of a few nanoseconds [1–4]. This transition induces abrupt changes in its electronic, optical and structural properties, making it a preferred candidate for many applications: optical switching devices, phase change memory, smart mirrors, infrared sensors.

At low temperatures ($\theta < 68$ °C), VO_2 adopts a monoclinic structure (M-phase) and behaves like a semiconductor, with a bandgap of about 0.7 eV [5, 6]. At high temperatures ($\theta > 68$ °C), it changes to a rutilated tetragonal structure (R-phase), becoming a metal with a significantly increased conductivity. This structural transition is accompanied by profound changes in the structure of electronic bands.

Previous studies, including those by Mossanek et al. [7, 8], have highlighted the characteristic electronic transitions in massive VO_2. For incident photons with energy $\hbar\omega > 0.7$ eV, interband transitions occur between the $3d_{//}$ band and the $3d\pi$ band. When the energy of the photons exceeds 2.5 eV, transitions from the $2d\pi$ band of oxygen to the $3d\pi^*$ band of vanadium are also observed.

In contrast, in the spectral region where $\hbar\omega < 0.7$ eV—below the VO_2 gap—there is not enough energy to excite an electron towards the conduction band. In this situation, incident light can generate excited electron–hole bound states, called excitons. In particular, the fundamental exciton of type 1S is of great interest, as it contributes significantly to the optical properties of the material in this energy range.

In this study, we propose to model and simulate, using the Maple software, the evolution of the reflectivity (R) and transmittivity (T) of massive VO_2 as a function of the E energy of the 1S exciton. This analysis will be conducted under the assumption of a weakly absorbent semiconductor state, taking into account the effective mass approximation and the variational solutions of the Schrödinger equation for excited states. The dynamics of this variation, especially in the vicinity of the phase transition, occur on the scale of a few nanoseconds, which gives this phenomenon a strong potential for ultrafast technologies.

2 Method

In our study of the optical properties of vanadium dioxide (VO_2) in the massive state, we seek to determine the binding energy of the fundamental exciton 1S, i.e. the bound state resulting from the electrostatic interaction between an excited electron and a hole in the valence band.

To do this, we use the variational method, a powerful quantum mechanical technique that allows us to estimate the energy of the fundamental level of a system from a judicious choice of test function. This method is well suited to excitons in semiconductors, especially in the context of effective mass approximation, where exciton is equated with a hydrogen system modified by the properties of the material.

The exciton Hamiltonian in an isotropic semiconductor can be written as follows [9–13]:

$$H = \frac{p_e^2}{2m_e^*} + \frac{p_h^2}{2m_h^*} - \frac{e^2}{\varepsilon r} \tag{1}$$

where ε: the relative dielectric constant of the material. m_e^*: the effective mass of the electron. m_h^*: the effective mass of the hole. a_0: the effective three-dimensional Bohr radius of the 1S exciton.

$$r = |\vec{r}_e - \vec{r}_h|; \quad \vec{K} = \vec{k}_e - \vec{k}_h; \quad \vec{P} = \vec{p}_e - \vec{p}_h$$

The Schrödinger equation describing the relative dynamics of this pair (electron–hole) is therefore of the form:

$$H\Psi(\vec{r}_e, \vec{r}_h) = \left(E - E_g\right)\Psi(\vec{r}_e, \vec{r}_h) \tag{2}$$

$\vec{r} = \vec{r}_e - \vec{r}_h; \quad \left(m_e^* + m_h^*\right)\vec{R} = m_e^*\vec{r}_e + m_h^*\vec{r}_h$: center-of mass system.

$\mu = \frac{m_e^* m_h^*}{m_e^* + m_h^*}; \quad \mu$: the reduced mass of the exciton.

$m = m_e^* + m_h^*$; m the total mass of the exiton.

R_0^{ex}: Rydberg effective of the excitement.

$$R_0^{ex} \ll E_g \Rightarrow \begin{cases} P = -i\hbar\dfrac{\partial}{\partial R} \\ p = -i\hbar\dfrac{\partial}{\partial r} \end{cases}$$

$$\Psi\left(\vec{r}, \vec{R}\right) = \frac{1}{\sqrt{\Omega}}\exp\left(i\vec{K}\cdot\vec{R}\right)\varphi(\vec{r}); \quad \Omega : \text{crystal volume}$$

Substituting in Schrödinger's equation [14] we have:

$$E - E_g = \frac{\hbar^2 K^2}{2m} + \eta \tag{3}$$

$$\text{Où :} \quad \left(\frac{p^2}{2\mu} - \frac{e^2}{\varepsilon r}\right)\varphi(\vec{r}) = \eta\varphi(\vec{r}) \tag{4}$$

$$\varphi(\vec{r}) = \frac{1}{\sqrt{\pi a_0^3}} \exp\left(-\frac{\vec{r}}{a_0}\right) \tag{5}$$

$\varphi(\vec{r})$: the test function by making a calculation by variational method [15, 16] where a_0 is the variational parameter.

$$E = \frac{\langle\Psi|H|\Psi\rangle}{\langle\Psi|\Psi\rangle} \tag{6}$$

$$\frac{dE}{da_0} = 0 \rightarrow E_{min} = R_0^{ex} \tag{7}$$

$$\langle\Psi|\Psi\rangle = \int \Psi_{1S}\Psi_{1S}^* d\tau = \frac{1}{\pi a_0^3}\int \exp\left(-\frac{2r}{a_0}\right)4\pi r^2 dr$$

$$\langle\Psi|\Psi\rangle = 1$$

We pose:

$$I = \langle\Psi|H|\Psi\rangle. \tag{8}$$

$\varphi(\vec{r}) = \Psi_{1S}(\vec{r}) = \frac{1}{\sqrt{\pi (a_0^{ex})^3}} \exp\left(-\frac{\vec{r}}{a_0^{ex}}\right)$; $\varphi(\vec{r})$: the testing function.

We put $a_0^{ex} = a$ et $B = \frac{1}{\sqrt{\pi a^3}}$ to simply calculations.

We proceed by variational method for the calculation of η

$$p \rightarrow -i\hbar\frac{\partial}{\partial r}; \quad \varphi(\vec{r}) = B\exp\left(-\frac{r}{a}\right)$$

$$\text{either } J = \varphi(r)|H|\varphi(r)$$

$$p^2 \rightarrow \left(-i\hbar\frac{\partial}{\partial r}\right)\left(-i\hbar\frac{\partial}{\partial r}\right) = -\hbar^2\frac{\partial^2}{\partial r^2}$$

$$J = \frac{1}{2\mu}\varphi(r)| - \hbar^2\frac{\partial^2}{\partial r^2}|\varphi(r) - \frac{e^2}{\varepsilon}\varphi(r)\left|\frac{1}{r}\right|\varphi(r)$$

we pose J = L − G where:

$$L = -\frac{\hbar^2}{2\mu}\varphi(r)|\frac{\partial^2}{\partial r^2}|\varphi(r);\quad G = \frac{e^2}{\varepsilon}\varphi(r)\left|\frac{1}{r}\right|\varphi(r)$$

Calculation of L

$$L = -\frac{\hbar^2}{2\mu}\frac{B^2}{a^2}\int_a^0 \exp\left(-\frac{r}{a}\right)\exp\left(-\frac{r}{a}\right)4\pi r^2 dr$$

$$L = -\frac{\hbar^2}{2\mu}\frac{B^2 4\pi}{a^2}\int_a^0 r^2 \exp\left(-\frac{2r}{a}\right) dr$$

$$L = -\frac{4\pi B^2\hbar^2}{2\mu a^2}\int_a^0 r^2 \exp\left(-\frac{2r}{a}\right) dr$$

$$L = -\frac{4\pi B^2\hbar^2}{2\mu a^2}\cdot I \quad \text{Où} \quad I = \int_a^0 r^2 \exp\left(-\frac{2r}{a}\right)dr$$

Calculation of I: we proceed with an integration by part

We pose:

$$\begin{cases} u = r^2 \\ dv = \exp\left(-\frac{2r}{a}\right)dr \end{cases} \Rightarrow \begin{cases} du = 2rdr \\ v = -\frac{a}{2}\exp\left(-\frac{2r}{a}\right) \end{cases}$$

$$I = \left[-\frac{a}{2}r^2 \exp\left(-\frac{2r}{a}\right)\right]_a^0 + \int_a^0 a\, r \exp\left(-\frac{2}{a}r\right) dr$$

$$I = \frac{a}{2}\left(a^2 \exp(-2) - 0\right) + \frac{a}{2}\cdot 2\int_a^0 r \exp\left(-\frac{2}{a}r\right) dr$$

$$I = \frac{a^3}{2}\exp(-2) + a\Delta \quad \text{Où} \quad \Delta = \int_a^0 r \exp\left(-\frac{2}{a}r\right)dr$$

Calculation of Δ

Let us pose:

$$\begin{cases} u = r \\ dv = \exp\left(-\frac{2r}{a}\right)dr \end{cases} \Rightarrow \begin{cases} du = dr \\ v = -\frac{a}{2}\exp\left(-\frac{2r}{a}\right) \end{cases}$$

$$\Delta = \left[-\frac{a}{2}r\exp\left(-\frac{2}{a}r\right)\right]_a^0 + \frac{a}{2}\int_a^0 \exp\left(-\frac{2}{a}r\right)dr$$

$$\Delta = \frac{a^2}{2}\exp(-2) + \frac{a}{2}\left[-\frac{a}{2}\exp\left(-\frac{2}{a}r\right)\right]_a^0$$

$$\Delta = \frac{a^2}{2}\exp(-2) + \frac{a}{2}\left[-\frac{a}{2}(-\exp(-2) + 1)\right]$$

$$\Delta = \frac{a^2}{2}\exp(-2) - \frac{a^2}{4}[(1 - \exp(-2))]$$

$$\Delta = \frac{a^2}{2}\left(\exp(-2) + \frac{\exp(-2)}{2} - \frac{1}{2}\right)$$

$$\Delta = \frac{a^2}{2}\left(\frac{3}{2}\exp(-2) - \frac{1}{2}\right)$$

$$\Delta = \frac{a^2}{4}(3\exp(-2) - 1)$$

$$I = \frac{a^3}{2}\exp(-2) + a\Delta$$

$$I = \frac{a^3}{2}\exp(-2) + \frac{a^3}{2}\left(\frac{3}{2}\exp(-2) - \frac{1}{2}\right)$$

$$I = \frac{a^3}{2}\left[\exp(-2) + \frac{3}{2}\exp(-2) - \frac{1}{2}\right]$$

$$I = \frac{a^3}{2}\left[\frac{5}{2}\exp(-2) - \frac{1}{2}\right]$$

$$\text{So:}\quad L = -\frac{4\pi B^2\hbar^2}{2\mu a^2}\frac{a^3}{2}\left[\frac{5}{2}\exp(-2) - \frac{1}{2}\right]$$

$$L = -\frac{\pi\hbar^2}{\mu a^2}\frac{1}{\pi a^3}\cdot a^3\left[\frac{5}{2}\exp(-2) - \frac{1}{2}\right]$$

$$L = -\frac{\hbar^2}{\mu a^2}\left[\frac{5}{2}\exp(-2) - \frac{1}{2}\right]$$

Calculation of G

$$G = \frac{e^2}{\varepsilon}\varphi(r)\left|\frac{1}{r}\right|\varphi(r)$$

$$G = \frac{e^2}{\varepsilon}\int_a^0 B^2 \exp\left(-\frac{2r}{a}\right)\cdot\frac{1}{r}\cdot 4\pi r^2\,dr$$

$$G = \frac{e^2 B^2 4\pi}{\varepsilon}\int_a^0 r\exp\left(-\frac{2r}{a}\right)dr$$

$$G = \frac{e^2 B^2 4\pi}{\varepsilon}\Delta;\quad \text{where}\quad \Delta = \int_a^0 r\exp\left(-\frac{2r}{a}\right)dr$$

We have $\Delta = \frac{a^2}{4}(3\exp(-2) - 1)$.
Therefore:

$$G = \frac{e^2 B^2 4\pi}{\varepsilon}\frac{a^2}{4}(3\exp(-2) - 1)$$

$$G = \frac{e^2 4\pi}{\varepsilon}\frac{1}{\pi a^3}\frac{a^2}{4}(3\exp(-2) - 1)$$

$$G = \frac{e^2}{\varepsilon a}[3\exp(-2) - 1]$$

We have $J = L - G$

$$J = -\frac{\hbar^2}{2\mu\, a^2}[5\exp(-2) - 1] - \frac{e^2}{\varepsilon a}[3\exp(-2) - 1] \tag{9}$$

We minimize J with respect to the variational parameter a, We set:

$$J = \alpha a^{-2} + \beta a^{-1};\quad \frac{dJ}{da} = 0 \tag{10}$$

$$\text{avec}\quad \alpha = -\frac{\hbar^2}{2\mu}[5\exp(-2) - 1];\quad \beta = -\frac{e^2}{\varepsilon}[3\exp(-2) - 1]$$

$$\alpha \cong 0.16\frac{\hbar^2}{\mu}; \quad \beta \cong 0.59\frac{e^2}{\varepsilon}$$

$$\frac{dJ}{da} = -2\alpha a^{-3} - \beta a^{-2} = 0 \Rightarrow a^{-2}\left(\beta + 2\alpha a^{-1}\right) = 0$$

$$a^{-1} = -\frac{\beta}{2\alpha} \Rightarrow a = -\frac{2\alpha}{\beta}; \quad a^{-2} \neq 0$$

$$a = -\frac{\varepsilon\hbar^2}{\mu\, e^2}\frac{[-1 + 5\exp(-2)]}{[-1 + 3\exp(-2)]}$$

$$a \cong -2\left(\frac{0.16\frac{\hbar^2}{\mu}}{0.59\frac{e^2}{\varepsilon}}\right) \cong -0.54\frac{\hbar^2\varepsilon}{e^2\mu}$$

Let us replace a in the expression of J:

$$J = \alpha a^{-2} + \beta a^{-1}; \quad a = -\frac{2\alpha}{\beta}; \quad a^{-1} = -\frac{\beta}{2\alpha}$$

$$J = \alpha\left(-\frac{\beta}{2\alpha}\right)^2 + \beta\left(-\frac{\beta}{2\alpha}\right)$$

$$J = \frac{\alpha\beta^2}{4\alpha^2} - \frac{\beta^2}{2\alpha} = \frac{\beta^2}{4\alpha} - \frac{\beta^2}{2\alpha}$$

$$J = -\frac{\beta^2}{4\alpha}$$

$$\text{Hence} \quad J = \frac{-\frac{e^4}{\varepsilon^2}[3\exp(-2) - 1]^2}{-\frac{4\hbar^2}{2\mu}[5\exp(-2) - 1]}$$

$$J = \frac{\mu\, e^4}{2\,\hbar^2}\frac{[-1 + 3\exp(-2)]^2}{[-1 + 5\exp(-2)]} \cdot \frac{1}{\varepsilon^2} \tag{11}$$

$$\textit{so} \quad E = E_g + \frac{\hbar^2 K^2}{2m} + \frac{\mu\, e^4}{2\,\hbar^2}\frac{[-1 + 3\exp(-2)]^2}{[-1 + 5\exp(-2)]} \cdot \frac{1}{\varepsilon^2} \tag{12}$$

E_g: The gap energy of vanadium dioxide in the semiconductor state.

$$E_g = 0.7\text{ eV}; \quad \frac{[-1 + 3\exp(-2)]^2}{2[-1 + 5\exp(-2)]} \cong -0.54$$

$$E \cong E_g + \frac{\hbar^2 K^2}{2m} - 0.54 \frac{\mu\, e^4}{\hbar^2} \cdot \frac{1}{\varepsilon^2}$$

$$\text{therefore} \quad J \cong -0.54 \frac{\mu\, e^4}{\hbar^2} \cdot \frac{1}{\varepsilon^2}$$

E: the energy of a 1S exciton.

Around $k \to 0$ (in the vicinity of the band gap):

$$E \cong E_g - 0.54 \frac{\mu e^4}{\hbar^2} \cdot \frac{1}{\varepsilon^2} \tag{13}$$

$$E = E_g + \frac{\mu\, e^4}{2\hbar^2} \frac{\left[-1 + 3\exp(-2)\right]^2}{\left[-1 + 5\exp(-2)\right]} \cdot \frac{1}{\varepsilon^2}$$

Therefore, knowing the gap energy E_g of this material, the reduced mass of an exciton 1S and around $k \to 0$, we can trace $\varepsilon = f(E)$ and extract by simulation by means of the Maple software the refractive index $n = \sqrt{\varepsilon}$ of this material in the semiconductor state (SC), which can be verified by the model of the dielectric function at low temperature, as well as the T and R coefficients.

In the vicinity of the bandgap B.I: ($k \to 0$)

$$E - E_g \cong -0.54 \frac{\mu\, e^4}{\hbar^2} \cdot \frac{1}{\varepsilon^2}$$

$$\frac{1}{\varepsilon^2} \cong -(E - E_g) 0.54 \frac{\mu\, e^4}{\hbar^2}$$

$$\frac{1}{\varepsilon} = \frac{\hbar}{e^2} \frac{1}{(\mu 0.54)^{\frac{1}{2}}} (-E + E_g)^{\frac{1}{2}}$$

$$\varepsilon = (0.54\mu)^{\frac{1}{2}} \frac{e^2}{\hbar} (E_g - E)^{-\frac{1}{2}}$$

$$\varepsilon \cong 9 \times 10^9 \cdot (0.54 \cdot \mu)^{\frac{1}{2}} \cdot \frac{e^2}{\hbar} \cdot (E_g - E)^{-\frac{1}{2}} \quad \text{(in S.I)} \tag{14}$$

And using the following data in the system (S.I):

$$\mu = 4 \cdot 9.1 \times 10^{-31}\,\text{kg}\,;\ \hbar = 1.05457 \times 10^{-34}\ \text{J} \cdot \text{s};$$
$$E_g = 1.12 \times 10^{-19}\ \text{J}\,;\ e = 1.6 \times 10^{-19}\ \text{C}$$

$$0.84 \times 10^{-19} < E(\text{J}) < 1.12 \times 10^{-19}$$

ε: the relative dielectric constant of the VO_2 material.

μ: la masse reduite de l'exciton IS.

ħ: Planck's constant (reduced).

E_g: the gap energy of vanadium dioxide in the semiconductor state.

n: the refractive index of the material VO_2.

$n = \sqrt{\varepsilon}$; $n^2 = \varepsilon$; $R \cong \frac{(n-1)^2}{(n+1)^2}$; $T \cong \frac{(1-R)^2}{1-R^2}$; considering a weakly absorbing state.

$$n = (\mu 0.54)^{\frac{1}{4}} \frac{e}{\sqrt{\hbar}} (-E + E_g)^{-\frac{1}{4}} \tag{15}$$

At the edge of the first Brillouin [7]**: (k ≠ 0)**

$$\varepsilon \cong 9 \times 10^9 \cdot (0.54 \cdot \mu)^{\frac{1}{2}} \cdot \frac{e^2}{\hbar} \cdot \left(E_g - E + \frac{\hbar^2 K^2}{2m} \right)^{-\frac{1}{2}} \tag{16}$$

And using the following data in the system (S.I):

$$\mu = 4 \cdot 9.1 \times 10^{-31} \text{kg} ; \hbar = 1.05457 \times 10^{-34} \text{ J} \cdot \text{s} ;$$
$$E_g = 1.12 \times 10^{-19} \text{ J}; \; e = 1.6 \times 10^{-19} \text{ C} ;$$

$$m = 173 \times 10^{-31} \text{kg}$$

$$k = \frac{2\pi}{5.75 \times 10^{-10}}; \quad m = m_e^* + m_h^*$$
$$= 4 \cdot m_e + 15 \cdot m_e = 19 \cdot m_e; \mu = 4 \cdot m_e$$

$$0.84 \times 10^{-19} < E(\text{J}) < 1.12 \times 10^{-19}$$

In each case we have simulated the optical functions: ε, n, T, R as a function of the energy E(J) of the exciton 1S in the approximation of the effective mass of the massive Vanadium dioxide.

2.1 Results

We integrate by part, we have obtained the results:

$$L = \frac{-\hbar^2}{\mu a^2} \cdot \left[\frac{5}{2} \cdot e^{-2} - \frac{1}{2} \right]; \; G = \frac{e^2}{\varepsilon a} \cdot \left[3 \cdot e^{-2} - 1 \right];$$

$$J = \frac{-\hbar^2}{2\mu a^2} \cdot \left[5 \cdot e^{-2} - 1\right] - \frac{e^2}{\varepsilon a} \cdot \left[3 \cdot e^{-2} - 1\right]$$

Let us pose:

$$J = \alpha \cdot a^{-2} + \beta \cdot a^{-1}; \alpha = \frac{-\hbar^2}{2\mu} \cdot \left[5 \cdot e^{-2} - 1\right] \cong 0.16\frac{\hbar^2}{\mu};$$
$$\beta = -\frac{e^2}{\varepsilon} \cdot \left[3 \cdot e^{-2} - 1\right] \cong 0.59\frac{e^2}{\varepsilon}$$

Solving the equation $\frac{dJ}{da} = 0$, we get:

$$a = \frac{-\varepsilon\hbar^2}{\mu e^2} \cdot \frac{\left[5 \cdot e^{-2} - 1\right]}{\left[3 \cdot e^{-2} - 1\right]} \cong -0.54 \cdot \frac{\varepsilon\hbar^2}{\mu e^2}$$

$$J = \frac{\mu e^4}{2\hbar^2} \cdot \frac{\left[3 \cdot e^{-2} - 1\right]^2}{\left[5 \cdot e^{-2} - 1\right]} \cdot \frac{1}{\varepsilon^2}$$

So the expression of the energy E of a 1S exciton:

$$E = E_g + \frac{{}^2K^2}{2m} + \frac{\mu e^4}{2\hbar^2} \cdot \frac{\left[3 \cdot e^{-2} - 1\right]^2}{\left[5 \cdot e^{-2} - 1\right]} \cdot \frac{1}{\varepsilon^2}$$

$E_g = 0.7$ eV : the gap energy of VO_2 in the semiconductor state. e: the elementary electric charge.

$$E \cong E_g + \frac{{}^2K^2}{2m} - \frac{0.54 \cdot \mu e^4}{\hbar^2} \cdot \frac{1}{\varepsilon^2} \quad (18)$$

In the vicinity of the band gap ($k \to 0$) we have:

$$\varepsilon \cong (0.54 \cdot \mu)^{\frac{1}{2}} \cdot \frac{e^2}{\hbar} \cdot \left(E_g - E\right)^{-\frac{1}{2}} \quad (19)$$

Let n be the refractive index of VO_2 that we considered to be weakly absorbent: $n = \sqrt{\varepsilon}$.

$$n \cong (0.54 \cdot \mu)^{\frac{1}{4}} \cdot \frac{e}{\sqrt{\hbar}} \cdot \left(E_g - E\right)^{-\frac{1}{4}} \quad (20)$$

Let's use the effective mass approximation:

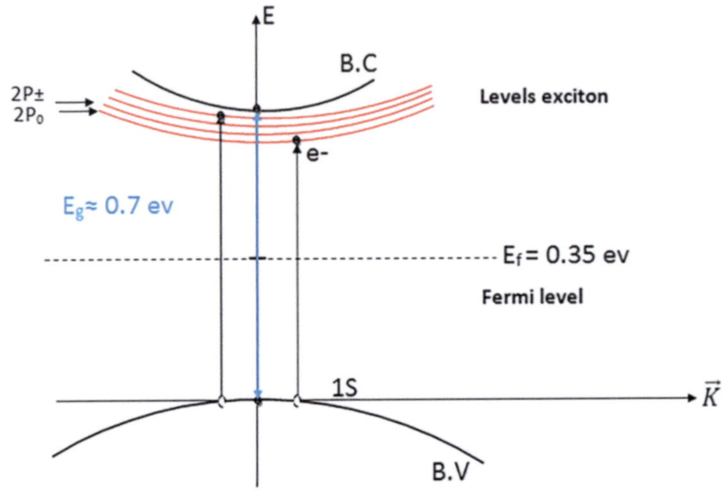

Diagram: levels of the 1S exciton in the approximation of the effective mass

We simulate using the Maple software for r and n as a function of the energy E of the exciton in the vicinity of the bandgap (k → 0) and for (k ≠ 0 at the edge of the first Brillouin zone):

For k → 0 we got the formula

$$\varepsilon \cong 9 \times 10^9 \cdot (0.54 \cdot \mu)^{\frac{1}{2}} \cdot \frac{e^2}{\hbar} \cdot \left(E_g - E\right)^{-\frac{1}{2}};$$
$$R \cong \frac{(n-1)^2}{(n+1)^2}; \quad T \cong \frac{(1-R)^2}{1-R^2}; \quad n = \sqrt{\varepsilon} \tag{21}$$

and Figs. 1, 2, 3, and 4 using the following data in the system (S.I):

$$\mu = 4 \cdot 9.1 \times 10^{-31} \text{kg} \,; \hbar = 1.05457 \times 10^{-34}\ \text{J} \cdot \text{s} \,;$$
$$E_g = 1.12 \times 10^{-19}\ \text{J} \,; \ e = 1.6 \times 10^{-19}\ \text{C}$$

For k ≠ 0 we have obtained the formula

$$\varepsilon \cong 9 \times 10^9 \cdot (0.54 \cdot \mu)^{\frac{1}{2}} \cdot \frac{e^2}{\hbar} \cdot \left(E_g - E + \frac{{}^2K^2}{2m}\right)^{-\frac{1}{2}} \tag{22}$$

and Figs. 5, 6, 7 and 8 using the following data in the system (S.I):

$$\mu = 4 \cdot 9.1 \times 10^{-31}\ \text{kg} \,; E_g = 1.12 \times 10^{-19}\ \text{J} \,;$$
$$m = 173 \times 10^{-31}\ \text{kg} \,; \ k = \frac{2\pi}{5.75 \times 10^{-10}}\ \text{m}^{-1}$$

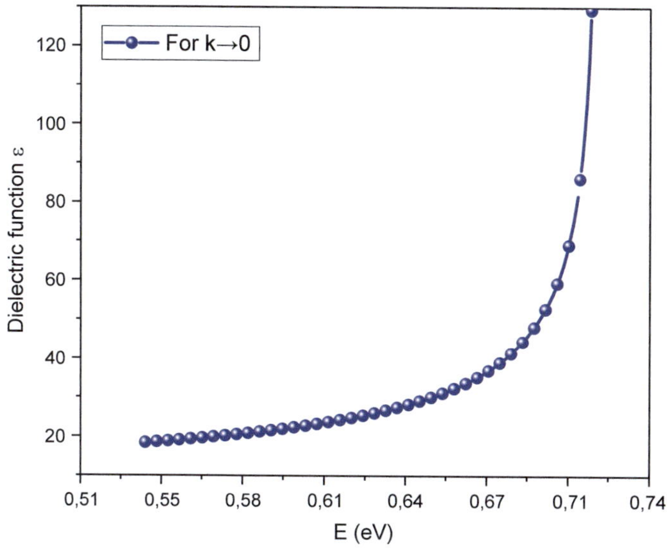

Fig. 1 The variation of the dielectric constant of VO_2 as a function of the energy E of the exciton 1S for $k \rightarrow 0$

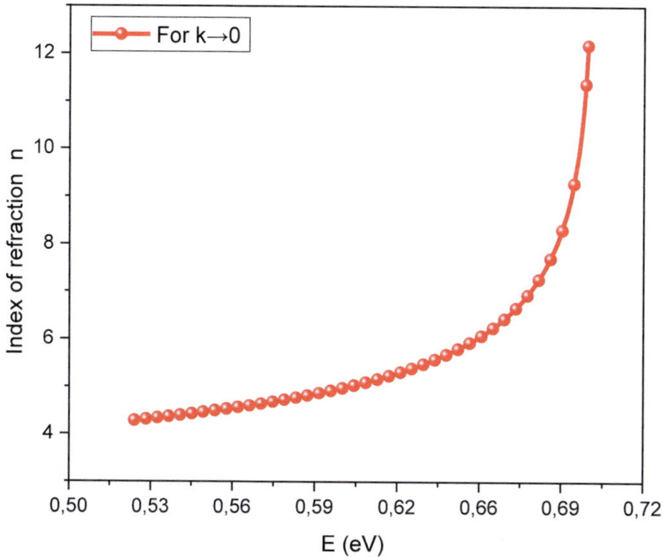

Fig. 2 The variation of the refractive index n of VO_2 as a function of the energy E of the exciton 1S for $k \rightarrow 0$

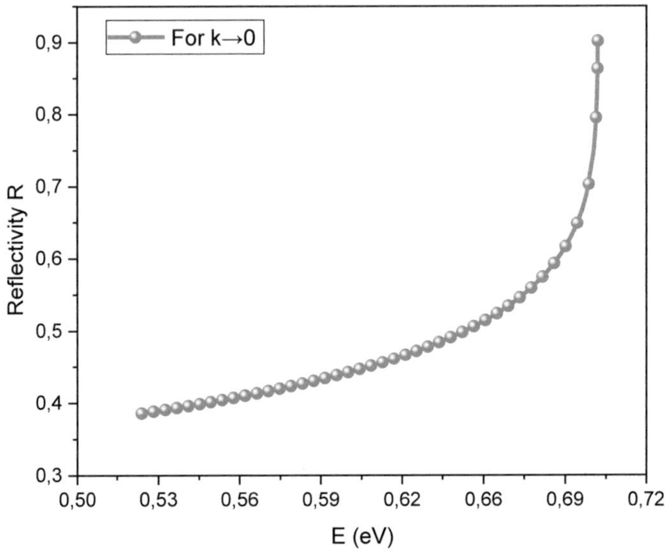

Fig. 3 The variation of the reflectivity R of VO_2 as a function of the energy E of the exciton 1S for $k \rightarrow 0$

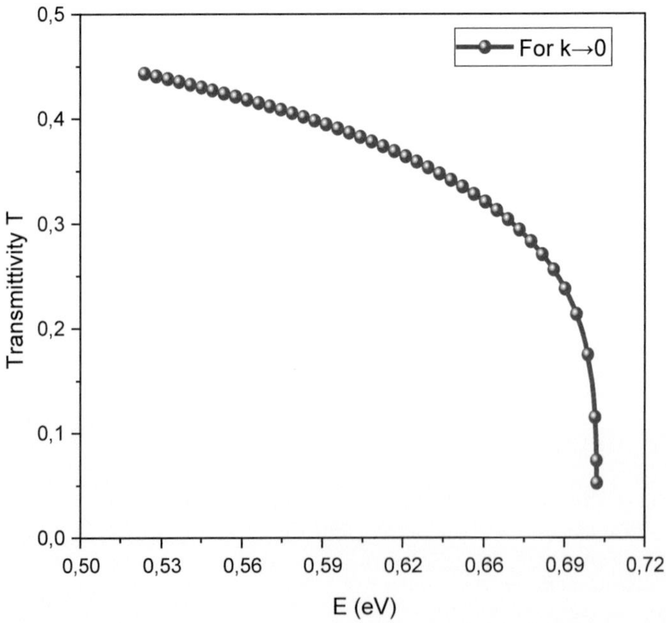

Fig. 4 The variation of the T transmittivity of VO_2 as a function of the energy E of the 1S exciton for $k \rightarrow 0$

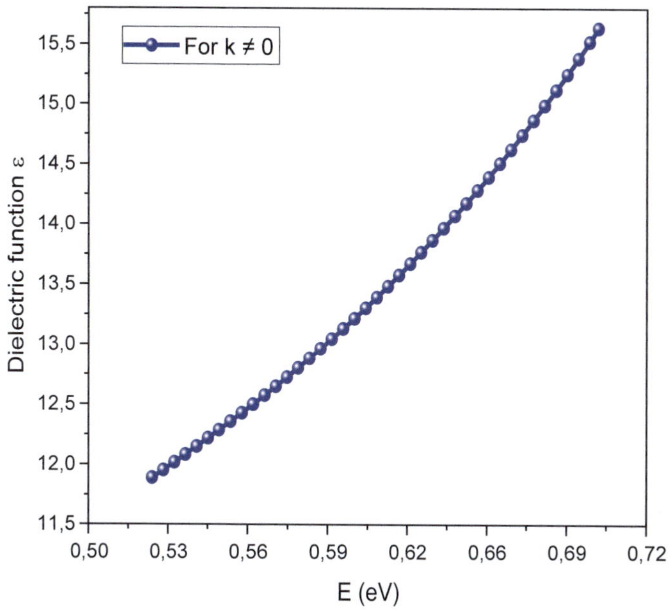

Fig. 5 The variation of the dielectric constant ε of VO_2 as a function of the energy E of the exciton 1S for $k \neq 0$

$$m = m_e^* + m_h^* = 4 \cdot m_e + 15 \cdot m_e = 19 \cdot m_e; \mu = 4 \cdot m_e$$

2.2 *Discussions*

The Hamiltonian H of an exciton 1S is given in the CGS system and for the energy of the exciton we have used the system (S.I) in our simulation of curves by making the necessary conversion. The creation of exciton in this material affects the variation of its optical properties like the dielectric function and the refractive index n around the bandgap and at the edge of the first Brillouin zone of the massive vanadium dioxide where the Coulomb interaction between the electron and the hole of the 1S exciton plays an important role in our study. The variation of the dielectric constant for this energy interval becomes important when we approach the gap energy E_g of this material because the Coulomb interaction energy decreases abruptly and the electron becomes almost free, it will be able to break the bandgap because the binding energy of the exciton decreases sharply and it loses its stability. For the lowest energy levels of these excitons, the variation of these optical constants r and n are slow around the bandgap ($k \rightarrow 0$); whereas for ($k \neq 0$) and at the edge of the first Brillouin zone, this variation becomes almost linear for $\frac{k}{2\pi}$, because of the kinetic energy $\frac{\hbar^2 K^2}{2m}$ of

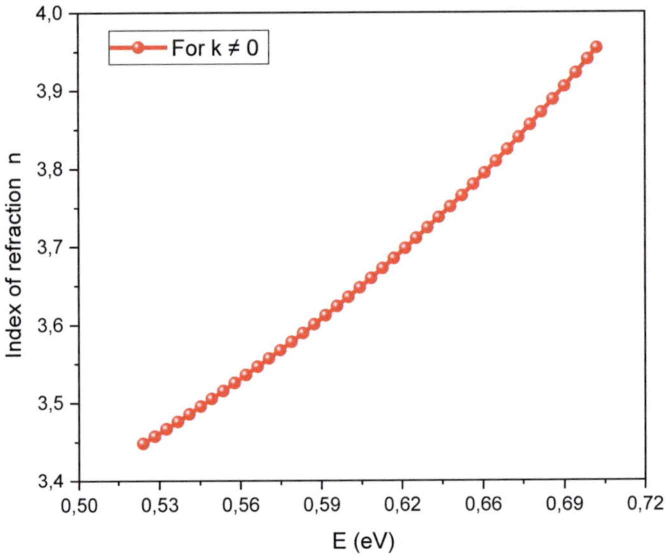

Fig. 6 The variation of the refractive index n of VO_2 as a function of the energy E of the exciton 1S for $k \neq 0$

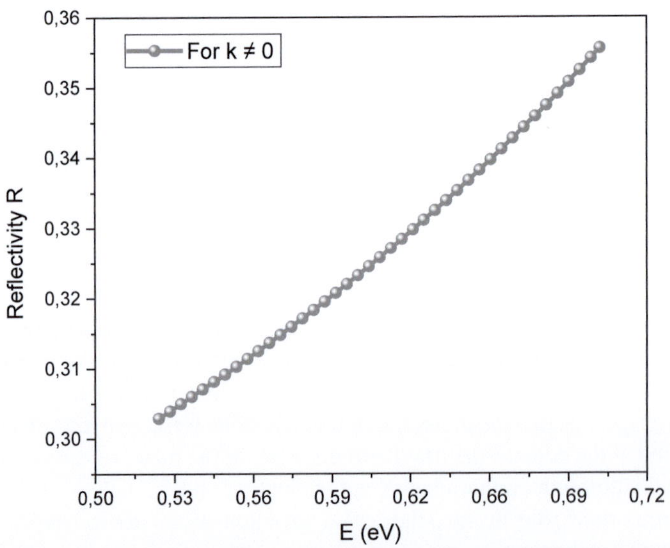

Fig. 7 The variation of the R reflectivity of VO_2 as a function of the energy E of the 1S exciton for $k \neq 0$

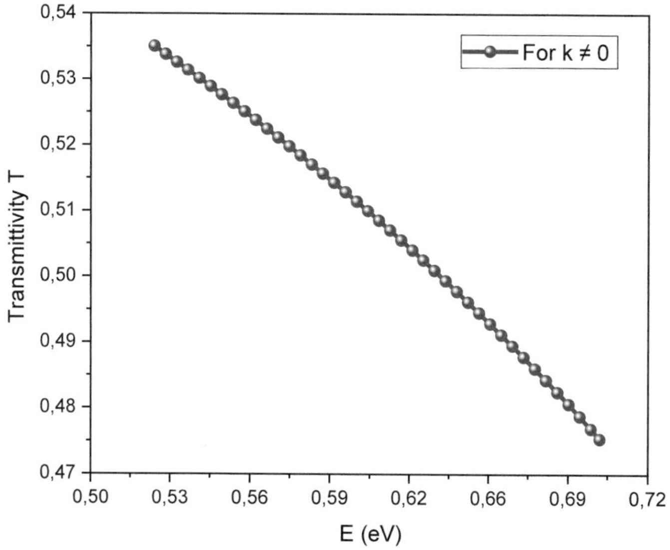

Fig. 8 The variation of the T transmittivity of VO_2 as a function of the energy E of the 1S exciton for $k \neq 0$

the exciton which becomes important with respect to the energy term $E_g - E$, and consequently its momentum P increases. Our results are based on the calculation of the energy E of a 1S exciton by the purely quantum variational method, using the approximation of the effective mass, the values of the reduced mass and the total mass previously assigned to the exciton [17]. The variations in the refractive index n of this material resemble those in the literature for the massive state. The exciton energy interval that we have proposed is highly probable in our study, based on the exciton levels allowed using the rules for selecting the excitonic transitions allowed and their positioning relative to the bottom of the conduction band (B.C) and the Fermi level E_f for the semiconductor state of this material. The expression of the energy E of the exciton 1S which we have obtained by our working method is an approximate result for excitons, because the Schrödinger equation obtained can be solved exactly by resembling the exciton S to the three-dimensional hydrogen atom. The lifetime of an exciton is of the order of a few nanoseconds, during this very low period we have shown that the optical response of massive vanadium dioxide is not constant. Indeed, we have found that from this work the dielectric function and the refractive index vary according to the energy of the exciton created by incident photons.

Around the band gap ($k \to 0$)

Reflectivity $R \cong 0.442$ and transmittivity $T \cong 0.386$, for the energy of the exciton $E = 8.4 \times 10^{-20}$ J. As the energy level E of the exciton approaches the conduction band, we find that $R \cong 0.90$ and $T \cong 0.051$; the material becomes highly reflective.

We therefore note that the reflectivity R increases slowly and the transmittivity T decreases for vanadium dioxide in the massive semiconductor state, if the energy E of the exciton increases. These variations occur within a few nanoseconds.

At the edge of the first area of Brillouin ($k \neq 0$)

For the lowest energy levels E of the 1S exciton, we find that $R \cong 0.30$ if $E = 8.4 \times 10^{-20}$ J. The reflectivity R slowly increases until it reaches a value of $R \cong 0.36$ if $E = 1.12 \times 10^{-19}$ J. While transmittivity $T \cong 0.54$ if $E = 8.4 \times 10^{-20}$ J. The transmittivity of this material decreases slowly as the energy E of the exciton increases; it is about 0.475 if $E = 1.12 \times 10^{-19}$ J.

We see that the variations of R and T are simultaneously opposite. Transmittivity T decreases while reflectivity R increases as the energy E of the exciton approaches the conduction band. This result is logical because the Coulomb interaction of the exciton decreases and the electron becomes almost free.

These simulated results clearly prove that the creation of exciton in massive vanadium dioxide in the semiconductor state by incident photons influences the R reflectivity and the T transmittivity of VO_2.

The interest of this optoquantum study concerning excitons is to be able to control and exploit the variation of the optical constants of this material [18, 19] at the nanosecond scale for various nanotechnological applications at very fast speed such as optical detectors and photovoltaic equipment. We are thinking of doing a similar study for the thin films of vanadium dioxide and comparing the two results.

2.3 *Conclusion*

The creation of 1S excitons in the massive VO_2 material by incident photons of ω energy lower than its gap energy $E_g = 0.7\,\text{eV}$ has a definite influence on the optical properties of VO_2 such as the refractive index n, the dielectric constant, on the one hand around the bandgap of its semiconductor state ($k \rightarrow 0$) and at the edge of the first Brillouin zone ($k \neq 0$), This is confirmed in our study. Our results are consistent with those of Mossaneck et al. We know in the literature that the presence of an exciton increases the α absorption coefficient of the material. We chose to study the influence of a 1S exciton on other optical constants such as the dielectric function Γ, the refractive index n, the transmittivity T and the reflectivity R, we arrived at a confirmation. This makes this study necessary to be able to control the optical response of this material in time intervals of the order of a few nanoseconds.

This exciton-based study allows us to put this smart material at the heart of scalable nano-technological applications such as VO_2-based solar air conditioners that can replace electric air conditioning, which remains expensive in comparison.

References

1. Paradis, S.: Doped vanadium dioxide with enhanced infrared modulation. DRDC Valcartier TM 2007-002
2. Cavalleri, A., Dekorsy, T., Chong, H.H.W., Kieffer, J.C., Schoenlein. R.W.: Evidence for a structurally-driven insulator- to-metal transition in VO_2: a view from the ultrafast timescale. Phys. Rev. B, **70** (2004)
3. Morin, F.J.: Oxides which show a metal-to-insulator transition at the neel temperature. Phys. Rev. Lett. Am. Phys. Soc. (APS). **3**(1), 34–36 (1959)
4. Vanadium oxides section. Gmelin Handbook, vol. 16, p. 482 (1990)
5. Doctoral thesis of the University of Paris-Sud XI: by Fanny Rodolakis "Metal-insulator transition in Vanadium oxides". December 4, 2009
6. Yang, Z., Ko, C., Ramanathan, S.: Metal-insulator transition characteristics of VO2 thin films grown on Ge(100) single crystals. J. Appl. Phys. **108**(7):073708 (2010)
7. Mossanek, R.J.O., Abbate, M.: Optical response of metallic and insulating VO_2 calculated with the LDA approach. J. Phys.: Condens. Matter. **19**(34), 346225 (2007)
8. Gatti, M., Bruneval, F., Olevano, V., Reining, L.: Understanding correlations in vanadium dioxide from first principles. Phys. Rev. Lett. **99**(26) (2007)
9. Elliott, R.J.: Intensity of optical absorption by excitons. Phys. Rev. **108**(6), 1384–1389 (1957)
10. Yu, P.Y., Cardona, M.: Fundamentals of Semi-Conductors. Springer, Berlin (1999)
11. Wooten, F.: Optical Properties of Solids. Academic Press (1972)
12. Kittel, C.: Physics of the Solid State. Dunod Edition, $7^{ém}$ (1983)
13. Pankove, J.: Optical Processes in Semiconductors. Dover, New York (1975)
14. Cohen, M.M.: Introduction to the Quantum Theory of Semiconductors. Gordon and Breach (1972)
15. Messiah, A.: Quantum Mechanics, vol. 1, pp. 114–117, 269–270. Dunod (1995)
16. Cohen-Tannoudji, C., Diuand, B., Laloé, F.: Mécanique quantique II. Edition Hermann (1986)
17. Kana, J.B.K., Ndjaka, J.M., Vignaud, G., Gibaud, A., Maaza, M.: Thermally tunable optical constants of vanadium dioxide thin films measured by spectroscopic ellipsometry. Opt. Commun. **284**(3), 807–812 (2011)
18. Ben Chaib, A., Zouini, M.: Modulating the optical behaviour of metallic vanadium dioxide by adjusting the number of Drude oscillators. Pramana - J. Phys. **97**, 201 (2023). https://doi.org/10.1007/s12043-023-02685-z
19. Ben Chaib, A., Zouini, M., Tahiri, A.: Drude-Lorentz oscillators' effect on the optical response of avanadium dioxide semiconductor. Opt. Mater. **133**, 112895 (2022). https://doi.org/10.1016/j.optmat.2022.112895

MIX
Papier aus verantwortungsvollen Quellen
Paper from responsible sources
FSC® C105338

If you have any concerns about our products,
you can contact us on
ProductSafety@springernature.com

In case Publisher is established outside the EU,
the EU authorized representative is:
Springer Nature Customer Service Center GmbH
Europaplatz 3, 69115 Heidelberg, Germany

Printed by Libri Plureos GmbH
in Hamburg, Germany